# Lecture Notes in Bioinformatics    11183

Subseries of Lecture Notes in Computer Science

More information about this series at http://www.springer.com/series/5381

Mathieu Blanchette · Aïda Ouangraoua (Eds.)

# Comparative Genomics

16th International Conference, RECOMB-CG 2018
Magog-Orford, QC, Canada, October 9–12, 2018
Proceedings

 Springer

*Editors*
Mathieu Blanchette (iD)
McGill University
Montréal, QC
Canada

Aïda Ouangraoua (iD)
Université de Sherbrooke
Sherbrooke, QC
Canada

ISSN 0302-9743          ISSN 1611-3349   (electronic)
Lecture Notes in Bioinformatics
ISBN 978-3-030-00833-8      ISBN 978-3-030-00834-5   (eBook)
https://doi.org/10.1007/978-3-030-00834-5

Library of Congress Control Number: 2018954769

LNCS Sublibrary: SL8 – Bioinformatics

This Springer imprint is published by the registered company Springer Nature Switzerland AG
The registered company address is: Gewerbestrasse 11, 6330 Cham, Switzerland

# Preface

This volume contains the papers presented at RECOMBCG-18: the 16th RECOMB Comparative Genomics Satellite Workshop, held during October 9–12, 2018, in Magog-Orford, Québec, Canada.

There were 30 submissions. Each submission was reviewed by at least 4 reviewers, and some received up to 6 papers to review. The reviews were conducted by members of the Program Committee (PC) as well as by additional reviewers who were sought based on their expertise for specific papers. The PC decided to accept 18 papers. The program also included 6 invited talks and 2 poster sessions.

We thank the members of the Program Committee as well as the additional reviewers for their diligent work reviewing the manuscripts and conducting thorough discussions on the manuscripts and their reviews that informed the decision process.

We also thank the members of the Steering Committee, Marlia Braga, Dannie Durand, Jens Lagergren, Aoife McLysaght, Luay Nakhleh, and David Sankoff, for their guidance.

Special thanks go to the members of the Local Organizing Committee at the Université de Sherbrooke, Sarah Belhamiti, Alan Cohen, Ali Fotouhi, Jean-Michel Garant, Pierre-Étienne Jacques, Safa Jammali, Esaie Kuitche, Manuel Lafond, Lynn Lebrun, Jean-Pierre Perreault, Sébastien Rodrigue, Michelle Scott, Fanny Thuriot, Yves Tremblay, Anais Vannutelli, and Shengrui Wang, the students in particular, for all their efforts and generosity in organizing the conference.

Last but not least, we would like to thank the six keynote speakers who agreed to speak at the conference despite their very busy schedules: Belinda Chang, Dannie Durand, Daniel Durocher, Christian Landry, Gwenaël Piganeau, and Xavier Roucou.

The workshop would not have been possible without the generous contribution of our sponsors, whose support is greatly appreciated:

- Faculté de médecine et des sciences de la santé - Université de Sherbrooke
- Faculté des sciences - Université de Sherbrooke
- Programme de soutien à la tenue de colloques scientifiques étudiants - Université de Sherbrooke
- Fonds institutionnel de soutien aux activités étudiantes - Université de Sherbrooke
- Département d'informatique - Université de Sherbrooke
- Département de biochimie - Université de Sherbrooke
- Département de biologie - Université de Sherbrooke
- Calcul Québec

We used the EasyChair system for submissions, reviews, and proceedings formatting.

August 2018

Mathieu Blanchette
Aïda Ouangraoua

# Organization

## Program Chairs

Mathieu Blanchette   McGill University, Canada
Aida Ouangraoua   University of Sherbrooke, Canada

## Steering Committee

Marília Braga    Bielefeld University, Germany
Dannie Durand    Carnegie Mellon University, USA
Jens Lagergren    Stockholm University, Sweden
Aoife McLysaght   Trinity College Dublin, Ireland
Luay Nakhleh    Rice University, USA
David Sankoff    University of Ottawa, Canada

## Program Committee Members

Max Alekseyev    George Washington University, USA
Lars Arvestad    Stockholm University, Sweden
Sèverine Bérard    Université de Montpellier, France
Anne Bergeron    Université du Québec à Montréal, Canada
Marilia Braga    Bielefeld University, Germany
Alessandra Carbone   Sorbonne University, France
Cedric Chauve    Simon Fraser University, Canada
Leonid Chindelevitch   Simon Fraser University, Canada
Miklos Csuros    University of Montreal, Canada
Daniel Doerr    Bielefeld University, Germany
Dannie Durand    Carnegie Mellon University, USA
Ingo Ebersberger   Goethe University, Germany
Nadia El-Mabrouk   University of Montreal, Canada
Oliver Eulenstein   Iowa State University, USA
Guillaume Fertin   CNRS, University of Nantes, France
Pawel Gorecki    University of Warsaw, Poland
Michael Hallett    Concordia University, Canada
Katharina Jahn    ETH Zurich, Switzerland
Asif Javed    Genome Institute of Singapore (A-STAR), Singapore
Manuel Lafond    University of Sherbrooke, Canada
Jens Lagergren    KTH Royal Institute of Technology, Sweden
Kevin Liu    Michigan State University, USA
Fábio H. V. Martinez   Universidade Federal de Mato Grosso do Sul, Brazil
Joao Meidanis    University of Campinas and Scylla Bioinformatics, Brazil
István Miklós    Rényi Institute, Hungary

| | |
|---|---|
| Luay Nakhleh | Rice University, USA |
| David Sankoff | University of Ottawa, Canada |
| Eric Tannier | Inria, France |
| Glenn Tesler | University of California San Diego, USA |
| Olivier Tremblay-Savard | University of Manitoba, Canada |
| Jens Stoye | Bielefeld University, Germany |
| Krister Swenson | CNRS and Université de Montpellier, France |
| Tamir Tuller | Tel Aviv University, Israel |
| Jean-Stéphane Varré | University of Lille, France |
| Lusheng Wang | City University of Hong Kong, SAR China |
| Tandy Warnow | University of Illinois at Urbana-Champaign, USA |
| Louxin Zhang | National University of Singapore, Singapore |
| Jie Zheng | ShanghaiTech University, China |

Also a big thank you to our sub-reviewers: Sergey Aganezov, Eloi Araujo, Maria Atamanova, Pavel Avdeyev, Annie Chateau, Sarah Christensen, Riccardo Dondi, Philippe Gambette, Hussein Hejase, Géraldine Jean, Yu Lin, Bernard Moret, Agnieszka Mykowiecka, Anton Nekhai, Fabio Pardi, Jarek Paszek, Diego P. Rubert, Celine Scornavacca, Marco Stefanes, Riccardo Vicedomini, Roland Wittler, and João Paulo Pereira Zanetti.

## Local Organizing Committee

| | |
|---|---|
| Mathieu Blanchette | McGill University, Canada |
| Aïda Ouangraoua | University of Sherbrooke, Canada |
| Sarah Belhamiti | University of Sherbrooke, Canada |
| Alan Cohen | University of Sherbrooke, Canada |
| Ali Fotouhi | University of Sherbrooke, Canada |
| Jean-Michel Garant | University of Sherbrooke, Canada |
| Pierre-Étienne Jacques | University of Sherbrooke, Canada |
| Safa Jammali | University of Sherbrooke, Canada |
| Esaie Kuitche | University of Sherbrooke, Canada |
| Manuel Lafond | University of Ottawa and University of Sherbrooke, Canada |
| Jean-Pierre Perreault | University of Sherbrooke, Canada |
| Sébastien Rodrigue | University of Sherbrooke, Canada |
| Michelle Scott | University of Sherbrooke, Canada |
| Fanny Thuriot | University of Sherbrooke, Canada |
| Anais Vannutelli | University of Sherbrooke, Canada |
| Shengrui Wang | University of Sherbrooke, Canada |

# Contents

## Phylogenetics

# Genome Rearrangements

# A Cubic Algorithm for the Generalized Rank Median of Three Genomes

Leonid Chindelevitch[1]([✉]) and Joao Meidanis[2]

[1] School of Computing Science, Simon Fraser University, Burnaby, Canada
leonid@sfu.ca
[2] Institute of Computing, University of Campinas, Campinas, Brazil

**Abstract.** The area of genome rearrangements has given rise to a number of interesting biological, mathematical and algorithmic problems. Among these, one of the most intractable ones has been that of finding the median of three genomes, a special case of the ancestral reconstruction problem. In this work we re-examine our recently proposed way of measuring genome rearrangement distance, namely, the rank distance between the matrix representations of the corresponding genomes, and show that the median of three genomes can be computed exactly in polynomial time $O(n^{\omega})$, where $\omega \leq 3$, with respect to this distance, when the median is allowed to be an arbitrary orthogonal matrix.

We define the five fundamental subspaces depending on three input genomes, and use their properties to show that a particular action on each of these subspaces produces a median. In the process we introduce the notion of $M$-stable subspaces. We also show that the median found by our algorithm is always orthogonal, symmetric, and conserves any adjacencies or telomeres present in at least 2 out of 3 input genomes.

We test our method on both simulated and real data. We find that the majority of the realistic inputs result in genomic outputs, and for those that do not, our two heuristics perform well in terms of reconstructing a genomic matrix attaining a score close to the lower bound, while running in a reasonable amount of time. We conclude that the rank distance is not only theoretically intriguing, but also practically useful for median-finding, and potentially ancestral genome reconstruction.

**Keywords:** Comparative genomics · Ancestral genome reconstruction
Phylogenetics · Rank distance

## 1 Introduction

The genome median problem consists of computing a genome $M$ that minimizes the sum $d(A, M) + d(B, M) + d(C, M)$, where $A$, $B$, and $C$ are three given genomes and $d(\cdot, \cdot)$ is a distance metric that measures how far apart two genomes are, and is commonly chosen to correlate with evolutionary time. In this paper, we present a polynomial-time algorithm for the computation of a median for the rank distance. We call it a generalized median because, despite attaining a lower

© Springer Nature Switzerland AG 2018
M. Blanchette and A. Ouangraoua (Eds.): RECOMB-CG 2018, LNBI 11183, pp. 3–27, 2018.
https://doi.org/10.1007/978-3-030-00834-5_1

bound on the best score with respect to the rank distance, it may not be a genome in all cases. However, we report on experiments that show that the median is genomic in the majority of the cases we examined, including real genomes and artificial genomes created by simulation, and when it is not, a genome close to the median can be found via an efficient post-processing heuristic.

This result is a significant improvement on the first algorithm for generalized medians with respect to the rank distance, which makes it fast enough to be used on real genomes, with thousands of genes. Our experiments deal with genomes with up to 1000 genes, but the measured running times of the algorithm and their extrapolation suggest that reaching tens of thousands of genes is feasible.

Our work builds upon a recent result from our group that shows the first polynomial-time algorithm for rank medians of orthogonal matrices [1], delivering an alternative specific to genomes which avoids any floating-point convergence issues, guarantees the desirable properties of symmetry and majority adjacency/telomere conservation, and provides a speed-up from $\Theta(n^{1+\omega})$ to $\Theta(n^\omega)$ in the worst case, where $\omega$ is the exponent of matrix multiplication known to be less than 2.38 [2], but close to 3 on practical instances. Prior to this result, there were fast, polynomial-time median algorithms for simpler distances, such as the breakpoint distance [3] and the SCJ distance [4]. In contrast, for more sophisticated distances such as the inversion distance [5] and the DCJ distance [3], the median problem is NP-hard, meaning that it is very unlikely that fast algorithms for it exist. The rank distance is equal to twice the algebraic distance [6], which in turn is very close to the widely used DCJ distance [7]. More specifically, it assigns a weight of 1 to cuts and joins, and a weight of 2 to double swaps; it is known that the rank distance equals the total weight of the smallest sequence of operations transforming one genome into another under this weighting scheme [8]. Therefore, it is fair to place the rank distance among the more sophisticated distances, that take into account rearrangements such as inversions, translocations, and transpositions, with weights that correlate with their relative frequency.

A more complete distance will also take into account content-changing events, such as duplications, gene gain and loss, etc. We hope that our contribution provides significant insight towards studies of more complex genome distances.

## 1.1 Definitions

Let $n \in \mathbb{N}$ be an integer and let $\mathbb{R}^{n \times n}$ be the set of $n \times n$ matrices with entries in $\mathbb{R}$. Following [6], we say that a matrix $M$ is *genomic* when it is:

- *binary*, i.e. $M_{ij} \in \{0,1\} \; \forall \, i,j$
- *orthogonal*, i.e. $M^T = M^{-1}$ (so the columns of $M$ are pairwise orthogonal)
- *symmetric*, i.e. $M^T = M$ (so $M_{ij} = M_{ji} \; \forall \, i,j$)

Strictly speaking, $n$ must be even for a genomic matrix, because $n$ is the number of gene extremities, and each gene contributes two extremities, its head and its tail [6]. However, most of our results apply equally well to all integers $n$.

A genomic matrix $M$ defines a permutation $\pi$ via the relationship

$$\pi(i) = j \iff M_{i,j} = 1.$$

It is easy to see that the permutation $\pi$ corresponding to a genomic matrix is a product of disjoint cycles of length 1 and 2. The cycles of length 1 correspond to *telomeres* while the cycles of length 2 correspond to *adjacencies*. The correspondence between a genome $G$ and a genomic matrix $M$ is defined by

$$M_{i,j} = 1 \iff i \neq j \text{ and } (i,j) \text{ is an adjacency in } G, \text{ or}$$
$$i = j \text{ and } i \text{ is a telomere in } G.$$

## 1.2 Rank Distance

The rank distance $d(\cdot, \cdot)$ [9] is defined on $\mathbb{R}^{n \times n}$ via

$$d(A, B) = r(A - B),$$

where $r(X)$ is the *rank* of the matrix $X$, defined as the dimension of the *image* (or column space) of $X$ and denoted $\text{im}(X)$. This distance is a metric and is equivalent to the Cayley distance between the corresponding permutations when $A$ and $B$ are both permutation matrices [1,6].

The relevance of the rank distance for genome comparison stems from the fact that some of the most frequent genome rearrangements occurring in genome evolution, such as inversions, transpositions, translocations, fissions and fusions, correspond to a perturbation of a very low rank (between 1 and 4, depending on the operation) of the starting genomic matrix. This suggests that the rank distance may be a good indicator of the amount of evolution that separates two genomic matrices. We previously reviewed its relationship to other distances [1].

## 1.3 The Median Problem and Invariants

Given three matrices $A, B, C$, the median $M$ is defined as a global minimizer of the *score* function $d(M; A, B, C) := d(A, M) + d(B, M) + d(C, M)$.

In previous work we identified three important invariants for the median-of-three problem. The first invariant is defined as:

$$\beta(A, B, C) := \frac{1}{2}[d(A, B) + d(B, C) + d(C, A)].$$

This invariant is known to be integral if $A$, $B$, and $C$ are orthogonal matrices, which include genomic matrices and permutation matrices as special cases [1].

The first invariant is also a lower bound for the score: $d(M; A, B, C) \geq \beta(A, B, C)$, with equality if and only if

$$d(X, M) + d(M, Y) = d(X, Y) \text{ for any distinct } X, Y \in \{A, B, C\}. \tag{1}$$

The second invariant is the dimension of the "triple agreement" subspace [1]:

$$\alpha(A, B, C) := \dim(V_1), \text{ where } V_1 := \{x \in \mathbb{R}^n | Ax = Bx = Cx\}. \tag{2}$$

Finally, the third invariant combines the first two with the dimension $n$:

$$\delta(A, B, C) := \alpha(A, B, C) + \beta(A, B, C) - n. \tag{3}$$

This invariant is known to be non-negative if $A$, $B$, and $C$ are orthogonal [1]. We therefore call it the *deficiency* of $A, B$ and $C$, by analogy with the deficiency of a chemical reaction network defined in the work of Horn, Jackson and Feinberg [10]. We recall here our "deficiency zero theorem" for medians of permutations [1].

**Theorem 1 (Deficiency Zero Theorem).** *Let $A, B, C$ be permutations with $\delta(A, B, C) = 0$. Then the median is unique, and can be found in $O(n^2)$ time.*

## 1.4   The Five Subspaces and Their Dimensions

The inputs of a median-of-three problem partition $\mathbb{R}^n$ into five subspaces [6], which we describe in this section.

The "triple agreement" subspace $V_1 = V(.A.B.C.)$ is defined in Eq. (2), and is the subspace of all vectors on which all three matrices agree. Its dimension is $\alpha(A, B, C)$, by definition.

The subspace $V_2 := V(.AB.C.) \cap V_1^\perp$ is defined via $V_1$ and the subspace

$$V(.AB.C) := \{x \in \mathbb{R}^n | Ax = Bx\}.$$

The dimension of $V(.AB.C)$ is precisely $c(\rho^{-1}\sigma)$, where $\rho$ and $\sigma$ are the permutations corresponding to $A$ and $B$, respectively, and $c(\pi)$ is the number of cycles (including fixed points) in a permutation $\pi$. This follows from this observation:

$$Ax = Bx \iff A^{-1}Bx = x \iff x \text{ is constant on every cycle of } \rho^{-1}\sigma. \tag{4}$$

Since $V_1 \subseteq V(.AB.C)$, it follows that a basis of $V_1$ can be extended to a basis of $V(.AB.C)$ with vectors orthogonal to those spanning $V_1$, so that

$$\dim(V_2) = \dim(V(.AB.C.) \cap V_1^\perp) = \dim(V(.AB.C.) - \dim(V_1) = c(\rho^{-1}\sigma) - \alpha.$$

We can apply a similar reasoning to the subspaces $V_3 := V(.A.BC.) \cap V_1^\perp$ and $V_4 := V(.AC.B) \cap V_1^\perp$, where $V(.A.BC.) := \{x \in \mathbb{R}^n | Bx = Cx\}$ and $V(.AC.B) := \{x \in \mathbb{R}^n | Cx = Ax\}$, to get

$$\dim(V_2) = c(\rho^{-1}\sigma) - \alpha; \ \dim(V_3) = c(\sigma^{-1}\tau) - \alpha; \ \dim(V_4) = c(\tau^{-1}\rho) - \alpha,$$

where $\tau$ is the permutation corresponding to $C$.

It was shown by Pereira Zanetti et al. [6] that

$$\mathbb{R}^n = V_1 \oplus V_2 \oplus V_3 \oplus V_4 \oplus V_5, \tag{5}$$

where $V_5$ is the subspace orthogonal to the sum of the other four subspaces, and the $\oplus$ notation represents a direct sum, i.e. $V_i \cap V_j = \{0\}$ whenever $1 \leq i < j \leq 5$. For each $1 \leq j \leq 5$, we also define the projector $P_j$, as the projector onto $V_j$ along $\oplus_{i \neq j} V_i$. After that Eq. (5) can also be equivalently written as $\sum_{j=1}^{5} P_j = I$.

Since $V_5$ is the last term in the direct sum decomposition of $\mathbb{R}^n$, we get that

$$\dim(V_5) = n - \sum_{i=1}^{4} \dim(V_i) = n + 2\alpha - (c(\rho^{-1}\sigma) + c(\sigma^{-1}\tau) + c(\tau^{-1}\rho))$$

$$= n + 2\alpha(A, B, C) - (3n - 2\beta(A, B, C)) = 2(\alpha + \beta - n) = 2\delta(A, B, C).$$

## 1.5 A Specific Example

Let us now look at a specific example (which is one of our simulated inputs). To save space, we write a genome as a permutation in cycle notation, with singletons omitted, and use hexadecimal notation for numbers exceeding a single digit. Let

$$A = (24)(39)(68)(ab), B = (27)(38)(45)(69)(ab), C = (23)(45)(67)(89)(ab).$$

We use $n = 12$ although $c$ is a singleton in all inputs. First note that $AB = (2745)(36)(89)$, $BC = (286)(379)$, and $CA = (25438769)$, so $\alpha(A, B, C) = 5$ because the triple agreement space is spanned by the indicator vectors of the sets $\{1\}, \{2, 3, 4, 5, 6, 7, 8, 9\}, \{a\}, \{b\}, \{c\}$. Furthermore, by counting the cycles in the products above we get $d(A, B) = 5, d(B, C) = 4, d(C, A) = 7$, so $\beta(A, B, C) = 8$ and $\delta(A, B, C) = 1$. The dimensions of the subspaces $V_1$ through $V_5$ are thus 5, 2, 3, 0, and 2.

We note that we can ignore the common telomeres 1 and $c$ as well as the common adjacency $(ab)$ because we can assume they will be present in a median (see Theorem 1 in [6]). Thus, we can simplify our example by adding the known adjacencies and telomeres to the median and removing them from the input. After renumbering the remaining extremities from 1 to 8, the input becomes

$$A' = (13)(28)(57), B' = (16)(27)(34)(58), C' = (12)(34)(56)(78).$$

Now the invariants get reduced to $\alpha(A', B', C') = 1, \beta(A', B', C') = 8, \delta(A', B', C') = 1$ and the subspace dimensions become 1, 2, 3, 0, and 2, respectively.

## 1.6 Highlights for Small $n$

To gain insight into the median problem, we scrutinized the problem of computing the median for all genomic matrices for $n = 3$ to $n = 8$. For each $n$, we classified the input matrices in a number of equivalent cases. For $n = 3$ and $n = 4$, we computed all the medians for all cases. For $n = 5$ and higher, we concentrated on the cases with positive deficiency $\delta$, given that cases with $\delta = 0$ are easy (Theorem 1). We tested an algorithm, which we call algorithm $\mathcal{A}$, that is a modification of the algorithm in [6] where $M$ agrees with the corresponding input on the 4 "agreement subspaces", but mimics the identity matrix on the subspace $V_5$. More specifically, Algorithm $\mathcal{A}$, given genomic matrices $A$, $B$, and $C$, returns matrix $M_I$ defined as follows:

$$M_I(v) = \begin{cases} Av \text{ if } v \in V_1 \\ Av \text{ if } v \in V_2 \\ Bv \text{ if } v \in V_3 \\ Cv \text{ if } v \in V_4 \\ v \ \text{ if } v \in V_5 \end{cases}$$

where the subspaces $V_1, \ldots, V_5$ were defined in Sect. 1.4.

We observed that in all cases we examined the result $M_I$ was an orthogonal matrix, and algorithm $\mathcal{A}$ was able to find a median attaining the lower bound $\beta(A, B, C)$. In cases where the median is not unique and we computed all the medians, we observed that all the medians $M$ satisfy an equation of the form

$$(M - O)(M - O)^T = R,$$

for suitable matrices $O$ and $R$ of size $n \times n$ depending only on $A$, $B$ and $C$, meaning that the medians lie on a "circle" in matrix space. We provide a detailed example in the Appendix and conjecture that such relationships hold for all triplets of genomic matrices $A, B$ and $C$.

## 2   $M_I$ and Its Computation

Following our experiments with algorithm $\mathcal{A}$, we conjectured — and proved — that it always produces a median when the inputs are genomic matrices. Furthermore, we proved that this median is always orthogonal, symmetric, and has rows and columns that add up to 1. It also contains only rational entries, and in our experiments, these entries are 0 and 1 most of the time, meaning that the median produced by algorithm $\mathcal{A}$ is actually genomic. For the few cases when this property does not hold, we introduce two heuristics in the next section.

The rest of this section is organized as follows: we begin by defining $M_I$, the output of algorithm $\mathcal{A}$, and provide sufficient conditions for its optimality in Sect. 2.1. We prove its symmetry in Sect. 2.2 and its orthogonality in Sect. 2.3. We sketch the proof of its optimality in Sect. 2.4, providing the complete version in the Appendix. We prove a result showing that $M_I$ contains any adjacencies and telomeres common to at least two of the three input genomes in Sect. 2.5. Lastly, we discuss how to compute $M_I$ efficiently in Sect. 2.6.

### 2.1   Definition of $M_I$ and Sufficient Conditions for Optimality

We start with a general result on matrices that mimic the majority of inputs in $V_1$ through $V_4$, and mimic a certain matrix $Z$ in $V_5$.

**Definition 1.** *Let $A, B, C$ be permutation matrices of size $n$, and let $Z$ be a fixed matrix of size $n$. As above, let $V_1$ through $V_5$ be the 5 subspaces in the direct sum decomposition of $\mathbb{R}^n$ induced by $A, B, C$, and let $P_j$ be the projector onto $V_j$ for $1 \leq j \leq 5$. We define $M_Z := AP_1 + AP_2 + BP_3 + CP_4 + ZP_5$ as the matrix that agrees with the corresponding inputs on the "agreement spaces" $V_1, V_2, V_3, V_4$ and acts by the operator $Z$ on the "disagreement space" $V_5$.*

**Definition 2.** *Let $A, B, C$ be permutation matrices, and let $Z$ be a fixed matrix, and let $V_1$ through $V_5$ be the 5 subspaces in the direct sum decomposition of $\mathbb{R}^n$ induced by $A, B, C$. We define $V_Z^A := \{x + y | x \in V_3, y \in V_5, A(x+y) = Bx + Zy\}$, and similarly, $V_Z^B := \{x + y | x \in V_4, y \in V_5, B(x+y) = Cx + Zy\}$ and $V_Z^C := \{x + y | x \in V_2, y \in V_5, C(x+y) = Ax + Zy\}$.*

**Lemma 1.** *Let $M_Z$ be the matrix in Definition 1 and let $V_Z^A$, $V_Z^B$, $V_Z^C$ be the subspaces in Definition 2. Then the score of $M_Z$ with respect to $A, B, C$ is $s(M_Z) := \beta(A, B, C) + 3\delta(A, B, C) - (\dim(V_Z^A) + \dim(V_Z^B) + \dim(V_Z^C))$.*

*Proof.* Recall Eq. (5): $\mathbb{R}^n = \bigoplus_{i=1}^5 V_i$. By construction, $M_Z$ agrees with $A$ on the subspaces $V_1, V_2, V_4$ so those do not contribute to the rank of $M_Z - A$. Therefore, by the rank plus nullity theorem,

$$d(M_Z, A) = \dim(V_3) + \dim(V_5) - \dim\{z \in V_3 + V_5 | Az = M_Z z\}.$$

However, the space whose dimension is subtracted can also be rewritten as

$$\{z = x + y | x \in V_3, y \in V_5, A(x+y) = Bx + Zy\} =: V_Z^A,$$

since $M_Z$ acts by $B$ on $V_3$ and by $Z$ on $V_5$, by Definition 1. We combine this result with similar results for $B$ and $C$ to deduce that

$$d(M_Z, A) = \dim(V_3) + \dim(V_5) - \dim(V_Z^A); \tag{6}$$
$$d(M_Z, B) = \dim(V_4) + \dim(V_5) - \dim(V_Z^B); \tag{7}$$
$$d(M_Z, C) = \dim(V_2) + \dim(V_5) - \dim(V_Z^C). \tag{8}$$

By adding these up and using the fact that $\dim(V_5) = 2\delta(A, B, C)$ and $\dim(V_2) + \dim(V_3) + \dim(V_4) = n - \dim(V_5) - \alpha(A, B, C)$ we obtain the desired conclusion.

**Lemma 2.** *The median candidate $M_Z$ from lemma 1 attains the lower bound if and only if $\dim(V_Z^A) = \dim(V_Z^B) = \dim(V_Z^C) = \delta(A, B, C)$.*

*Proof.* We start by considering Eq. (6) in the proof of lemma 1, since the other two are analogous. By the necessary conditions for optimality in Eq. (1),

$$d(M_Z, A) = \beta(A, B, C) - d(B, C) = \beta(A, B, C) - (n - c(\sigma^{-1}\tau)). \tag{9}$$

On the other hand, we have $\dim(V_3) = c(\sigma^{-1}\tau) - \alpha(A, B, C)$ and $\dim(V_5) = 2\delta(A, B, C)$, so by combining Eq. (6) with Eq. (9) we obtain

$$\dim(V_Z^A) = \dim(V_3) + \dim(V_5) - d(M_Z, A)$$
$$= \beta(A, B, C) + \alpha(A, B, C) - n$$
$$= \delta(A, B, C).$$

For the sufficiency, it is enough to check that when all three spaces have this dimension, then $s(M_Z) = \beta(A, B, C)$, which follows immediately from lemma 1.

## 2.2   Symmetry of $M_I$

We first define a new term that we call an $M$-stable subspace; this is closely related to the notion of an $M$-invariant subspace [11], which is a subspace $V$ such that $MV \subseteq V$, but with the additional specification that the dimensions are preserved. More specifically, we propose the following

**Definition 3.** *Let $M$ be an invertible $n \times n$ matrix and let $V$ be a subspace of $\mathbb{R}^n$. Then $V$ is an $M$-stable subspace if and only if $MV = V$.*

We have the following properties that we prove in the Appendix:

**Theorem 2.** *Let $M$ and $N$ be invertible matrices. Then*

a. *If $V, W$ are two $M$-stable subspaces, then so are $V \cap W$ and $V + W$.*
b. *If $M$ is symmetric and $V$ is an $M$-stable subspace, then so is $V^\perp$.*
c. *If $M^2 = I = N^2$ then the subspace $\{x | Mx = Nx\}$ is $M$-stable and $N$-stable.*

An easy but useful consequence of this theorem is the following

**Lemma 3.** *Let $A, B, C$ be involutions. Then the subspace $V_1$ is $A$-stable, $B$-stable and $C$-stable; the subspace $V_2$ is $A$-stable and $B$-stable; the subspace $V_3$ is $B$-stable and $C$-stable; and the subspace $V_4$ is $A$-stable and $C$-stable.*

*Proof.* We begin by showing that $V_1$ is $A$-stable. Indeed, $V_1 = \{x | Ax = Bx = Cx\} = \{x | Ax = Bx\} \cap \{x | Ax = Cx\}$ is the intersection of two subspaces, each of which is $A$-stable by part c of Theorem 2, and therefore is itself $A$-stable by part a. The fact that it is also $B$-stable and $C$-stable follows by symmetry.

Similarly, $V_2 = \{x | Ax = Bx\} \cap V_1^\perp$ is the intersection of two subspaces that are $A$-stable by parts c and b of Theorem 2, respectively, and so is $A$-stable itself by part a. By symmetry, $V_2$ is also $B$-stable, and the same reasoning applied to $V_3$ and $V_4$ shows that they are stable for the two involutions defining them.

**Theorem 3.** *$M_I$ is always symmetric for involutions $A$, $B$ and $C$.*

*Proof.* To prove the symmetry of an $n \times n$ matrix $M$, it is sufficient to show that

$$x^T My = y^T Mx \; \forall \; x, y \in \mathbb{R}^n \tag{10}$$

By linearity, it is enough to show this for a set of basis vectors of $\mathbb{R}^n$. We choose the basis of $\mathbb{R}^n$ to be the union of the bases for the subspaces $V_i$ for $i = 1$ to $i = 5$. Now Lemma 3 shows that for any of these subspaces, $x \in V_i$ implies $M_I x \in V_i$. Indeed, this is clear for $i = 1$ to $i = 4$, since the corresponding vector gets projected into its own subspace $V_i$ and then acted on by an involution that fixes $V_i$. This is also clear for $i = 5$ since any vector in $V_5$ is fixed by $M_I$.

Suppose first that $x, y$ be two vectors from different subspaces, say $x \in V_i, y \in V_j$, with $i < j$ without loss of generality; then we have three cases to consider.

Case (A) $i = 1$ and $j \in \{2, 3, 4, 5\}$; since $V_1$ and $V_j$ are mutually orthogonal, we have $x^T M_I y = 0 = y^T M_I x$, since $M_I x \in V_1$ and $M_I y \in V_j$ by the result above.

Case (B) $i \in \{2, 3, 4\}$ and $j = 5$; since $V_i$ and $V_5$ are mutually orthogonal, we have $x^T M_I y = 0 = y^T M_I x$, since $M_I x \in V_i$ and $M_I y \in V_5$ by the result above.

Case (C) $i \in \{2, 3, 4\}$ and $j \in \{2, 3, 4\} - \{i\}$; we consider the case $i = 2$ and $j = 3$, as the others follow by symmetry. Since $M_I = B$ on both $V_2$ as well as $V_3$,

$$x^T(M_I y) = x^T(By) = x^T B^T y = (Bx)^T y = \langle Bx, y \rangle = y^T(Bx) = y^T(M_I x).$$

Now, suppose that $x, y$ are two vectors from the same subspace, say $x, y \in V_i$. In this case, the matrix $M_I$ acts on $V_i$ via a symmetric matrix, and the same argument as in the previous equation shows equality, proving the desired result.

## 2.3   Orthogonality of $M_I$

**Theorem 4.** $M_I$ *is always orthogonal for involutions* $A$, $B$, *and* $C$.

The proof proceeds along very similar lines to the proof that $M_I$ is symmetric, and is provided in the Appendix.

## 2.4   Optimality of $M_I$

To show the optimality of $M_I$, it suffices to show that $\dim(V_I^C) \geq \delta(A, B, C)$, since symmetry implies that the same holds for $\dim(V_I^A)$ and $\dim(V_I^B)$, and then Lemma 1 shows that $M_I$ is a median because it achieves the lower bound.

Recall that the definition of $V_I^C$ asks for vectors $x + y$ such that $x$ is in $V_2$, $y$ is in $V_5$, and $C(x + y) = Ax + y$, or $(C - A)x + (C - I)y = 0$. The main idea is to show that it is enough to restrict ourselves to vectors $x$ such that $(A - I)x = 0$, meaning that the equation simply becomes $(C - I)(x + y) = 0$. The full details are provided in the Appendix.

## 2.5   Conservation of Common Adjacencies and Telomeres

We say that an adjacency $i, j$ is *present* in a matrix $M$ if $M_{ij} = 1 = M_{ji}$, $M_{kj} = 0 = M_{jk}$ for any $k \neq i$, and $M_{ik} = 0 = M_{ki}$ for any $k \neq j$. Similarly, we say that a telomere $i$ is *present* in a matrix $M$ if $M_{ii} = 1$ and $M_{ik} = 0 = M_{ki}$ for any $k \neq i$. In other words, the association of $i$ to $j$ (for an adjacency) or to $i$ (for a telomere) is unambiguous according to $M$. We now show that any adjacencies or telomeres common to 2 of 3 input genomes are present in any orthogonal median of three genomes, including $M_I$.

**Theorem 5.** *Let* $A, B, C$ *be three genomic matrices with median* $M$. *If* $A_{ij} = 1 = B_{ij}$ *for some* $i, j$, *then* $M_{ij} = 1 = M_{ji}$, $M_{kj} = 0 \; \forall \; k \neq i$, *and* $M_{ki} = 0 \; \forall \; k \neq j$.

*Proof.* By optimality of $M_I$ shown in the previous section, any median $M$ of three genomes attains the lower bound $\beta(A, B, C)$ on the score. Hence, by Eq. (1) it must satisfy $d(A, M) + d(M, B) = d(A, B)$. By corollary 1 in [1] it follows that for any vector $x$ with $Ax = Bx$, we also have $Mx = Ax$. We have two cases:

Case (A) $i = j$; then, taking $x = e_i$, the $i$-th standard basis vector, we get that $Ax = Bx = x$, so $Mx = x$ as well. It follows that the $i$-th column of $M$ is $e_i$, so that $M_{ij} = M_{ii} = M_{ji} = 1$ and $M_{kj} = M_{ki} = 0 \ \forall \ k \neq i$, as required.

Case (B) $i \neq j$; then taking $x = e_i + e_j$ and $y = e_i - e_j$, we get that $Ax = Bx = x$ and $Ay = By = -y$, so that $Mx = x$ and $My = -y$ as well. By linearity, we take the half-sum and half-difference of these equations to get $Me_i = e_j$ and $Me_j = e_i$. The first of these implies that $M_{ij} = 1$ and $M_{kj} = 0 \ \forall \ k \neq i$, while the second one implies that $M_{ji} = 1$ and $M_{ki} = 0 \ \forall \ k \neq j$, as required.

**Corollary 1.** *If $M$ is an orthogonal median of genomic matrices $A, B, C$, and $A_{ij} = 1 = B_{ij}$ for some pair $i, j$, then $M_{jk} = 0 \ \forall \ k \neq i$. In particular, any adjacency or telomere common to 2 out of 3 input genomes is present in $M_I$.*

*Proof.* The first statement follows immediately from Theorem 5 and orthogonality. The second statement is clear for telomeres, and follows for adjacencies since an adjacency $i, j$ is common to $A$ and $B$ if and only if $A_{ij} = B_{ij} = 1 = B_{ji} = A_{ji}$.

## 2.6   Computation of $M_I$

In order to compute $M_I$ we need the projection matrices $P_j$, which require a basis matrix $B_j$ for each of the spaces $V_j$, for $1 \leq j \leq 5$, as well as a nullspace matrix $N_j$ for $2 \leq j \leq 4$ [6]. However, it turns out that we can dispense with the nullspace matrices altogether and bypass the computation of $B_5$, which tends to be complicated, by using column-wise matrix concatenation $[\cdot, \cdot]$ and the following formula:

$$M_I = I + ([AB_1, AB_2, BB_3, CB_4] - B_{14})(B_{14}^T B_{14})^{-1} B_{14}^T, \qquad (11)$$

where $B_{14} := [B_1, B_2, B_3, B_4]$.

To verify this equation, it suffices to check that the right-hand side agrees with $M_I$ on the basis vectors of each subspace $V_j$, for $1 \leq j \leq 5$. This is clear for $V_5$ since $B_{14}^T x = 0 \ \forall \ x \in V_5$, and is also true for the basis vectors of $V_j$ for $1 \leq j \leq 4$ since Eq. (11) implies that $M_I B_{14} = [AB_1, AB_2, BB_3, CB_4]$.

It is easy to compute a basis $B_1$ for the triple agreement space $V_1$. Indeed, we note that, by Eq. (4),

$$x \in V_1 \iff Ax = Bx = Cx$$
$$\iff x \text{ is constant on the cycles of } \rho^{-1}\sigma \text{ and } \sigma^{-1}\tau,$$

where $\rho, \sigma, \tau$ are the permutations corresponding to $A, B, C$, respectively. The computation of $\rho^{-1}\sigma$ and $\sigma^{-1}\tau$ takes $O(n)$ time, and $V_1$ is spanned by the indicator vectors of the weakly connected components of the union of their graph representations (the graph representation of a permutation $\pi \in S_n$ has a vertex for each $i$ for $1 \le i \le n$, and a directed edge from $i$ to $\pi(i)$ for each $i$). Note that the basis vectors in $B_1$ are orthogonal because their supports are disjoint. We refer to this basis as the *standard basis* of $V_1$.

Likewise, by Eq. (4), a basis $B_2$ for the space $V_2$ can be computed by determining the cycles of $\rho^{-1}\sigma$ and subtracting the orthogonal projection onto the $\alpha(A, B, C)$ standard basis vectors of $B_1$ from the indicator vector $\chi(C)$ of each cycle $C$. We refer to the resulting basis as the *standard basis* of $V_2$.

The same construction can be applied to $B_3$ and $B_4$, and the overall computation of $B_1$ through $B_4$ takes $O(n^2)$ time. Thus, the most time-consuming step is inverting $B_{14}^T B_{14}$ in (11), which requires $O(n^\omega)$ time, or $O(n^3)$ in practice.

In our running example, with $A' = (13)(28)(57), B' = (16)(27)(34)(58), C' = (12)(34)(56)(78)$, using the notation $e_i$ for the $i$th standard basis and $e$ for the vector of all 1's, we end up with the bases $B_1 = \{e\}, B_2 = \{e_2 + e_5 - e/4, e_7 + e_8 - e/4\}, B_3 = \{e_1 + e_5 + e_7 - 3e/8, e_3 - e/8, e_4 - e/8\}, B_4 = \{0\}$, so by (11),

$$
M_I = \frac{1}{6}
\begin{pmatrix}
4 & 2 & 0 & 0 & -2 & 2 & -2 & 2 \\
2 & 1 & 0 & 0 & -1 & -2 & 5 & 1 \\
0 & 0 & 0 & 6 & 0 & 0 & 0 & 0 \\
0 & 0 & 6 & 0 & 0 & 0 & 0 & 0 \\
-2 & -1 & 0 & 0 & 1 & 2 & 1 & 5 \\
2 & -2 & 0 & 0 & 2 & 4 & 2 & -2 \\
-2 & 5 & 0 & 0 & 1 & 2 & 1 & -1 \\
2 & 1 & 0 & 0 & 5 & -2 & -1 & 1
\end{pmatrix}.
$$

$M_I$ it is both symmetric, in agreement with Theorem 3, and orthogonal, in agreement with Theorem 4, although it is certainly not genomic. Furthermore, it contains the adjacency (34) common to $B'$ and $C'$, in agreement with Corollary 1. The process of turning it into a genome is the subject of the following section.

## 3   From Matrices Back to Genomes

In this section we describe the two heuristics for extracting back a genome from a symmetric median, in cases when this median is not itself a genomic matrix. The first one is an improvement of the one proposed by Pereira Zanetti et al. [6], while the second one is a brute-force approach only applicable in certain cases.

### 3.1   The First Heuristic: Maximum-Weight Matching

Let $M$ be a symmetric median to be transformed back into a genome. Since a genome can also be seen as a matching on the extremities of the genes involved, we can construct a weighted graph $H$ with a weight of $|M_{ij}| + |M_{ji}| = 2|M_{ij}|$ on the edge from $i$ to $j$, provided this weight exceeds $\epsilon = 10^{-6}$, a bound introduced to avoid numerically insignificant values. We modify this by also adding self-loops to $H$ with weight $|M_{ii}|$, so that those extremities $i$ with a high value of

$|M_{ii}|$ can be encouraged to form a telomere. We then extract a maximum-weight matching of $H$ by using an implementation of the Blossom algorithm [12]. More specifically, we used the NetworkX package [15] in Python [14], which in turn is based on a detailed paper by Galil [13]. This implementation runs in $O(mn \log n)$ time for a graph with $n$ nodes and $m$ edges, or in $O(n^3)$ time for dense graphs.

In our running example, the maximum-weight matching is obvious by inspection (in fact, the greedy algorithm yields the optimum matching), and is $M = (34)(27)(58)$. Unfortunately, its score, 10, exceeds the lower bound $\beta = 8$.

## 3.2   The Second Heuristic: The Closest Genome by Rank Distance

Let $R$ be the set of rows of a symmetric, orthogonal median $M$ that contain at least one non-integer entry; by symmetry, this is the same as the set of columns that contain at least one non-integer entry. Note that $M$ cannot contain a $-1$ value since otherwise, we would have the rest of the row equal to 0 by orthogonality, and its sum would then be -1 instead of 1 (as it must be in order to satisfy the lower bound: $A\mathbf{1} = B\mathbf{1} = 1$, so $M\mathbf{1} = 1$ as well, by corollary 1 in [1]). Hence, $M$ must be binary outside of the rows and columns indexed by $R$.

We consider the matrix $M^R := M[R, R]$, i.e. the square submatrix of $M$ with rows and columns indexed by $R$. We would like to find the genomic matrix $G$ closest to $M^R$ in rank distance and replace $M^R$ with $G$ to obtain a candidate genome (since the rest of $M$ contains only integers, and $M$ is symmetric, the closest genome to all of $M$ will agreed with $M$ there).

We create an auxiliary graph $H$ with a node for each element of $R$ and an undirected edge between $i$ and $j$ if and only if $M_{ij}^R \neq 0$. Let $C_1, \ldots, C_k$ denote the connected components of $H$. Our heuristic consists in restricting the search to block-diagonal genomes with blocks determined by $C_1, \ldots, C_k$. This can be done in an exhaustive manner if each block has size at most $n = 10$, in which case there are only $9,496$ genomes to check. This can be done reasonably fast, in fact, under a second on a modern laptop running R [17]; larger sizes, such as $n = 12$ with over $140,000$ genomes to check, already take substantially longer.

In our running example, we take $R = [1, 2, 5, 6, 7, 8]$. There is a single block. We compute that, out of the 76 possible genomes with $n = 6$, only one is at rank distance 1 from $M^R$, namely, $M = (14)(25)(36)$, which, after renumbering it according to $R$ and adding back the adjacency $(34)$, gives us $(16)(27)(34)(58)$, which happens to be $B'$. It gets a score of 9 with the reduced inputs $A', B', C'$. Although this still exceeds the lower bound $\beta = 8$, an exhaustive check reveals that $M$ is one of the three best-scoring genomes, the other two being $M' = (16)(28)(34)(57)$ and $M'' = (16)(25)(34)(78)$. Thus, in this example our second heuristic works better than the first one and, in fact, finds a genomic median.

## 4   Experiments

We tested our algorithm $\mathcal{A}$, as well as the two heuristics described in the previous section, on simulated and real data. For our simulations, we started from a random genome with $n$ genes, for $n$ varying from 12 to 1000, and applied $rn$ random rearrangement operations to obtain the three input genomes, with $r$ ranging from 0.05 to 0.3, and the rearrangement operations were chosen to be either SCJ (single cut-or-join) [4] or DCJ (double cut-and-join) [16] operations. In both cases the operations are chosen uniformly at random among the possible ones, as described in previous work [6]. For each combination of $n$ and $r$ we generated 10 samples, for a total of 600 samples for each of SCJ and DCJ.

For the real data, we selected a dataset containing 13 plants from the *Campanulaceæ* family, with the gene order for $n = 210$ gene extremities (i.e. 105 genes) each, and created all possible triples for a total of 286 inputs. We present a summary of our results in the next subsections.

## 4.1 Results on the SCJ Samples

Perhaps because the SCJ rearrangements involve smaller rank distances, the SCJ samples turned out to be particularly easy to process. It turned out that all but 19 (or $\approx 3\%$) of them actually had $\delta = 0$, and all but 5 (or $\approx 1\%$) of them had a median $M_I$ that was genomic. Of these 5 cases, 4 had a submatrix $M^R$ of size $n = 4$ with all the entries equal to $\pm\frac{1}{2}$, and one had a submatrix $M^R$ of size $n = 6$ with $\frac{2}{3}$ in each diagonal entry and $\pm\frac{1}{3}$ in each off-diagonal entry.

For those 5 inputs, both the maximum matching as well as the closest genome heuristics resulted in the same conclusion, namely, that all possible genomes had the exact same distance from $M^R$, equal to the size of $R$ (i.e. the maximum possible rank), and all matchings had the same score. Nevertheless, the solution produced by the maximum matching heuristic (picked arbitrarily among many possible matchings), namely, the one in which every element of $R$ was a telomere, always scored $\beta + 1$ with the original inputs, which was the best possible score among all genomes in every case.

## 4.2 Results on the DCJ Samples

The situation was more complex with the DCJ samples, as 424 out of 600 samples, or more than 70%, had $\delta > 0$, and for 337 out of 600, or more than 56%, $M_I$ had some fractional entries. Unsurprisingly, there was an increasing trend for the proportion of medians $M_I$ with fractional entries as a function of both $n$ and $r$. The matching heuristic did not produce very good results, with the score of the resulting genome exceeding the lower bound $\beta$ by a value in the range from 1 to 173, with a mean of 19.

The submatrices $M^R$ varied in size from 4 to 354, with a mean size of 64. Nevertheless, over 40% all the fractional cases (135 out of 337) had the largest connected component of size at most 10, so the closest genome heuristic was applicable to them. For those that it was applicable to, the closest genome heuristic produced relatively good results, with the score of the resulting genome exceeding the lower bound $\beta$ by a value in the range from 0 to 21, including one exact match, with a mean of just under 3. It appears that the closest genome heuristic generally exhibits a better performance than the maximum matching heuristic, but is applicable in a smaller number of cases.

### 4.3 Results on the *Campanulaceæ* Dataset

We construct all 286 possible distinct triples of the 13 genomes on $n = 210$ extremities present in our dataset. Out of these, 189 (or 66%) have $\delta = 0$ and 165 (or 58%) have a genomic median $M_I$. For the remaining ones we apply the two heuristics to determine the best one in terms of the score.

The matching heuristic produced reasonable results this time, with deviations from $\beta$ ranging from 1 to 12, and a mean of just over 4. The submatrices $M^R$ varied in size from 4 to 22, with a mean size of 9. Nearly two-thirds of them (79/121) had the largest connected component of size at most 10, so the closest genome heuristic was applicable to them. Among those, the deviations from $\beta$ ranged from 1 to 4, with a mean of just over 2. Once again, the closest genome heuristic performed better, but was applicable to a smaller number of cases.

### 4.4 Running Times

The average running time for DCJ samples with $\delta > 0$ of size 100, 300 and 1000, respectively was 0.04, 0.07 and 0.45 s, suggesting a slightly sub-cubic running time; indeed, the best-fitting power law function of the form $f(x) = ax^b$ had $b \approx 2.97$. Both post-processing heuristics were similarly fast to apply, taking an average of 0.5 s for the closest genome and 0.7 s for the maximum matching per instance of the largest size, $n = 1000$. The computations were even faster for SCJ samples and real data. By extrapolating these running times, we expect that even much larger instances, with, $n \approx 10^4$, would still run in minutes. We performed all our experiments in the R computing language [17] on a single Mac laptop with a 2.8 GHz Intel Core i7 processor and 16 GB of memory.

## 5   Conclusions

In this work we presented the first polynomial-time exact solution of the median-of-three problem for genomes under the rank distance. Although the resulting median is only guaranteed to be symmetric and orthogonal, not binary, we observed that it frequently happens to be binary (i.e. genomic) with both simulated and real data. For the cases when it is not, we presented two effective heuristics for trying to find the genome closest to the median, and showed that they tend to produce good results in practice.

Despite this important step forward, the fundamental problem of finding the genomic median of three genomic matrices, or, more generally, the permutation median of three permutation matrices, remains open. The additional question of discovering a faster algorithm for the generalized rank median of three genomes (i.e. when there are no restrictions on it being binary) is also open - we conjecture that it is possible to do it in $O(n^2)$.

In future work, we plan to explore the relationships between the rank distance and other well-studied genome rearrangement distances such as the breakpoint distance, DCJ, and SCJ. In addition, we intend to test the suitability of the rank distance for phylogenetic inference, ancestral genome reconstruction, and orthology assignment. Lastly, it would be very interesting to establish the computational complexity of finding the genomic rank median of three genomes.

**Acknowledgments.** The authors would like to thank Cedric Chauve, Pedro Feijão, Yann Ponty, and David Sankoff for helpful discussions. LC would like to acknowledge financial support from NSERC, CIHR, Genome Canada and the Sloan Foundation. JM would like to acknowledge financial support from FAPESP (Fundacão de Amparo a Pesquisa do Estado de São Paulo), grant 2016/01511-7.

# Appendix

## All the Medians for a Small Example

Let $n = 3$ and take the only triplet of distinct genomes for which $\delta(A, B, C) > 0$, namely, $A = (12), B = (13), C = (23)$, i.e. each of the genomes contains a single adjacency as well as a telomere. Note that we identify a permutation with its corresponding matrix in this section. It is easy to see that the identity $I$ is a rank median, as are the two 3-cycles $K = (123)$ and $L = (132)$. Using the Maple software [18] we found that all the medians can be written as a subset of the linear combinations of these three "basic" solutions. More precisely,

$$\mathcal{M} := \{aI + bK + cL \,|\, a + b + c = 1 = a^2 + b^2 + c^2\}$$

is the exact description of all the rank medians. It is easy to see (from the properties of the corresponding permutations) that

$$I^2 = I^T = I, \ L^2 = L^T = K, \ K^2 = K^T = L, \ KL = LK = I. \tag{12}$$

Now let $J := \frac{1}{3}[1,1,1]^T[1,1,1]$ be the normalized matrix of all 1's and let $N := I - J$. Note that $J^2 = J^T = J$ and $N^2 = N^T = N$; in fact, these two matrices are complementary orthogonal projections. Then it is easy to check, using Eq. (12), that the set of medians $\mathcal{M}$ is a subset of the matrices satisfying the equation

$$(M - J)(M - J)^T = N,$$

which is indeed the equation of a circle in matrix space. Note, however, that this equation is also satisfied by non-median matrices, including all those in the set

$$\mathcal{N} := \{aA + bB + cC \,|\, a + b + c = 1 = a^2 + b^2 + c^2\}.$$

## Proof of Theorem 2

*Proof.* Note that, because of the invertibility of $M$, to prove that $V$ is $M$-stable it is sufficient to show that $MV \subseteq V$.

a. If $V, W$ are two $M$-stable subspaces, let $u \in V \cap W$. Then $u \in V$ and $u \in W$, so $Mu \in V$ and $Mu \in W$, and therefore $Mu \in V \cap W$. Hence $M(V \cap W) \subseteq V \cap W$, and $V \cap W$ is $M$-stable.
Similarly, let $u \in V + W$. Then $u = v + w$ with $v \in V, w \in W$, so $Mu = Mv + Mw \in V + W$, so $M(V + W) \subseteq V + W$, and $V + W$ is $M$-stable.

b. Suppose $M$ is symmetric and $V$ is an $M$-stable subspace. Let $u \in V^\perp$, so that $u^T v = 0$ for any $v \in V$. Let $w = Mv$; by hypothesis, $w \in V$, so that

$$(Mu)^T v = u^T M^T v = u^T M v = u^T w = 0$$

since $w \in V$. However, $v \in V$ was chosen arbitrarily, and therefore $Mu \in V^\perp \ \forall \ u \in V^\perp$, meaning that $MV^\perp \subseteq V^\perp$, and $V^\perp$ is indeed $M$-stable.

c. If $M^2 = I = N^2$, let $x$ be such that $Mx = Nx$. Then

$$M(Mx) = Ix = x = N(Nx) = N(Mx),$$

so that $Mx$ is also in the desired subspace $\{x | Mx = Nx\}$, meaning that it is $M$-stable. By symmetry, it is also $N$-stable, completing the proof.

## Proof that $M_I$ is Orthogonal for Genomes $A, B, C$

*Proof.* First, we recall that a matrix $M$ is orthogonal if and only if

$$(Mx)^T(My) = x^T y \ \forall \ x, y \in \mathbb{R}^n. \tag{13}$$

Second, it is sufficient to prove that Eq. (13) holds for any pair of vectors in a basis $\mathbb{B} = \{v_1, \ldots, v_n\}$ of $\mathbb{R}^n$. We take $\mathbb{B}$ to be the union of the bases for the subspaces $V_i$ for $i = 1$ to $i = 5$, and consider different cases, once again using the fact that $M_I$ maps vectors in each $V_i$ into other vectors in $V_i$, which follows from Lemma 3 and the fact that $M_I$ fixes each vector in $V_5$. If $x \in V_i, y \in V_j$ with $i \neq j$, without loss of generality $i < j$, then there are three cases to consider.

Case (A) $i = 1$ and $j \in \{2, 3, 4, 5\}$; since $V_1$ and $V_j$ are mutually orthogonal, we have $(M_I x)^T(M_I y) = 0 = x^T y$, since $M_I x \in V_1$ and $M_I y \in V_j$.

Case (B) $i \in \{2, 3, 4\}$ and $j = 5$; since $V_i$ and $V_5$ are mutually orthogonal, we have $(M_I x)^T(M_I y) = 0 = x^T y$, since $M_I x \in V_i$ and $M_I y \in V_5$.

Case (C) $i \in \{2, 3, 4\}$ and $j \in \{2, 3, 4\} - \{i\}$; we consider the case $i = 2$ and $j = 3$, as the others follow by symmetry. Since $M_I = B$ on both $V_2$ as well as $V_3$

$$(M_I x)^T(M_I y) = (Bx)^T(By) = x^T B^T B y = x^T I y = x^T y.$$

Now, suppose that $x, y$ are two vectors from the same subspace, say $x, y \in V_i$. In this case, the matrix $M_I$ acts on $V_i$ via an orthogonal matrix, and the same argument as in the previous equation shows equality, proving the desired result.

## Proof that $M_I$ is a Median for Genomes $A, B, C$

We begin with the following three lemmas, which will be useful in the proof.

**Lemma 4.** *If $V$ is a vector subspace of $\mathbb{R}^n$ of dimension $k$ and $M$ is a square matrix of size $n$, then $MV := \{Mx | x \in V\}$ is a vector subspace of $\mathbb{R}^n$ of dimension $k - d$, where $d := \dim(\ker(M) \cap V)$. Furthermore, for any two subspaces $V$ and $W$ of $\mathbb{R}^n$ and $M$ a square matrix of size $n$, $M(V + W) = MV + MW$.*

*Proof.* The first part of the statement, the fact that $MV$ is a vector subspace of $\mathbb{R}^n$, is true because

$$\alpha_1 M(v_1) + \alpha_2 M(v_2) = M(\alpha_1 v_1 + \alpha_2 v_2)$$

for any scalars $\alpha_1$ and $\alpha_2$ in $\mathbb{R}$ and vectors $v_1$ and $v_2$ in $V$.

The second part can be proven as follows. Let $v_1, \ldots, v_d$ be a basis of $\ker(M) \cap V$, and let us extend it to a basis of $V$ by adding the vectors $v_{d+1}, \ldots, v_k$. Clearly, $Mv_i = 0$ for each $1 \le i \le d$, since $Mx = 0$ for any $x \in \ker(M)$. Furthermore, the $Mv_j$ for $d + 1 \le j \le k$ are linearly independent since

$$\sum_{j>d} \alpha_j M v_j = M\left(\sum_{j>d} \alpha_j v_j\right) = 0 \iff \sum_{j>d} \alpha_j v_j \in \ker(M) \cap V \iff \alpha_j = 0 \,\forall\, j,$$

where the last conclusion follows from the linear independence of the basis vectors $v_1, \ldots, v_k$ and the fact that the first $d$ of those form a basis of $\ker(M) \cap V$. Therefore, the space $MV$ is spanned by $\{Mv_j\}_{j=d+1}^{j=k}$, and its dimension is $k - d$.

For the last part, we note that

$$x \in M(V + W) \iff \exists v \in V, w \in W \text{ with } x = M(v + w) \iff$$
$$\iff \exists v \in V, w \in W \text{ with } x = Mv + Mw \iff x \in MV + MW.$$

**Lemma 5.** *$A$ is an involution on the standard basis $B_1$ of $V_1$ for genomes $A, B, C$.*

*Proof.* Consider the graph $G$ containing the union of the graph representations of the permutations $AB$ and $CA$. The standard basis $B_1$ of $V_1$ contains the indicator vectors of the connected components of $G$. We will show that these basis vectors are either fixed or interchanged in pairs by $A$.

By Lemma 3, $AV_1 = V_1$. Now let $C_t$ be a component of $G$, and let $\chi(C_t)$ be its indicator vector. since $\chi(C_t) \in V_1$, the same is true of $\chi(AC_t) := A\chi(C_t)$ by the $A$-stability of $V_1$. However, since $A$ is a permutation, $\chi(AC_t)$ is a vector with $|C_t|$ entries equal to 1 and $n - |C_t|$ entries equal to 0. It follows that $AC_t$, the image of the elements of $C_t$ under $A$, is a disjoint union of components of $G$.

Now we show that this disjoint union in fact contains a single component of $G$. Indeed, note that the $A$-stability of $V_1$ means that

$$(x_i = x_j \,\forall\, x \in V_1) \iff (x_{\rho(i)} = (Ax)_i = (Ax)_j = x_{\rho(j)} \,\forall\, x \in V_1). \tag{14}$$

This shows that whenever $i, j$ belong to the same component of $G$, then so do $\rho(i), \rho(j)$. Therefore, $AC_t$ must be a single component of $G$ for any $t$, and $A$ permutes the set of components of $G$ by its action, so it is an involution on $B_1$.

**Lemma 6.** *A is an involution on the standard basis $B_2$ of $V_2$ for genomes $A, B, C$.*

*Proof.* Consider the cycles of the permutation $AB$. The standard basis vectors of $V_2$ are the indicator vectors of these cycles, from which we subtract the orthogonal projections onto each of the vectors in $V_1$. We will show that these basis vectors are either fixed or interchanged in pairs by $A$, meaning that $A$ is indeed an involution on them.

By Lemma 3, $AV_2 = V_2$. Now let $C_t$ be a cycle of $AB$, and let $\chi(C_t)$ be its indicator vector; the corresponding basis vector of $B_2$ will be given by

$$v := \chi(C_t) - \sum_{i=1}^{\alpha} \frac{|C_t \cap C_i|}{|C_i|} \chi(C_i), \tag{15}$$

where the $C_i$ are the components of the graph $G$ defining $V_1$. It follows that $Av$ is given by

$$Av = \chi(AC_t) - \sum_{i=1}^{\alpha} \frac{|C_t \cap C_i|}{|C_i|} \chi(AC_i).$$

From the proof of Lemma 5, we have $|AC_i| = |C_i| \; \forall \, i$. Furthermore, we have

$$|AC_t \cap AC_i| = |A(C_t \cap C_i)| = |C_t \cap C_i|,$$

since $A$ is a permutation. It finally follows that

$$Av = \chi(AC_t) - \sum_{i=1}^{\alpha} \frac{|AC_t \cap AC_i|}{|AC_i|} \chi(AC_i) = \chi(AC_t) - \sum_{j=1}^{\alpha} \frac{|AC_t \cap C_j|}{|C_j|} \chi(C_j) \tag{16}$$

where the second equality follows from the fact, shown in the proof of Lemma 5, that $A$ permutes the standard basis $B_1$ of $V_1$. Also analogously to the proof of Lemma 5 we can show that $AC_t$ is a single cycle of $AB$. Indeed, it suffices to consider Eq. (14) with $V_1$ replaced by $V_1 + V_2$, which is also $A$-stable.

By combining this fact with Eqs. (15) and (16) we see that the vector $Av$ is the basis vector of $B_2$ defined by the single cycle $AC_t$. In fact, $C_t$ and $AC_t$ are either both equally-sized parts of an even cycle in the graph union of the representations of $A$ and $B$, or coincide and correspond to a path in that graph.

**Corollary 2.** *Both $A$ and $B$ are involutions on the standard basis $B_2$ of $V_2$. Similarly, both $B$ and $C$ are involutions on the standard basis $B_3$ of $V_3$, and both $A$ and $C$ are involutions on the standard basis $B_4$ of $V_4$. These results also hold for the subspaces $\ker(A - B) = V_1 + V_2$ with basis $B_1 \cup B_2$, $\ker(B - C) = V_1 + V_3$ with basis $B_1 \cup B_3$, and $\ker(C - A) = V_1 + V_4$ with basis $B_1 \cup B_4$.*

We will need two additional definitions and three additional simple lemmas.

**Definition 4.** *Let $A$ be a permutation on $n$ elements. We denote by $f(A)$ the number of fixed points of $A$.*

**Lemma 7.** *Let $A$ be a permutation on $n$ elements, let $f(A)$ be as in Definition 4, and let $c(A)$ be the number of cycles of $A$. Then*

$$f(A) \geq 2c(A) - n,$$

*with equality if and only if $A$ is an involution.*

*Proof.* The cycles counted by $c(A)$ can be trivial (fixed points) or non-trivial (size at least 2). There are $c(A) - f(A)$ non-trivial cycles, and they involve $n - f(A)$ elements. It follows that

$$2(c(A) - f(A)) \leq n - f(A) \iff f(A) \geq 2c(A) - n,$$

with equality if and only if each non-trivial cycle has size exactly 2, i.e. $A$ is an involution.

**Definition 5.** *Let $A$ and $B$ be two involutions. Let $G(A, B)$ be the graph union of the representations of $A$ and $B$, which contains paths and even cycles. We define $p(AB)$ to be the number of paths in $G(A, B)$.*

**Lemma 8.** *Let $A$ and $B$ be two involutions. Then*

$$p(AB) = \frac{f(A) + f(B)}{2}.$$

*Proof.* Let $P$ be an arbitrary path in $G(A, B)$. Then the endpoints of $P$ are two fixed points, one at either end. Since all the fixed points of $A$ and $B$ form the endpoints of some path, the result follows.

**Lemma 9.** *Let $A, B, C$ be three involutions, and let $\ker(A - B) = V_1 + V_2$ have the basis $B_1 \cup B_2$. Then the number of pairs of distinct basis vectors of $B_1 \cup B_2$ that are exchanged by $A$ (or $B$) is precisely $\frac{c(AB) - p(AB)}{2}$.*

*Proof.* We start by showing that this number is independent of the chosen basis. Note that each pair of vectors $(v, w)$ that are exchanged by $A$ yield an eigenvalue 1 for $v + w$ and an eigenvalue of $-1$ for $v - w$, while any vector $u$ that is fixed by $A$ yields an eigenvalue 1. Thus, we can diagonalize $A$ with respect to any basis on which it is an involution, to get a number of $-1$ eigenvalues equal to the number of exchanged pairs. But the algebraic multiplicity of an eigenvalue is invariant under similarity (similar matrices have the same characteristic equation) [11], so this number, the number of exchanged pairs, is independent of the chosen basis.

Now consider the union graph $G(A, B)$. Each connected component in it is either a path or an even cycle. Each path creates a single cycle in the product $AB$ which is fixed by $A$ (and $B$). On the other hand, each even cycle splits into a pair of equal-sized cycles in the product $AB$, and those are exchanged by $A$ (or $B$). Therefore, if we use the basis of $\ker(A - B)$ consisting of the indicator vectors of the cycles of $AB$, the desired number of pairs is indeed $\frac{c(AB) - p(AB)}{2}$.

We are now ready to prove our main result. We begin by proving it for the case $\alpha = 1$, and then generalize it to arbitrary $\alpha$.

**Theorem 6.** *The matrix $M_I$ is a median of genomes $A, B, C$ if $\alpha(A, B, C) = 1$.*

Let us first define $V_{12}$ to be the restriction of $V_1 + V_2$ to those vectors which are fixed by $A$ (equivalently, $B$). In other words, let $V_{12} := (V_1 + V_2) \cap \ker(A - I)$.

We begin with the decomposition of $\mathbb{R}^n$ from Zanetti et al. [6], to which we apply $(C - I)$:

$$\mathbb{R}^n = V_1 + V_3 + V_1 + V_4 + V_1 + V_2 + V_5 \supseteq (V_1 + V_3) + (V_1 + V_4) + (V_{12} + V_5);$$
$$(C - I)\mathbb{R}^n \supseteq (C - I)(V_1 + V_3) + (C - I)(V_1 + V_4) + (C - I)(V_{12} + V_5). \quad (17)$$

We will show that the sum on the right-hand side of Eq. (17) is direct. We will then compute the dimension of each term to reach the desired conclusion.

First, we show that $(C - I)(V_1 + V_3)$ and $(C - I)(V_1 + V_4)$ are disjoint subspaces, so that the sum of the first two terms is direct.

**Lemma 10**
$$(C - I)(V_1 + V_3) \cap (C - I)(V_1 + V_4) = \{0\}.$$

*Proof.* We reason as follows.

$$x \in (C - I)(V_1 + V_3) \cap (C - I)(V_1 + V_4) \iff$$
$$\iff \exists v \in \ker(B - C), w \in \ker(C - A) \text{ s.t. } (C - I)v = x = (C - I)w \iff$$
$$\iff (B - I)v = x = (A - I)w.$$

Now, by Lemma 3, $(B - I)\ker(B - C) \subseteq B\ker(B - C) - \ker(B - C) \subseteq \ker(B - C)$ by the $B$-stability of $\ker(B - C)$, and similarly, $(A - I)\ker(C - A) \subseteq \ker(C - A)$ by the $A$-stability of $\ker(C - A)$. Since $x$ is in their intersection, we get $x \in V_1$.

However, since $\mathbf{1}^T x = \mathbf{1}^T (B - I)v = 0^T v = 0$, it follows that $x = 0$ because when $\alpha = 1$, $V_1$ is spanned by $\mathbf{1}$, meaning that the subspaces are indeed disjoint.

We now show that the addition of the third term in Eq. (17) keeps the sum direct.

By the same reasoning as in the proof of Lemma 10, we see that $C - I$ maps both $V_1 + V_3 = \ker(B - C)$ and $V_1 + V_4 = \ker(C - A)$ into themselves.

Since $V_1 + V_2 = \ker(A - B)$, we get

$$V_{12} \subseteq \ker(A - B) \cap \ker(A - I) = \ker(A - I) \cap \ker(B - I).$$

We will now show that $(C - I)V_{12} \subseteq \text{im}(C - A)$. Indeed, we have

$$y \in (C - I)V_{12} \implies y = (C - I)x, x \in \ker(A - I) \cap \ker(B - I)$$
$$\implies Ax = x = Bx$$
$$\implies y = Cx - x = CAx - AAx = (C - A)Ax \in \text{im}(C - A).$$

By the same reasoning, $(C - I)V_{12} \subseteq \text{im}(B - C)$.

Furthermore, we have $V_5 \subseteq \text{im}(B - C) \cap \text{im}(C - A)$, and both $\text{im}(B - C) = \ker(B - C)^\perp$ as well as $\text{im}(C - A) = \ker(C - A)^\perp$ are $C$-stable by parts b and c of Theorem 2, and their intersection is also $C$-stable by part a of this theorem. It follows that

$$(C - I)V_5 \subseteq CV_5 - V_5 \subseteq \text{im}(B - C) \cap \text{im}(C - A).$$

By combining this with the previous results on $(C - I)V_{12}$, we conclude that

$$(C - I)(V_{12} + V_5) \subseteq (C - I)V_{12} + (C - I)V_5 \subseteq \text{im}(B - C) \cap \text{im}(C - A).$$

Since $\text{im}(B - C) \cap \text{im}(C - A)$ is orthogonal to the sum of $V_1 + V_3$ and $V_1 + V_4$, which equals $\ker(B - C) + \ker(C - A)$, it follows *a fortiori* that $(C - I)(V_{12} + V_5)$ is disjoint from the sum of these subspaces, so the sum in Eq. (17) is direct.

We now consider the dimension of each of the terms in Eq. (17).

Since $C$ permutes the basis vectors of $V_1$, $V_3$ and $V_4$ by Lemmas 5 and 6, the dimension of $\ker(C - I) \cap (V_1 + V_3)$ equals the number of those basis vectors that $C$ maps into themselves, plus the number of pairs of basis vectors that get swapped by $C$. It follows by Lemmas 4 and 9 that

$$\dim((C - I)(V_1 + V_3)) = \dim(V_1 + V_3) - \dim(\ker(C - I) \cap (V_1 + V_3))$$
$$= c(BC) - p(BC) - \frac{c(BC) - p(BC)}{2} = \frac{c(BC) - p(BC)}{2}.$$

In the same way, we get

$$\dim((C - I)(V_1 + V_4)) = \frac{c(CA) - p(CA)}{2}.$$

Analogously, by using Lemmas 4 and 9 once again, we have

$$\dim(V_{12}) = \dim((V_1 + V_2) \cap \ker(A - I))$$
$$= \dim(V_1 + V_2) - \dim((A - I)(V_1 + V_2))$$
$$= n_2 + \alpha(A, B, C) - \frac{c(AB) - p(AB)}{2},$$

where $n_2 := \dim(V_2)$.

Lastly, by Lemma 4 the dimension of $\text{im}(C - I) = (C - I)\mathbb{R}^n$ equals

$$\dim(\mathbb{R}^n) - \dim(\ker(C - I) \cap \mathbb{R}^n) = n - c(C).$$

From the directness of the sum in the second part of Eq. (17), we have

$$n - c(C) \geq \dim((C - I)(V_{12} + V_5)) + \dim((C - I)(V_1 + V_3)) + \dim((C - I)(V_1 + V_4))$$
$$= \dim((C - I)(V_{12} + V_5)) + \frac{c(BC) - p(BC)}{2} + \frac{c(CA) - p(CA)}{2} \implies$$
$$\implies \dim((C - I)(V_{12} + V_5)) \leq n - c(C) - \frac{c(BC) - p(BC)}{2} - \frac{c(CA) - p(CA)}{2}.$$

By using Lemmas 7 and 8, the definition of $n_2$, and the invariants $\alpha(A, B, C)$, $\beta(A, B, C)$, and $\delta(A, B, C)$ we can rewrite the right-hand side above to obtain

$$\dim((C - I)(V_{12} + V_5)) \le n - c(C) - \frac{c(BC) + c(CA)}{2} + \frac{p(BC) + p(CA)}{2}$$

$$= n - c(C) - \frac{c(AB) + c(BC) + c(CA)}{2} + \frac{c(AB)}{2} + \frac{f(A) + f(B)}{4} + \frac{2c(C) - n}{2}$$

$$= \frac{n + c(AB)}{2} - \frac{3n - 2\beta(A, B, C)}{2} + \frac{f(A) + f(B)}{4}$$

$$= \beta(A, B, C) - n + \frac{c(AB)}{2} + \frac{p(AB)}{2} = c(AB) - \frac{c(AB) - p(AB)}{2} + \beta(A, B, C) - n$$

$$= n_2 + \alpha(A, B, C) + \beta(A, B, C) - n - \frac{c(AB) - p(AB)}{2} = n_2 + \delta - \frac{c(AB) - p(AB)}{2}.$$

And now we use Lemma 4 and the fact that $\dim(V_5) = 2\delta$ to obtain

$$\dim(V_I^C) = \dim(\{x + y | x \in V_2, y \in V_5, C(x + y) = Ax + y\})$$

$$\ge \dim(\{x + y | x \in V_{12}, y \in V_5, C(x + y) = Ax + y\}) - 1$$

$$= \dim(\{x + y | x \in V_{12}, y \in V_5, C(x + y) = x + y\}) - 1$$

$$= \dim(\ker(C - I) \cap (V_{12} + V_5)) - 1 = \dim(V_{12} + V_5) - \dim((C - I)(V_{12} + V_5)) - 1$$

$$\ge n_2 + \alpha(A, B, C) - \frac{c(AB) - p(AB)}{2} + 2\delta - \left(n_2 + \delta - \frac{c(AB) - p(AB)}{2}\right) - 1 = \delta.$$

Therefore, all the intermediate inequalities are equalities as well. This proves that $M_I$ is always a median for three involutions provided $\alpha(A, B, C) = 1$. Note that we subtract 1 in the first step above to account for the fact that any multiple of the vector $\mathbf{1}$ can be added to any solution of the set of equations defining $V_I^C$.

### Proof that $M_I$ is a Median for General $\alpha$

This time we use a slightly different decomposition of $\mathbb{R}^n$ because the intersection of $(C - I)(V_1 + V_3)$ and $(C - I)(V_1 + V_4)$ may be non-trivial. Namely, we replace Eq. (17) with

$$(C - I)\mathbb{R}^n \supseteq (C - I)V_1 + (C - I)V_3 + (C - I)V_4 + (C - I)(V_{12} + V_5). \quad (18)$$

We will show that the resulting sum is direct.

First, we note that, because of the $C$-stability of $V_1, V_3, V_1 + V_3$, and $V_4$, we have that $(C - I)V_1 \cap (C - I)V_3 \subseteq V_1 \cap V_3 = \{0\}$, and furthermore, $((C - I)V_1 + (C - I)V_3) \cap (C - I)V_4 = (C - I)(V_1 + V_3) \cap (C - I)V_4 \subseteq (V_1 + V_3) \cap V_4 = \{0\}$, where we used the last part of Lemma 4 in the second step.

Second, by the last part of Lemma 4, we have that $(C - I)V_1 + (C - I)V_3 + (C - I)V_4 = ((C - I)V_1 + (C - I)V_3) + ((C - I)V_1 + (C - I)V_4) = (C - I)(V_1 + V_3) + (C - I)(V_1 + V_4)$.

We already showed in the previous section that the intersection of the sum $(C - I)(V_1 + V_3) + (C - I)(V_1 + V_4)$ with $(C - I)(V_{12} + V_5)$ is trivial. It follows that the sum in Eq. (18) is indeed direct.

Now we consider the dimension of each term. Let us define $q$ as the dimension of $(C-I)V_1$ (it is not simple to express in terms of other basic quantities, but we will see that it cancels out at the end). By the directness of the sum in Eq. (18), and reasoning in the same way we did in the previous section, we have

$$\dim((C - I)V_1) + \dim((C - I)V_3) = \dim((C - I)V_1 + (C - I)V_3)$$
$$= \dim((C - I)(V_1 + V_3)) = \frac{c(BC) - p(BC)}{2},$$

and similarly,

$$\dim((C - I)V_1) + \dim((C - I)V_4) = \dim((C - I)(V_1 + V_4)) = \frac{c(CA) - p(CA)}{2}.$$

Therefore

$$\dim((C-I)V_1 + (C-I)V_3 + (C-I)V_4) = \frac{c(BC) - p(BC)}{2} + \frac{c(CA) - p(CA)}{2} - q.$$

By repeating the calculation in the previous subsection, but carrying the extra $q$ term throughout, we now obtain the upper bound

$$\dim((C - I)(V_{12} + V_5)) \leq n_2 + \delta - \frac{c(AB) - p(AB)}{2} + q.$$

And now, we have to carefully estimate the number of degrees of freedom gained by going from $V_I^C := \{x + y | x \in V_2, y \in V_5, C(x + y) = Ax + y\}$ to the potentially larger subspace $\{x + y | x \in V_{12}, y \in V_5, C(x + y) = Ax + y\}$ (this was simple in the previous section since there was at most 1 extra dimension when $\dim(V_1) = \alpha = 1$).

We first restrict the space $V_I^C$ to allow only those vectors $x$ for which $Ax = x$, i.e. we replace it with

$$\{x + y | x \in V_2 \cap \ker(A - I), y \in V_5, C(x + y) = Ax + y\} \tag{19}$$

This restriction clearly does not increase its dimension.

Second, we go from this subspace to the subspace

$$\{x + y | x \in V_{12}, y \in V_5, C(x + y) = Ax + y\}. \tag{20}$$

Recall that $V_{12} := (V_1 + V_2) \cap \ker(A - I)$. By Lemmas 5 and 6, $A$ is an involution on the standard bases of both $V_1$ and $V_2$, and these bases can be altered so that each pair of basis vectors $v$ and $w$ permuted by $A$ is replaced by $v + w$ and $v - w$, of which the first one is in $\ker(A - I)$ and the second one is not. Together with the vectors $u$ fixed by $A$, which are also in $\ker(A - I)$, the resulting bases will contain sub-bases for the intersection of the corresponding vector space with $\ker(A - I)$. It follows that $V_{12} = (V_1 \cap \ker(A - I)) + (V_2 \cap \ker(A - I))$.

We note that in general, for three finite-dimensional vector spaces $U, V, W$, we have $(U \cap W) + (V \cap W) \subseteq (U + V) \cap W$, and the inclusion can be strict; however, we have equality here thanks to the representation of $A$ on $V_1 + V_2$.

It is now easy to see from the foregoing discussion that the subspace in Eq. (20) differs from the one in Eq. (19) by the vectors in the subspace

$$V_1 \cap \ker(A - I) = \{x \in \mathbb{R}^n | x = Ax = Bx = Cx\} = V_1 \cap \ker(C - I),$$

whose dimension, by Lemma 4, is given by

$$\dim(V_1 \cap \ker(C - I)) = \dim(V_1) - \dim((C - I)V_1) = \alpha - q.$$

The final calculation from the previous section (with some parallel intermediate steps omitted) now becomes

$$\dim(V_I^C) = \dim(\{x + y | x \in V_2, y \in V_5, C(x + y) = Ax + y\})$$
$$\geq \dim(\{x + y | x \in V_2 \cap \ker(A - I), y \in V_5, C(x + y) = Ax + y\})$$
$$\geq \dim(\{x + y | x \in V_{12}, y \in V_5, C(x + y) = Ax + y\}) - (\alpha - q)$$
$$\geq n_2 + \alpha - \frac{c(AB) - p(AB)}{2} + 2\delta - \left(n_2 + \delta - \frac{c(AB) - p(AB)}{2} + q\right) - (\alpha - q) = \delta,$$

which completes the proof.

# References

1. Chindelevitch, L., Zanetti, J.P.P., Meidanis, J.: On the Rank-Distance Median of 3 permutations. BMC Bioinform. **19**(Suppl. 6), 142 (2018). A preliminary version appeared. In: Meidanis, J., Nakhleh, L. (eds.) Proceedings of 15th RECOMB Comparative Genomics Satellite Workshop. LNCS. vol. 10562, pp. 256–276. Springer, Heidelberg (2017). https://doi.org/10.1007/978-3-319-67979-2_14
2. Coppersmith, D., Winograd, S.: Matrix multiplication via arithmetic progressions. J. Symbolic Comput. **9**(3), 251 (1990)
3. Tannier, E., Zheng, C., Sankoff, D.: Multichromosomal median and halving problems under different genomic distances. BMC Bioinform. **10**, 120 (2009)
4. Feijao, P., Meidanis, J.: SCJ: a breakpoint-like distance that simplifies several rearrangement problems. IEEE/ACM Trans. Comput. Biol. Bioinform. **8**(5), 1318–1329 (2011)
5. Caprara, A.: Formulations and hardness of multiple sorting by reversals. In: Proceedings of 3rd Annual International Conference on Research in Computational Molecular Biology, pp. 84–94. ACM Press, New York (1999)
6. Zanetti, J.P.P., Biller, P., Meidanis, J.: Median approximations for genomes modeled as matrices. Bull. Math. Biol. **78**, 786 (2016)
7. Feijao, P., Meidanis, J.: Extending the algebraic formalism for genome rearrangements to include linear chromosomes. IEEE/ACM Trans. Comput. Biol. Bioinform. **10**(4), 819–831 (2012)
8. Meidanis, J., Biller, P., Zanetti, J.P.P.: A Matrix-Based Theory for Genome Rearrangements. Technical Report, Institute of Computing, University of Campinas (2017)
9. Delsarte, P.: Bilinear forms over a finite field, with applications to coding theory. J. Comb. Theory A. **25**(3), 226–241 (1978)
10. Horn, F.: Necessary and sufficient conditions for complex balancing in chemical kinetics. Arch. Ration. Mech. Anal. **49**, 172–186 (1972)

11. Axler, S.: Linear Algebra Done Right. UTM. Springer, Cham (2015). https://doi.org/10.1007/978-3-319-11080-6
12. Edmonds, J.: Paths, trees, and flowers. Canad. J. Math. **17**, 449–467 (1965)
13. Galil, Z.: Efficient algorithms for finding maximum matching in graphs. ACM Comput. Surv. **18**(1), 23–38 (1986)
14. van Rossum, G. Python tutorial. Technical Report CS-R9526, Centrum voor Wiskunde en Informatica (CWI), Amsterdam (1995)
15. Hagberg, A.A., Schult, D.A., Swart, P.J.: Exploring network structure, dynamics, and function using NetworkX. In: Varoquaux, G., Vaught, T., Millman, J. (eds.) Proceedings of the 7th Python in Science Conference (SciPy2008), Pasadena, CA, USA, pp. 11–15 (2008)
16. Bergeron, A., Mixtacki, J., Stoye, J.: A unifying view of genome rearrangements. In: Moret, B. (ed.) Algorithms in Bioinformatics Proceedings of WABI (2006)
17. R Core Team. R: A Language and Environment for Statistical Computing. R Foundation for Statistical Computing, Vienna, Austria (2016). www.R-project.org/
18. Monagan, M.B., Geddes, K.O., Heal, K.M., Labahn, G., Vorkoetter, S.M.: Maple 10 Programming Guide. Maplesoft, Waterloo (2005)

# The Rooted SCJ Median with Single Gene Duplications

Aniket C. Mane[1], Manuel Lafond[2], Pedro Feijão[3], and Cedric Chauve[1(✉)]

[1] Department of Mathematics, Simon Fraser University, 8888 University Drive,
Burnaby, BC, Canada
cedric.chauve@sfu.ca
[2] Department of Computer Science, Université de Sherbrooke,
2500 Boul. de l'Université, Sherbrooke, Canada
[3] School of Computing Science, Simon Fraser University, 8888 University Drive,
Burnaby, BC, Canada

**Abstract.** The median problem is a classical problem in genome rearrangements. It aims to compute a gene order that minimizes the sum of the genomic distances to $k \geq 3$ given gene orders. This problem is intractable except in the related Single-Cut-or-Join and breakpoint rearrangement models. Here we consider the rooted median problem, where we assume one of the given genomes to be ancestral to the median, which is itself ancestral to the other genomes. We show that in the Single-Cut-or-Join model with single gene duplications, the rooted median problem is NP-hard. We also describe an Integer Linear Program for solving this problem, which we apply to simulated data, showing high accuracy of the reconstructed medians.

## 1 Introduction

Reconstructing the evolution of genomes at the level of large-scale genome rearrangements is an important problem in computational biology [17,19]. There are several computational problems related to rearrangements, ranging from the computation of pairwise distances in a given rearrangement model to the reconstruction of complete phylogenetic trees, often following a parsimony approach [12]. Among these problems, the reconstruction of ancestral gene orders given a species phylogeny has been considered in various frameworks, including the so-called Small Parsimony Problem (SPP), which aims at proposing gene orders at the internal nodes of the given species phylogeny while minimizing the sum of the genome rearrangement distances along its branches. The simplest instance of the SPP is the Median Problem, where the given phylogeny contains a single ancestral node whose gene order is to be reconstructed. In the present paper, we introduce novel results about the median problem, in a context where gene duplications are considered.

The median problem was introduced in 1996 [21], motivated by its application to iterative algorithms for solving the SPP [3]. Early results suggested that,

© Springer Nature Switzerland AG 2018
M. Blanchette and A. Ouangraoua (Eds.): RECOMB-CG 2018, LNBI 11183, pp. 28–48, 2018.
https://doi.org/10.1007/978-3-030-00834-5_2

even in the simple breakpoint distance model, computing a median gene order is intractable [20], and heuristics based on the Traveling Salesman Problem (TSP) were introduced to solve the breakpoint median problem [3,7]. However, in 2009, Tannier, Zheng and Sankoff proved that computing a median gene order that is allowed to contain an arbitrary mixture of linear and circular chromosomes was tractable in the breakpoint distance model, by using a reduction to the problem of computing a Maximum Weight Matching (MWM) [22]. This tractability result, the first of its kind in genome rearrangements, renewed the interest in gene order median problems, although most of the following work presented intractability results, even on variations of the breakpoint distance [5,9,14]. A notable exception was the Single-Cut-or-Join (SCJ) distance, introduced by Feijão and Meidanis [11], where it was shown that both the SCJ median problem and the SCJ SPP are tractable.

Gene duplication is another important evolutionary mechanism, ranging from single-gene duplication to whole-genome duplications (WGD) [13,15]. The first models of evolution by genome rearrangements considered the case of genomes with equal gene content, thus disregarding gene duplication and gene loss. When considered as a possible evolutionary event, gene duplication most often leads to intractability results, even for the simple pairwise gene order distance [1,4,6]. Notable exceptions include again variants of the SCJ distance. In [23] it was shown that in an evolutionary model including SCJ and whole-chromosome duplications, the pairwise distance problem is tractable. More recently, we introduced a variant of SCJ including single-gene duplications where the distance between an ancestral genome and a descendant genome can be computed, when orthology relations between the descendant and ancestral genes are provided [10]. We also showed that a directed median problem where the median is the ancestor of $k$ given genomes is tractable, again by reduction to a MWM problem. These results raised the question of tractability boundaries towards the SPP in a rearrangement model, including gene duplication.

In the present work, we show that a different median problem, which involves an additional given ancestral genome, is intractable. More precisely, we introduce the rooted median problem, where we are provided with $k + 1 \geq 3$ genomes, $A, D_1, \ldots, D_k$, such that $A$ is ancestral to $D_1, \ldots D_k$, and we are looking for a median $M$, whose gene content and orthology relation to the given genomes are provided, that minimizes the sum of the directed distances between $A$ and $M$, and $M$ and the $D_i$s, in the distance model defined in [10]. In Sect. 3, we prove that this median problem is NP-hard even when $k = 2$. In Sect. 4, we describe a simple Integer Linear Program (ILP) for this problem, based on a reduction to a colored MWM problem. We provide in Sect. 5 experimental results on simulated data.

## 2   Preliminaries

*Genes and Genomes.* A genome consists of a set of chromosomes, each being a linear or circular ordered set of oriented genes. Following the usual encoding of gene orders, we represent a genome by its *gene extremity adjacencies*. In this representation, a gene $g$ is represented using a pair of gene extremities $(g_t, g_h)$, $g_t$ denotes the tail of the gene $g$ and $g_h$ denotes its head, and an *adjacency* is a pair of gene extremities that are adjacent in a genome. If a gene $g_i$ is denoted with a subscript, we will denote the tail of $g_i$ by $g_{i,t}$ and its head by $g_{i,h}$. A gene extremity is *free* if it does not belong to an adjacency.

We assume that a given gene $g$ can have multiple copies in a genome, the number of copies being called its *copy number*. A genome in which every gene has copy number 1 is a *trivial genome*. A non-trivial genome sometimes cannot be represented unambiguously by its adjacencies, that can form a *multi-set*, unless we distinguish the copies of each gene, for example by denoting the copies of a gene $g$ with copy number $k$ by $g^1, \ldots, g^k$. Nevertheless, we identify a genome with its multi-set of gene extremity adjacencies, which we call adjacencies from now on. A *chromosome* is a maximal contiguous sequence of genes; a chromosome with $k$ genes can have either $k - 1$ adjacencies, in which case it is a *linear* chromosome, or $k$ adjacencies, in which case it is a *circular* chromosome.

*Evolutionary Model.* In this work, following [10], we consider a model of *directed evolution* in which, when comparing two genomes, we assume one, denoted by $A$, is a trivial genome and an ancestor of the other genome, denoted by $D$.

We now describe the evolutionary events defining our evolutionary model. Genome rearrangements are modeled by *Single-Cut-or-Join* (SCJ) operations, which either delete an adjacency from a genome (a cut) or join a pair of free gene extremities (a join), thus forming a new adjacency. For duplication events, we consider two types of duplications, both creating an extra copy of a single gene: *Tandem Duplications* (TD) and *Floating Duplications* (FD). A tandem duplication of an existing gene $g$ introduces an extra copy of $g$, say $g'$, by adding an adjacency $g_h g'_t$, and, if there was an adjacency $g_h x$ by replacing it by the adjacency $g'_h x$. A floating duplication introduces an extra copy $g'$ of a gene $g$ as a single-gene circular chromosome by adding the adjacency $g'_h g'_t$.

Given $A$ and $D$, we denote by gene family all copies of a given gene observed in $A$ and $D$. By definition, there is exactly one copy of the gene in $A$ and there might be several, paralogous, copies of the gene in $D$. We assume here that every gene in $A$ has at least one descendant gene in $D$ and conversely, every gene in $D$ has exactly one ancestral gene in $A$, so we do not consider gene gains or losses.

*Problem Statements.* In [10], Feijão *et al.* introduced the **directed SCJ-TD-FD (d-SCJ-TD-FD) distance problem** that asks to compute the minimum number of SCJ, TD and FD operations needed to transform $A$ into $D$, denoted by $d_{\text{DSCJ}}(A, D)$. They showed that this problem is tractable and that the distance can be computed using a simple set-theoretical formula, extending naturally the distance formula for the SCJ with no duplication model.

A first median problem was also introduced in [10], the **directed SCJ-TD-FD (d-SCJ-TD-FD) median problem**, defined as follows: given $D_1, \ldots, D_k$ $(k \geq 2)$ (possibly) non-trivial genomes, such that no gene family is absent from any $D_i$, compute a trivial genome $A$ on the same set of gene families, that

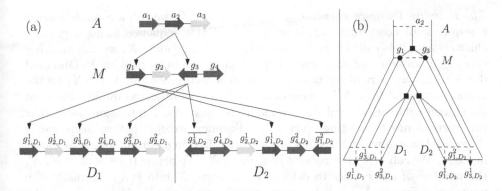

**Fig. 1.** In part (a), each color represents a gene family from $A$. Notice that each gene in $D_1$ and $D_2$ can be traced to a unique gene in $M$ whereas a gene from $A$ might have multiple daughters in $M$. Part (b) displays the gene tree of the gene family in blue (indicated by arrows in part (a)). Since the gene $a_2$ undergoes duplication (dark squares) to form $g_1$ and $g_3$ in $M$, $M$ is not trivial w.r.t $A$. (Color figure online)

minimizes $\sum_{i=1}^{k} d_{DSCJ}(A, D_i)$. It was shown that this median problem is also tractable through a simple reduction to a MWM problem.

In the present work, we introduce the **rooted SCJ-TD-FD (r-SCJ-TD-FD) median problem**. We are given $k + 1 \geq 3$ genomes, $A, D_1, \ldots, D_k$ such that $A$ is a trivial genome, ancestor to the $D_i$'s. The goal of the rooted median problem is to find a genome $M$ which is a descendant of $A$ and an ancestor of $D_1, \ldots, D_k$, minimizing the sum of its distance to $A$ and to the $D_i$'s. Following the approach introduced in [10], we assume we are given the *gene content* $\Gamma$ of $M$ and the *orthology relations* between $A$ and $M$, as well as between $M$ and the $D_i$'s. This implies that every gene of $M$ (resp. $D_1, \ldots, D_k$) has a unique ancestor in $A$ (resp. in $M$), so $M$ is a trivial genome compared to the $D_i$'s but might not be compared to $A$ (see Fig. 1 for an illustration). To formally handle this difference, we assume that all copies of a gene $g$ of $A$ in $M$ (i.e. the genes of $M$ whose ancestor in $A$ is gene $g$) are distinguishable (e.g. labeled, say $g_1, \ldots, g_k$) and, for a given gene $g_i$ of $M$, we denote its ancestor in $A$ by $a(g_i)$. Then for a given genome $M$ on $\Gamma$, we denote by $M_a$ the genome where every gene $g$ is relabeled by $a(g)$. The goal of the rooted median problem is to find a genome $M$ that minimizes the following function:

$$d_{DSCJ}(A, M_a) + \sum_{i=1}^{k} d_{DSCJ}(M, D_i). \tag{1}$$

*Remark 1.* If we assume there is no duplication from $A$ to $M$, i.e. both have the same gene content, then the MWM algorithm introduced in [10] for the directed median problem applies to the rooted median problem and the problem is thus tractable. So the difficulty in solving the rooted median problem is to account for duplications from $A$ to $M$.

*The Pairwise Distance Formula.* Given a gene $g \in \Gamma$, we call a *$g$-tandem array* a sequence of consecutive adjacencies $g_h g_t$; if this sequence forms a circular chromosome, it is called a *$g$-chromosome.* Given a genome $X$, we call an adjacency $g_h g_t$ an *observed duplication* if $g$ has more than one copy in $X$. Observed duplications are part of a $g$-tandem array or a $g$-chromosome. Let $r(X)$ be the genome obtained from $X$ by successively deleting an observed duplication from $X$, chosen arbitrarily, until there remains no observed duplication. Note that this corresponds to deleting every $g_h g_t$ adjacency, except that we keep one in the special case in which all copies of $g$ are organized in $g$-chromosomes, as shown in Fig. 2. We call $r(X)$ the *reduced* genome of $X$. We define $t(X) = |X - r(X)|$, the number of adjacencies to delete to transform $X$ into $r(X)$. Formally, the multi-set difference $X - Y$ between two multi-sets $X$ and $Y$ of adjacencies is the multi-set obtained as follows: it contains $k$ copies of a given adjacency if and only if $X$ contains exactly $k$ more occurrences of this adjacency than $Y$ (with $k = 0$ being possible).

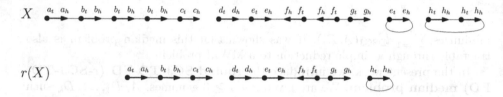

**Fig. 2.** An example of the reduced genome $r(X)$, of the genome $X$. Note that an instance of $h_h h_t$ is retained so that $r(X)$ contains at least one representative of gene family $h$. All observed duplications are removed in $r(X)$. Here, $t(X) = |X - r(X)| = 5$.

The directed SCJ-TD-FD distance between an ancestral genome $A$ and a descendant genome $D$ is given by [10]:

$$d_{\mathrm{DSCJ}}(A, D) = |A - r(D)| + |r(D) - A| + 2\delta(A, r(D)) + t(D) \qquad (2)$$

where $\delta(A, r(D))$ is the difference between the number of genes of $r(D)$ and the number of genes of $A$ (i.e. the number of duplications from $A$ to $r(D)$). We introduce[1] now a slightly different formulation of $d_{\mathrm{DSCJ}}$ that will be useful in our hardness proof:

$$d_{\mathrm{DSCJ}}(A, D) = |A - r(D)| + |r(D) - A| + 2\delta(A, D) - t(D) \qquad (3)$$

*Remark 2.* For $d_{\mathrm{DSCJ}}(M, D_i)$, the value of $t(D_i)$ does not depend on our choice of $M$, for $i = 1, \dots, k$. We will therefore assume that the $D_i's$ are reduced (hence we may refer to $r(D_i)$ as simply $D_i$ instead). However $t(M_a)$ has an impact on $d_{\mathrm{DSCJ}}(A, M_a)$, and so we will not assume that $M$ is reduced.

---

[1] The proof is given in the Appendix.

## 3   The Rooted Median Problem Is NP-hard

We show that finding the optimal gene order for $M$ is NP-hard even for $k = 2$, by reduction from the 2P2N-3SAT problem [2][2]. In 2P2N-3SAT, we are given $n$ variables $x_1, \ldots, x_n$ and $m$ clauses $C_1, \ldots, C_m$, each containing exactly 3 literals. Each $x_i$ variable appears as a positive literal in exactly 2 clauses, and as a negative literal in exactly 2 clauses. Note that since each variable occurs in exactly 4 clauses and each clause has 3 literals, $m = 4n/3$. An example of a 2P2N-3SAT instance is shown in Fig. 3.

We now describe how we transform the $x_i$ variables and $C_j$ clauses into an instance of the rooted median. The genes of $M$ are

$$\Gamma = \{g_1^+, \gamma_1^+, g_1^-, \gamma_1^-, \ldots, g_n^+, \gamma_n^+, g_n^-, \gamma_n^-, c_1, \ldots, c_m, \alpha_1, \ldots, \alpha_{2n-m}\}$$

The genes $g_i^+, \gamma_i^+, g_i^-, \gamma_i^-$ correspond to the $x_i$ variable, and $c_j$ to the clause $C_j$. The purpose of the $2n - m = 2n/3$ special $\alpha_i$ genes will become apparent later.

To simplify matters, every adjacency in our reduction is between the tails of two genes. Hence, the heads of each gene of $A, D_1$ and $D_2$ are telomeres (linear chromosomes extremities), so that all chromosomes are linear and have at most 2 genes. From now, we will omit the $t$ subscript from the extremities for these adjacencies, with the understanding that every adjacency is between tails; for instance, we may write $g_i^+ \gamma_i^+$ for the adjacency $g_{i,t}^+ \gamma_{i,t}^+$.

We can now describe $A$, $D_1$ and $D_2$. The genes of $A$ are $g_1', \gamma_1', \ldots, g_n', \gamma_n', c_1', \ldots, c_m', \alpha_1', \ldots, \alpha_{2n-m}'$. The genes $g_i^+$ and $g_i^-$ (resp. $\gamma_i^+$ and $\gamma_i^-$) are duplicates of $g_i'$ (resp. $\gamma_i'$), and there are no other duplications in $M$ compared to $A$. Formally, for each $i \in [n]$, put $a(g_i^+) = a(g_i^-) = g_i'$, $a(\gamma_i^+) = a(\gamma_i^-) = \gamma_i'$ and for each $j \in [m]$, put $a(c_j) = c_j'$. Finally, for each $i \in [2n - m]$, put $a(\alpha_i) = \alpha_i'$. The adjacencies of $A$ are $\{g_i'\gamma_i' : i \in [n]\}$.

The genomes $D_1$ and $D_2$ are identical, i.e. they contain the same set of genes and of adjacencies. We simply describe the set of adjacencies of $D_1$ and $D_2$ with the understanding that if an extremity, say $x$, appears in two adjacencies $xy$ and $xz$, then the two $x$ are the tails of two distinct copies of the same gene on two distinct chromosomes. The adjacencies of $D_1$ and $D_2$ are described as follows.

- For each $i \in [n]$, add to $D_1$ and $D_2$ the adjacencies $g_i^+ \gamma_i^+$ and $g_i^- \gamma_i^-$.
- For each $i \in [n]$, let $C_{j_1}, C_{j_2}$ be the two clauses in which $x_i$ occurs positively and let $C_{k_1}, C_{k_2}$ be the two clauses in which $x_i$ occurs negatively. Add to $D_1$ and $D_2$ the adjacencies $g_i^+ c_{j_1}$ and $\gamma_i^+ c_{j_2}$. Similarly, add to $D_1$ and $D_2$ the adjacencies $g_i^- c_{k_1}$ and $\gamma_i^- c_{k_2}$[3].
- Finally, for each $i \in [n]$ and each $j \in [2n - m]$, add to $D_1$ and $D_2$ the adjacencies $g_i^+ \alpha_j, g_i^- \alpha_j, \gamma_i^+ \alpha_j$ and $\gamma_i^- \alpha_j$.

---

[2] This problem is sometimes called the (3,B2)-SAT problem, where B2 indicates that the literals are *balanced* with two occurrences each.

[3] Intuitively, these adjacencies represent using a literal to satisfy a specific clause. For instance, the adjacency $g_i^+ c_{j_1}$ represents "setting $x_i$ to true and satisfying $C_{j_1}$".

This completes our construction. The intuition behind our hardness proof is that for each $i \in [n]$, we need to pick one of $g_i^+ \gamma_i^+$ or $g_i^- \gamma_i^-$ in $M$, as we will show. Simultaneously, we would like to include as many adjacencies which are in both $D_1$ and $D_2$. It will possible to choose the positive and negative adjacencies *and* match all the $c_j$ and $\alpha_j$ if and only if the 2P2N-3SAT instance is satisfiable.

It will be useful to think of $D_1$ (and $D_2$) as the set of adjacencies which are allowed to belong to $M$, as stated in the following.

**Lemma 1.** *Let $a$ be an adjacency in $M$, such that $a \notin D_1$ (equivalently, $a \notin D_2$). Then $M - \{a\}$ achieves a smaller total distance to $A$, $D_1$ and $D_2$ than $M$.*

*Proof.* By cutting $a$, we increase the distance to $A$ by at most 1, but decrease the distance to $D_1$ and $D_2$ by 1 each. This is because $|(M - \{a\}) - D_1| + |D_1 - (M - \{a\})| = |M - D_1| - 1 + |D_1 - M|$, the value of $\delta(M, D_1)$ is unchanged and $t(D_1) = 0$ by assumption (and the same holds for $D_2$). Therefore removing $a$ from $M$ yields a better median genome. □

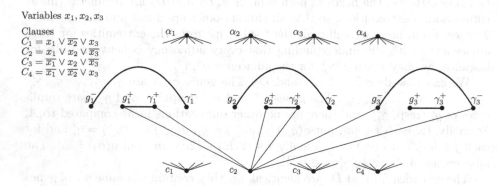

**Fig. 3.** An example of a 2P2N-3SAT instance, with an illustration of the genes of $M$ (only the gene tails are shown) and the adjacencies that are allowed by $D_1$ and $D_2$. The fat edges represent pairs of adjacencies of which at least one must be present according to Lemma 2. Among the $c_j$ extremities, only the adjacencies for $c_2$ are shown.

Therefore, we may assume that every adjacency of a median $M$ belongs to $D_1$ and $D_2$. Note that this implies that $M$ contains no observed duplications (with respect to $A$), as no such adjacency is in $D_1$ and $D_2$. Thus we will ignore the $t(M_a) = 0$ term in $d_{\mathrm{DSCJ}}(A, M_a)$ (Eq. (3)), and we will not make a distinction between $M_a$ and $r(M_a)$, as these are equal.

Another property of $M$ is that it must contain at least one "positive" or one "negative" adjacency for each $i \in [n]$.

**Lemma 2.** *For $i \in [n]$, $M$ contains at least one of $g_i^+ \gamma_i^+$ and $g_i^- \gamma_i^-$.*

The proof of this lemma is provided in the Appendix.

We now formally prove the hardness of computing the SCJTDFD median.

**Theorem 1.** *The rooted SCJ-TD-FD median problem is NP-hard.*

*Proof.* Let $x_1, \ldots, x_n$ and $C_1, \ldots, C_m$ be a 2P2N-3SAT-instance, and let $A, D_1, D_2$ and the genes $\Gamma$ of $M$ be the corresponding instance of the r-SCJ-TD-FD median genome problem. We will show that the given 2P2N-3SAT instance is satisfiable if and only if there exists a median genome $M$ satisfying

$$d_{\text{DSCJ}}(A, M_a) + d_{\text{DSCJ}}(M, D_1) + d_{\text{DSCJ}}(M, D_2) \leq 2|D_1| - 2n + 4\delta(M, D_1)$$

($\Rightarrow$) Suppose that the 2P2N-3SAT can be satisfied by an assignment of the $x_i$ variables to true or false. Construct a median genome using the following steps.

1. For each $i \in [n]$, if $x_i$ is set to true, then add $g_i^- \gamma_i^-$ to $M$, and if instead $x_i$ is set to false, add $g_i^+ \gamma_i^+$ to $M$.
2. Then, add to $M$ these adjacencies in an algorithmic fashion: for each $j = 1, 2, \ldots, m$, consider clause $C_j$ and let $x_i$ be any variable satisfying $C_j$.
   - If $x_i$ is set to true, then note that $g_i^+$ and $\gamma_i^+$ have not been matched in Step 1. Add $g_i^+ c_j$ to $M$ if $g_i^+$ is not part of an adjacency of $M$ yet, or add $\gamma_i^+ c_j$ to $M$ otherwise.
   - If instead $x_i$ is set to false, then $g_i^-$ and $\gamma_i^-$ have not been matched in Step 1. Add $g_i^- c_j$ if $g_i^-$ is not part of an adjacency in $M$ yet, or add $\gamma_i^- c_j$ to $M$ otherwise.
   Note that since each $x_i$ can satisfy at most two clauses, it will always be possible to find an extremity to match $c_j$ with.
3. Finally, observe that so far each of the $g_i^+, g_i^-, \gamma_i^+$ and $\gamma_i^-$ extremities are in an adjacency $M$, except $4n - 2n - m = 2n - m$ of them. Associate each such extremity $g$ with a distinct $\alpha_j$ extremity arbitrarily, and add each $g\alpha_j$ to $M$, noting that there are just enough $\alpha_j$ genes to do so.

Note that $M$ contains $n + m + 2n - m = 3n$ adjacencies in total, exactly $n$ of which correspond to an adjacency of $A$ (those included in Step 1). Also, every adjacency of $M$ occurs in both $D_1$ and $D_2$. We have

$$d_{\text{DSCJ}}(A, M_a) = |A - M_a| + |M_a - A| + 2\delta(A, M_a) - t(M_a)$$
$$= 0 + 2n + 2n - 0 = 4n$$

As for $D_1$ and $D_2$,

$$d_{\text{DSCJ}}(M, D_1) = d_{\text{DSCJ}}(M, D_2) = |D_1 - M| + |M - D_1| + 2\delta(M, D_1)$$
$$= |D_1| - 3n + 0 + 2\delta(M, D_1)$$

Therefore the total distance is $4n + 2(|D_1| - 3n + 2\delta(M, D_1)) = 2|D_1| - 2n + 4\delta(M, D_1)$, as we predicted.

($\Leftarrow$) Suppose that there exists a median genome $M$ of total distance at most $2|D_1| - 2n + 4\delta(M, D_1)$. By Lemma 1, we may assume that every adjacency of $M$ is present in both $D_1$ and $D_2$.

With the next two claims, we will prove that $M$ has exactly $3n$ adjacencies, of which exactly $n$ are adjacencies corresponding to those in $A$.

**Claim 1.** $|M| \leq 3n$, and $|M| = 3n$ only if every $c_j$ and $\alpha_j$ extremity is in some adjacency of $M$.

For the rest of the proof, denote by $q$ the number of distinct adjacencies $ab \in A$ for which there exists $xy \in M$ such that $a(x)a(y) = ab$.

**Claim 2.** $|M| = 3n$ and $q = n$.

The proofs of both claims will be discussed in detail in the Appendix.

Because $q = n$, Claim 2 implies that for each $i \in [n]$, (at least) one of $g_i^+\gamma_i^+$ and $g_i^-\gamma_i^-$ is in $M$. This lets us define as assignment for our 2P2N-3SAT instance: for each $i \in [n]$, set $x_i$ to *true* if $g_i^-\gamma_i^-$ is in $M$, and otherwise set $x_i$ to *false*. We claim this this assignment satisfies every clause.

To see this, let $C_j$ be a clause and let $c_j$ be its corresponding extremity in $M$. By Claim 2, every extremity that is part of some adjacency in $D_1$ must be part of an adjacency in $M$, including $c_j$. Thus there is some $e$ such that $c_j e \in M$. By Lemma 1, the adjacency $c_j e$ must also be in $D_1$, and by construction either (1) $e \in \{g_i^+, \gamma_i^+\}$ for some $x_i$ that occurs positively in $C_j$, or (2) $e \in \{g_i^-, \gamma_i^-\}$ for some $x_i$ that occurs negatively in $C_j$. Suppose that case (1) applies. Then $c_j g_i^+$ or $c_j \gamma_i^+$ being in $M$ means that $g_i^+\gamma_i^+ \notin M$, implying in turn that $g_i^-\gamma_i^-$ is in $M$. In this situation, we have set $x_i$ to *true* and we satisfy $C_j$. Suppose instead that case (2) applies. Then $g_i^-\gamma_i^- \notin M$, in which case we have set $x_i$ to false and satisfy $C_j$. As the argument applies to any clause $C_j$, this concludes the proof. □

*Remark 3.* In the reduction above, none of the considered genomes contain a $g$-tandem array or a $g$-chromosome. So our result also implies the hardness of the rooted median problem where the distance between two genomes $A$ and $D$, where $A$ is an ancestor of $D$, is computed in a simpler way as $|A-D|+|D-A|+2\delta(A, D)$, i.e. does not contain a term related to reducing the descendant genome.

## 4   An Integer Linear Program

We now describe a simple Integer Linear Program (ILP) to solve the rooted median problem. The key idea, already used in previous median problems [10, 22] is to convert the rooted median problem into an instance of a MWM problem, albeit with certain additional constraints. More precisely, in this approach we define a complete graph $G$ on the extremities $g_h$ and $g_t$ of every gene $g$ in $\Gamma$. A pair of distinct extremities defines an edge and thus a potential adjacency in $M$, which is thus defined by a matching in $G$. Each edge is assigned a weight that reflects the number of descendant genomes which contain the corresponding adjacency. Further, each edge is assigned a color that reflects its corresponding adjacency in the ancestral genome, if any, and the number of colors of the selected edges also contributes to the weight of the matching defining the median $M$.

*An Alternative Formulation for the Distance.* We first introduce an alternative formula to compute the directed distance, denoted by $d_{\mathrm{DSCJ}}(u, v)$, from an ancestor $u$ to a descendant $v$. For the rooted median problem, the pair $(u, v)$ can represent either the pair $(A, M_a)$ or any pair $(M, D_i)$. The new formulation is easier to handle in an ILP framework than Eq. (3). We denote by $n_v(g)$ the number of copies of gene $g$ in $v$, by $n_v(g_h g_t)$ the number of occurrences of adjacency $g_h g_t$ in $v$, and by $t_v(g)$ the number of observed duplications of gene $g$ in $v$. Note that $t_v(g) \in \{n_v(g_h g_t) - 1, n_v(g_h g_t)\}$, the case $t_v(g) = n_v(g_h g_t) - 1$ occurring when adjacencies $g_h g_t$ form only $g$-chromosomes. Further, let $t(v) = \sum_{g \in \Gamma_u} t_v(g)$ denote the total number of observed duplications in $v$, where $\Gamma_u$ is the set of genes of $u$ and also the alphabet of genes of $v$.

To rewrite $d_{\mathrm{DSCJ}}(u, v)$, we introduce an indicator variable $\alpha_{g,uv}$, where $\alpha_{g,uv} = 1$ if $g_h g_t$ is common to both $u$ and $v$, but all occurrences were removed while reducing $v$. Formally, $\alpha_{g,uv} = 1$ if $g_h g_t \in u \cap v$ and $g_h g_t \notin r(v)$; otherwise $\alpha_{g,uv} = 0$. It is then relatively straightforward to show[4] that

$$d_{\mathrm{DSCJ}}(u, v) = |u - v| + |v - u| + 2\delta(u, v) - 2t(v) + 2 \sum_{g \in \Gamma_u} \alpha_{g,uv} \qquad (4)$$

This formulation is interesting due to the fact it does not rely on the notion of a reduced genome. We will discuss later how variables $\alpha_{g,uv}$ and $t_v(g)$ can be handled simply in an ILP framework.

*Reformulating the Objective Function.* We now use Eq. (4) to reformulate the objective function of the rooted median problem[5].

**Claim 3.** *Minimizing the function Eq. (1) defining the evolutionary cost of a median $M$ is equivalent to maximizing the following expression:*

$$\sum_{i=1}^{k} \left( 2|M \cap D_i| - 2 \sum_{g \in \Gamma_M} \alpha_{g,MD_i} \right) + 2|A \cap M_a| + 2t(M_a) - 2 \sum_{g \in \Gamma_A} \alpha_{g,AM_a} - (k+1)|M| \quad (5)$$

*where $\Gamma_A$ and $\Gamma_M$ are the set of genes of $A$ and $M$, respectively, and so also the gene alphabets for $M$ and the $D_i$s, and variables $\alpha_{g,AM_a}$ and $\alpha_{g,MD_i}$ are defined as $\alpha_{g,uv}$ above.*

Such a reformulation of the objective function is inspired by [10]. This revision enables us to translate the problem as an instance of a *colored MWM problem*, as will be made clear in the subsequent paragraphs.

*An Interpretation as a Colored MWM Problem.* The terms $\alpha_{g,uv}$ and $t(M_a)$ in Eq. (5) account for the presence of observed duplications. In the absence of observed duplications however, solving the rooted median problem requires finding a matching in $G$ that maximizes the sum of the weight of the selected edges

---

[4] A proof is provided in the Appendix.
[5] The proof of this claim is discussed in the Appendix.

and of the number of colors represented by the matching edges. The matching edges weight is partly accounted for by the term $|M \cap D_i|$, while on the other hand, $|A \cap M_a|$ determines the number of colors used in the matching. Using the intersection terms in the objective function, we now interpret the notion of *weight* and *color* of an edge in terms of decision variables of an ILP.

In order to compute $|M \cap D_i|$, we introduce the variable $\gamma_i(e)$ denoting the existence of a potential adjacency $e$ of $M$ in a genome $D_i$: we put $\gamma_i(e) = |e \cap D_i|$, i.e. $\gamma_i(e) = 1$ if $e \in D_i$ and 0, otherwise. For each adjacency $e$ in the graph $G$, the weight $w(e)$ of $e$ is determined using the weight function $w : E(G) \to \mathbb{N}$:

$$ w(e) = 2 \left( \sum_{i=1}^{k} \gamma_i(e) \right) - (k+1) $$

Since $M$ is trivial w.r.t. every $D_i$, the weights for edges $e \in M$ will account for the term $\sum_{i=1}^{k} 2|M \cap D_i| - (k+1)|M|$ in Eq. (5). However, this principle does not work with $A$. Indeed, it is possible that $x_1 y_1 \in M$ and $x_2 y_2 \in M$ such that $a(x_1)a(y_1) = a(x_2)a(y_2) \in A$. In this situation, only one of $x_1 y_1$ or $x_2 y_2$ can contribute to $|A \cap M_a|$, but both $|x_1 y_1 \cap A|$ and $|x_2 y_2 \cap A|$ equal to 1. In other words, we cannot simply sum the adjacencies of $M_a$ which are in $A$.

To address this issue, we introduce the notion of a *color family*. Let $m_A$ be the number of adjacencies in $A$. Each number from the set $\{1, 2, ..., m_A\}$ represents a distinct color. We arbitrarily assign a distinct color from this set to each adjacency in $A$. If $E(G)$ is the edge set of $G$, representing all possible adjacencies in $M$, then every adjacency in $E(G)$ is assigned a color from $\{1, 2, ..., m_A\} \cup \{0\}$, consistent with the orthology relations: the adjacency $xy \in M$ receives color $i \neq 0$ if the adjacency $a(x)a(y)$ is present in $A$ and was assigned color $i$, and color 0 if $a(x)a(y)$ is not present in $A$. The set of adjacencies having the same color $i$ form a color family, represented by $E_i$. We denote by $C$ the coloring function $E(G) \to \{0, 1, ..., m_A\}$ defined as described above. Notice that a color $i$ contributes exactly once to the term $|A \cap M_a|$ if there exists at least one adjacency in $M$ that belongs to the color family $i$.

*Reducing the Size of the ILP.* The size of the ILP we are about to describe is polynomial in the sum of the considered genomes. As the total number of adjacencies is quadratic in the number of genes in $M$, it can reach large values when dealing with large genomes, thus making the ILP challenging to solve in practice. We show that the set of decision variables can be restricted to specific adjacencies, which we call *candidate adjacencies.* An adjacency $xy$ is a *candidate adjacency* for the median if at least $\lfloor \frac{k+1}{2} \rfloor + 1$ genomes from the set $\{A, D_1, D_2, ..., D_k\}$ contain $xy$ (where here $A$ contains $xy$ if $a(x)a(y) \in A$). Lemma 3, proved in the Appendix, shows that the number of adjacencies to consider in an ILP is linear in the sum of the sizes of the input genomes.

**Lemma 3.** *There exists an optimal median consisting of only candidate adjacencies. Furthermore, when $k$ is even, an adjacency which is not a candidate adjacency can not be a part of any optimal median.*

*Remark 4.* The difficulty of the rooted median problem stems from the fact that duplication from $M$ to the $D_i$s can create conflicting adjacencies, where a median gene extremity belongs to several candidate adjacencies. It is interesting to observe that this can happen only due to convergent evolution, i.e. the fact that the same adjacency is created independently in several $D_i$s. This suggests that in the practical context of a limited level of convergent evolution, the rooted median problem is easy to solve.

*The ILP for the Rooted Median Problem.* We can now provide the complete ILP formulation to solve the rooted SCJ-TD-FD median problem. Let $x(e)$ be a binary decision variable denoting the inclusion of edge (candidate adjacency) $e \in E(G)$ in $M$. Also, let $c_i$ be a binary decision variable indicating if at least one edge with color $i$ belongs to $M$. From the previous paragraph, one can write the objective function as

**Maximize:**

$$\sum_{e \in E(G)} w(e)x(e) + 2\sum_{i=1}^{m_A} c_i + 2t(M_a) - 2\sum_{g \in \Gamma_A} \alpha_{g,AM_a} - 2\sum_{i=1}^{k}\sum_{g \in \Gamma_M} \alpha_{g,MD_i}$$

We now describe the constraints of the ILP. The first set of constraints concern the *consistency* of the set of chosen adjacencies, that ensures that each gene extremity in $M$ belongs to at most one adjacency, or in other words that $M$ is a matching for the graph $G$ (these are the first two sets of constraints below). Next, we use an additional set of constraints to determine the values of $c_i$, $i = \{1, 2, ..., m_A\}$. If at least one adjacency of color $i$ is present in the median, $c_i = 1$, otherwise $c_i = 0$. The following inequalities define these color constraints:

$$\sum_{e=(y_h,z)} x(e) \leq 1 \qquad\qquad \forall y \in \Gamma_M \qquad\qquad (6)$$

$$\sum_{e=(y_t,z)} x(e) \leq 1 \qquad\qquad \forall y \in \Gamma_M \qquad\qquad (7)$$

$$c_i = \left\lceil \frac{\sum_{C(e)=i} x(e)}{|E_i|} \right\rceil \qquad\qquad \forall i \in \{1, 2, ..., m_A\} \qquad\qquad (8)$$

Note that for $c_i$ above, the constraints of the type $x = \lceil y \rceil$ are not linear, but if $x$ is restricted to be in $\{0, 1\}$, it can be replaced by the constraint $y \leq x \leq y+\epsilon$, where $\epsilon$ is very close to 1, say 0.999. A similar trick can be used for floor functions.

In order to compute $\alpha_{g,uv}$ for every pair $(u,v)$ – where either $u = A$, $v = M_a$ or $u = M, v = D_i$ for some $i$ – and every gene $g \in \Gamma_u$, we use some additional constraints. Let $p_v(e)$ be the binary variable denoting if the adjacency $e$ exists in $v$. We use an indicator variable $\lambda_{g,uv}$ such that $\lambda_{g,uv} = 1$ if and only if all copies of $g$ are involved in $g_h g_t$ adjacencies. Consequently, $\lambda_{g,uv} = 1$ ensures the existence of the $g_h g_t$ adjacency in $r(v)$. Thus, $\lambda_{g,uv} = \left\lfloor \frac{n_v(g_h g_t)}{n_v(g)} \right\rfloor$. Further, we use $\Lambda_{g,uv}$ to indicate if at least one instance of $g_h g_t$ has been observed in $v$. Thus, we can represent $\Lambda_{g,uv}$ as $\left\lceil \frac{n_v(g_h g_t)}{n_v(g)} \right\rceil$. Since we already know the gene orders of $A$ and each $D_i$, the values of $p_A(e)$ and $p_{D_i}(e)$ are known. Further, $p_M(e) = x(e)$. Thus, we obtain the following constraints for every gene $g$ and branch $(u, v)$:

$$\lambda_{g,uv} = \left\lfloor \frac{n_v(g_h g_t)}{n_v(g)} \right\rfloor \tag{9}$$

$$\Lambda_{g,uv} = \left\lceil \frac{n_v(g_h g_t)}{n_v(g)} \right\rceil \tag{10}$$

$$\alpha_{g,uv} = \min(p_u(g_h g_t), \Lambda_{g,uv} - \lambda_{g,uv}) \tag{11}$$

$$t_v(g) = n_v(g_h g_t) - \lambda_{g,uv} \tag{12}$$

We use the fact that if $g_h g_t \notin v$ for some $g$ then $g_h g_t \notin r(v)$. Thus, if $g_h g_t \notin v$, $\lambda_{g,uv} = 0$ thereby ensuring the correctness of constraints to find $\alpha_{g,uv}$. Again, note that the min function is not linear, but that a constraint $x = \min(y, z)$ can be replaced by $x \geq y$ and $x \geq z$, assuming that $x, y, z \in \{0, 1\}$.

## 5    Experimental Results

We ran experiments on simulated data in order to evaluate the ability of the ILP to correctly predict the gene order of the median genome. The input for the program, including gene orders for the ancestor genome $A$ and the descendant genomes $D_i$, along with the orthology relations, generated using the ZOMBI genome simulator [8]. The ILP was solved using the Gurobi solver.

*Simulations Parameters.* Our input genomes consisted of one ancestor $A$ and two descendants $D_1$ and $D_2$. We started with the ancestral genome $A$ as a single circular chromosome consisting of 1000 genes, belonging to different gene families (so without duplicate genes). The genome $A$ evolved into the median genome $M$ using duplications, inversions and translocations. The genome $M$ was further evolved along two independent branches to yield the descendant genomes, $D_1$ and $D_2$. The total number of rearrangements (inversions + translocations) from $A$ to $M$ and from $M$ to $D_i$ was varied from 100 to 500, in steps of 100. The parameter for duplication events was kept constant throughout the experiments. The average number of duplicated genes, over all three branches collectively, was found to be 362.8 with a standard deviation of 82 genes. Considering the number of duplication events, the mean and standard deviation of segmental duplications over the three branches was 72.6 and 15.8 respectively. The lengths of segmental duplications, inversions and translocations were controlled using specific extension rates. These extension rates (all between 0 and 1) are the parameters of a geometric distribution dictating the respective lengths. Thus, the length of the segment being acted upon would be 1 if the extension rate parameter is set to 1 and would increase as the parameter value reduces. In our experiments, the inversion, translocation and duplication extension rates were 0.05, 0.3 and 0.2 respectively. For each setting (number of rearrangements) we ran 40 simulations.

*Results.* For each simulation, we compared the optimal median according to the ILP to the actual median generated by the simulator. For each group, we measured the average precision and recall statistics. The ILP predicts the median genome in the form of its adjacency set. Thus, in this context, precision refers to the ratio of number of correctly predicted adjacencies to the total number of adjacencies in the computed optimal median. On the other hand, recall represents the ratio of the correctly predicted adjacencies to the total number of adjacencies in the actual median. For each instance, we measured the number of candidate adjacencies used in the ILP. Additionally, to evaluate the effectiveness of our approach, we also measured the number of adjacencies in the solution which were common to all genomes ($A, D_1$ and $D_2$) and those common to only two of the three.

An overview of the results is given in Table 1. The ILP rarely predicts an erroneous adjacency to be a part of the optimal median, with a near-perfect precision. This property is observed throughout the experiments irrespective of the number of rearrangement events. On the other hand, the ILP predicts more than 90% of the median for lower rates of rearrangement and a decreasing trend is observed as the number of rearrangement events increase. This can be partly attributed to the decrease in the number of candidate adjacencies. In general, the number of candidate adjacencies is lower than the true number of adjacencies in the median, as including other adjacencies may result in a non-optimal median. This, however, emphasizes the practicality of Lemma 3, as the number of adjacency variables is significantly reduced. It can also be observed that the number of adjacencies common to all genomes decreases with increase in rearrangements. These adjacencies will be preferred by the ILP on account of higher weight.

**Table 1.** Statistics of the ILP median experiment on simulated data.

| Events | Adj. in true median | Cand. adj. | Adj. in ILP median | Precision | Recall | % Adj. common to all genomes | % Adj. common to two genomes | No. of optimal solutions | Avg. time per run (in sec) |
|--------|------|------|------|--------|--------|--------|--------|--------|------|
| 100 | 1514 | 1503 | 1493 | 0.9998 | 0.9859 | 86.43 | 13.57 | 2.3 | 53 |
| 200 | 1107 | 1062 | 1044 | 0.9991 | 0.9428 | 69.49 | 30.51 | 15.8 | 29 |
| 300 | 1312 | 1192 | 1155 | 0.9985 | 0.8758 | 52.94 | 47.06 | 40.3 | 38 |
| 400 | 1151 | 985 | 961 | 0.9981 | 0.8329 | 49.44 | 50.56 | 393.7 | 51 |
| 500 | 1430 | 1174 | 1132 | 0.9972 | 0.7897 | 46.68 | 53.32 | 3682.6 | 84 |

Another notable observation is the increase in the number of optimal solutions with larger rates of rearrangement. This correlates naturally with the decrease in the number of adjacencies which are common to all genomes. For only 100 rearrangements, the ILP outputs a unique optimal median in most runs, with an overall average of 2.3 solutions. However, the average number of optimal solutions exceeded 3000 in case of 500 rearrangements. Despite a pool of optimal solutions, the SCJ distance between the actual median and an optimal median does not vary by much. If the SCJ distance between the actual median and a randomly chosen optimal median is $D$, then the distance between the actual median and any other optimal median was observed to stay within the range $(D - 2, D + 2)$. For most of our simulations, the ILP output an optimal median in under a minute, with the exception of the case with 500 rearrangement events.

# 6   Conclusion

In this chapter, we introduced the directed and rooted median problems and studied them under the SCJ-TD-FD model. We proved that computing the median with the most parsimonious directed distance for an ancestor $A$ and descendants $D_i$, $i = 1$ to $k$ is NP-hard by reduction from the 2P2N-3SAT problem. This contrasts with the directed median problem which does not involve an ancestral genome $A$. An interesting feature of our hardness proof is that it relies on two identical descendant genomes, showing a sharp tractability boundary between the directed pairwise distance problem and the rooted median of three genomes problem. Similarly to other SCJ-related median problems, our rooted median problem aims at selecting adjacencies among candidate adjacencies which are seen in a majority of the given input genomes; nevertheless the possibility of conflicting median adjacencies due to convergent evolution is at the heart of the intractability of the problem (Remark 4). To address this intractability, we provide a simple Integer Linear Program that computes an optimal median. Without surprise, we observe that our ILP outputs a more reliable estimate of the median in case of lower rates of rearrangements. Moreover, we observe that despite having many more optimal solutions for higher rates of rearrangement, the distance of a random solution from the actual median does not deviate by much.

Our work can be commented with regard to the Small Parsimony Problem under the directed SCJ-TD-FD model. The hardness result of the rooted median problem likely implies the corresponding SPP problem is also NP-hard. This motivates our current work about extending the rooted median ILP toward the SPP. It is worth noting that our median ILP can also be used to solve the SPP by iterative application from an initial assignment of ancestral gene orders, similarly to the early SPP solvers for genome rearrangements such as GRAPPA [18]. Considering the multiplicity of the solutions, it also remains to be investigated if the sampling and subsequent analysis of co-optimal evolutionary scenarios, in a similar manner as [16], is possible within this framework.

**Acknowledgments.** CC is supported by Natural Science and Engineering Research Council of Canada (NSERC) Discovery Grant RGPIN-2017-03986. CC and PF are supported by CIHR/Genome Canada Bioinformatics and Computational Biology grant B/CB 11106. Most computations were done on the Cedar system of Compute Canada through a resource allocation to CC.

## Appendix

***Proof of Eq. (3).*** We remind that the original pairwise distance formula (Eq. (2)) is

$$d_{\mathrm{DSCJ}}(A, D) = |A - r(D)| + |r(D) - A| + 2\delta(A, r(D)) + t(D)$$

and we want to prove it is equivalent to

$$d_{\mathrm{DSCJ}}(A, D) = |A - r(D)| + |r(D) - A| + 2\delta(A, D) - t(D).$$

Notice that the $2\delta(A, r(D))$ term from the original formula was switched for the $2\delta(A, D)$ term. Consider the difference in the number of genes from $D$ to $r(D)$. Each time we remove a $g_h g_t$ observed duplication from $D$ while reducing it, it corresponds to removing a copy of $g$ from $D$. Thus $D$ has $t(D)$ more genes than $r(D)$, so that $2\delta(A, D) = 2\delta(A, r(D)) + 2t(D)$. This implies $2\delta(A, D) - t(D) = 2\delta(A, r(D)) + t(D)$. $\qquad\square$

***Proof of Lemma 2.*** Suppose that for some $i$, $M$ contains none of $g_i^+ \gamma_i^+$ or $g_i^- \gamma_i^-$. Note that $M$ does not contain $g_i^+ \gamma_i^-$ nor $g_i^- \gamma_i^+$, by Lemma 1. This implies that $g_i' \gamma_i' \notin M_a$, as we have excluded all the four possibilities of having this adjacency in $M_a$.

Consider the median $M'$ obtained from $M$ by adding $g_i^+ \gamma_i^+$, cutting the adjacencies that $g_i^+$ and $\gamma_i^+$ were contained in, if needed. If $g_i^+$ and $\gamma_i^+$ are both telomeres in $M$, then it is easy to check that $M' = M + g_i^+ \gamma_i^+$ ($M$ augmented by the adjacency $g_i^+ \gamma_i^+$) attains a better distance than $M$ since $g_i^+ \gamma_i^+ \in D_1, D_2$ and $a(g_i^+) a(\gamma_i^+) = g_i' \gamma_i' \in A$ (this decreases the distance by 3).

Suppose that $g_i^+ x \in M$ for some $x$, and that $\gamma_i^+$ is a telomere in $M$. By Lemma 1, $g_i^+ x$ is in both $D_1$ and $D_2$, which implies that $x = c_j$ or $x = \alpha_j$ for some $j$. This implies in turn that $a(g_i^+) a(x) \notin A$. We can argue that $M' = M - g_i^+ x + g_i^+ \gamma_i^+$ is better. To see this, observe that $|M' - D_1| = |M - D_1|$ and $|D_1 - M'| = |D_1 - M|$ (and the same with $D_2$). On the other hand, recalling that $g_i' \gamma_i' \notin M_a$, we have $|M_a' - A| = |M_a - A| - 1$ (because $a(g_i^+) a(x) \notin A$ and $a(g_i^+) a(\gamma_i^+) \in A$) and $|A - M_a'| = |A - M_a| - 1$ (because $a(g_i^+) a(\gamma_i^+) \in A$). We have thus decreased the distance by 2. The same argument applies if $g_i^+$ is a telomere but $\gamma_i^+$ is not.

Finally, suppose that $g_i^+ x$ and $\gamma_i^+ y$ are adjacencies of $M$. As we argued above, $a(g_i^+) a(x) \notin A$ and $a(\gamma_i^+) a(y) \notin A$. Letting $M' = M - g_i^+ x - \gamma_i^+ y + g_i^+ \gamma_i^+$, we find that $|M' - D_1| = |M - D_1|$ and $|D_1 - M'| = |D_1 - M| + 1$. As the same holds with $D_2$, we have increased the distance to $D_1$ and $D_2$ by 2. On the other hand, $|A - M_a'| = |A - M_a| - 1$ and $|M_a' - A| = |M_a - A| - 2$. To sum up, the total distance decreases by 1. $\qquad\square$

***Proof of Claim 1.*** Call an extremity $e$ of a gene in $\Gamma$ *matchable* if there exists an adjacency of $D_1$ that contains $e$. By Lemma 1, the adjacencies of $M$ only contain matchable extremities. The $g_i^+, g_i^-, \gamma_i^+$ and $\gamma_i^-$ extremities account for $4n$ matchable extremities. The $c_j$ genes account for $m$ matchable extremities and the $\alpha_j$ genes for $2n - m$ matchable extremities. Thus there are $4n + m + 2n - m = 6n$

matchable extremities. Because an adjacency contains 2 extremities, there can be at most $3n$ adjacencies in $M$. The second part of the claim follows from the fact that we have to assume that every $c_j$ and $\alpha_j$ is matched to attain this bound. □

**Proof of Claim (2).** By the definition of $q$, we have $|A - M_a| = n - q$ and $|M_a - A| = |M| - q$. It follows that

$$d_{\mathrm{DSCJ}}(A, M_a) = |A - M_a| + |M_a - A| + 2\delta(A, M_a) - t(M_a)$$
$$= n - q + |M| - q + 2n - 0$$
$$= |M| + 3n - 2q$$

Using Lemma 1, we also have $d_{\mathrm{DSCJ}}(M, D_1) = |M - D_1| + |D_1 - M| + 2\delta(M, D_1) = 0 + |D_1| - |M| + 2\delta(M, D_1)$. Thus the sum of the 3 distances is

$$|M| + 3n - 2q + 2|D_1| - 2|M| + 4\delta(M, D_1) \leq 2|D_1| - 2n + 4\delta(M, D_1)$$

(this inequality is due to our initial assumption on the total distance of $M$). After simplifying, this gives $5n \leq |M| + 2q$. By Claim 1, $|M| \leq 3n$ and because $A$ has $n$ adjacencies, $q \leq n$. Hence, this inequality is only possible if $|M| = 3n$ and $q = n$. □

**Proof of Eq. (4).** From Eq. (3), we have $d_{DSCJ}(u, v) = |u - r(v)| + |r(v) - u| + 2\delta(u, v) - t(v)$. However, it is easier to express the distance without the reduced genome terms. Hence, we eliminate the need for computing the reduced genomes by replacing $|u - r(v)|$ and $|r(v) - u|$ by suitable expressions as follows. We show that (1) $|u - r(v)| = |u - v| + \sum_{g \in \Gamma_u} \alpha_g$, and (2) $|r(v) - u| = |v - u| - t(v) + \sum_{g \in \Gamma_u} \alpha_g$. Substituting the terms in Eq. (3) yield Eq. (4).

(1) Consider first the difference between $u - r(v)$ and $u - v$. Suppose that $xy \in u - v$ but $xy \notin u - r(v)$. Then $xy \in r(v)$ but $xy \notin v$, which is not possible. Thus the difference can only be due to some $xy \in u - r(v)$ such that $xy \notin u - v$. This means that $xy \notin r(v)$ and $xy \in v$, which only happens when $xy = g_h g_t$ for some gene $g$. As we have $xy = g_h g_t \in u \cap v$ and $g_h g_t \notin r(v)$, we also have $\alpha_g = 1$, by definition. Since only one such adjacency is possible for each gene $g$ (because $u$ is trivial), $u - r(v)$ and $u - v$ differ only by adjacencies on genes for which $\alpha_g = 1$. We have shown that $|u - r(v)| = |u - v| + \sum_{g \in \Gamma_u} \alpha_g$.

(2) Now consider the difference between $r(v) - u$ and $v - u$. Note that there are $t(v)$ adjacencies in $v$ not in $r(v)$, all observed duplications of the type $g_h g_t$. Let $g \in \Gamma_u$. If $g_h g_t \notin u$, then all of the $t(g)$ observed duplications in $g$ are counted in $v - u$ but not in $r(v) - u$. This is also true when $g_h g_t \in u$ and $g_h g_t \in r(v)$. In these cases, $\alpha_g = 0$. However when $g_h g_t \in u \cap v$ but $g_h g_t \notin r(v)$, there are $t(g) - 1$ of the $g_h g_t$ adjacencies counted in $v - u$ not counted in $r(v) - u$ (this is because exactly one $g_h g_t$ adjacency of $v$ can be matched with the $g_h g_t$ adjacency in $u$, and $r(v)$ has no such adjacency). This case occurs precisely when $\alpha_g = 1$. This shows that $|r(v) - u| = |v - u| - \sum_{g \in \Gamma_u}(t(g) - \alpha_g) = |v - u| - t(v) + \sum_{g \in \Gamma_u} \alpha_g$. □

**Proof of Claim (3).** By Eq. (4), we know that

$$d_{\text{DSCJ}}(A, M_a) = |A - M_a| + |M_a - A| + 2\delta(A, M_a) - 2t(M_a) + 2 \sum_{g \in \Gamma_A} \alpha_{g, AM_a}$$

$$d_{\text{DSCJ}}(M, D_i) = |M - D_i| + |D_i - M| + 2\delta(M, D_i) - 2t(D_i) + 2 \sum_{g \in \Gamma_M} \alpha_{g, MD_i}$$

where $\Gamma_A$ and $\Gamma_M$ are the set of genes in the gene orders of $A$ and $M$, respectively, and so also the genes alphabets for $M$ and the $D_i$s. Variables $\alpha_{g, AM_a}$ and $\alpha_{g, MD_i}$ are defined as $\alpha_{g, uv}$ above.

For any two adjacency sets $X$ and $Y$, we use the identity $|X - Y| + |Y - X| = |X| + |Y| - 2|X \cap Y|$ to obtain

$$d_{\text{DSCJ}}(A, M_a) = |A| + |M_a| - 2|A \cap M_a| + 2\delta(A, M_a) - 2t(M_a) + 2 \sum_{g \in \Gamma_A} \alpha_{g, AM_a},$$

$$d_{\text{DSCJ}}(M, D_i) = |M| + |D_i| - 2|M \cap D_i| + 2\delta(M, D_i) - 2t(D_i) + 2 \sum_{g \in \Gamma_M} \alpha_{g, MD_i}.$$

This eliminates the need to count the actual number of cut and join events along every branch. Instead, it suffices to compute the common adjacencies in the parent and child genomes (using the terms $|A \cap M_a|$ and $|M \cap D_i|$) for each branch $(A, M_a)$ and $(M, D_i)$.

For a median $M$, let $s(M) = d_{\text{DSCJ}}(A, M_a) + \sum_{i=1}^{k} d_{\text{DSCJ}}(M, D_i)$ be the *score* of $M$. It follows easily from above that

$$s(M) = \left[ |A| + 2\delta(A, M_a) + \sum_{i=1}^{k} (|D_i| + 2\delta(M, D_i)) \right]$$
$$- \left[ \sum_{i=1}^{k} \left( 2|M \cap D_i| + 2t(D_i) - 2 \sum_{g \in \Gamma_M} \alpha_{g, MD_i} \right) \right.$$
$$\left. + 2|A \cap M_a| + 2t(M_a) - 2 \sum_{g \in \Gamma_A} \alpha_{g, AM_a} - (k+1)|M| \right]$$

Let $N = |A| + 2\delta(A, M_a) + \sum_{i=1}^{k} \left( |D_i| + 2\delta(M, D_i) + 2t(D_i) \right)$. Given that $N$ depends only on $A$ and $D_i$ and not on $M$, it is constant (note that $\delta(A, M_a)$ and $\delta(M, D_i)$ are constant as the gene content of $M$ is an input to the problem). Thus in order to minimize the score $s(M)$, we only need to maximize the term:

$$\sum_{i=1}^{k} \left( 2|M \cap D_i| - 2 \sum_{g \in \Gamma_M} \alpha_{g, MD_i} \right) + 2|A \cap M_a| + 2t(M_a) - 2 \sum_{g \in \Gamma_A} \alpha_{g, AM_a} - (k+1)|M|$$

which is negated in $s(M)$, as required in Eq. (5).    □

***Proof of Lemma*** *(3).* To prove this lemma, we start with a median containing a non-candidate adjacency. For odd values of $k$, we prove that removing the non-candidate adjacency results in another median of the same cost whereas for even $k$, it is shown that the resultant median (on removing the non-candidate adjacency) is better. We temporarily ignore the influence of reduced genomes for this proof.

Consider an adjacency $xy$ that is not a candidate. Recall that since $xy$ is not a candidate it is present in at most $\lfloor \frac{k+1}{2} \rfloor$ genomes from $\{A, D_1, ..., D_k\}$. Assume that $M$ is a median genome and $xy$ is present in $M$. Further, assume that $M$ is optimal. Thus, the sum of the distances $d_{\mathrm{DSCJ}}(A, M_a) + \sum_{i=1}^{k} d_{\mathrm{DSCJ}}(M, D_i)$ should be the least over all medians. Let $M'$ be the genome obtained by removing $xy$ from $M$.

Let $D_{xy} \subseteq \{D_1, ..., D_k\}$ be the set of descendant genomes that contain $xy$, and let $\overline{D_{xy}}$ be the set of those that do not. For any $D_i \in D_{xy}$, the adjacency need not be cut along $(M, D_i)$, however it has to be added along $(M', D_i)$, introducing an extra cost of 1 to the total distance. Thus, $d_{\mathrm{DSCJ}}(M, D_i) = d_{\mathrm{DSCJ}}(M', D_i) - 1$, for all $D_i \in D_{xy}$. On the other hand, if $D_i \notin D_{xy}$, then it does not contain $xy$. Consequently, for all such $D_i$, the adjacency has to be cut along $(M, D_i)$ but not along $(M', D_i)$ (since $M'$ does not contain it in the first place). Thus, for all $D_i \notin D_{xy}$, $d_{\mathrm{DSCJ}}(M, D_i) = d_{\mathrm{DSCJ}}(M', D_i) + 1$.

Further if $A$ contains $a(x)a(y)$, it need not be cut along $(A, M_a)$ but may need to be cut along $(A, M_a')$ thereby introducing a possible extra cost of 1 (note here the possibility that some $x^*y^* \in M$ distinct from $xy$ such that $a(x^*)a(y^*) = a(x)a(y)$). Thus, $d_{\mathrm{DSCJ}}(A, M_a) \geq d_{\mathrm{DSCJ}}(A, M_a') - 1$. If instead, $A$ does not contain $xy$ then it has to be joined along $(A, M_a)$ and not along $(A, M_a')$. Unlike the previous case, the cost of the join is unavoidable. Hence, $d_{\mathrm{DSCJ}}(A, M_a) = d_{\mathrm{DSCJ}}(A, M_a') + 1$.

Case 1: $A$ contains $xy$. Then $|D_{xy}| \leq \lfloor \frac{k+1}{2} \rfloor - 1$.

$$d_{\mathrm{DSCJ}}(A, M_a) \geq d_{\mathrm{DSCJ}}(A, M_a') - 1$$
$$d_{\mathrm{DSCJ}}(M, D_i) = d_{\mathrm{DSCJ}}(M', D_i) - 1 \qquad \forall D_i \in D_{xy}$$
$$d_{\mathrm{DSCJ}}(M, D_i) = d_{\mathrm{DSCJ}}(M', D_i) + 1 \qquad \forall D_i \notin D_{xy}$$

Summing over all the input genomes, we get

$$d_{\mathrm{DSCJ}}(A, M_a) + \sum_{D_i \in D_{xy}} d_{\mathrm{DSCJ}}(M, D_i) \geq d_{\mathrm{DSCJ}}(A, M_a') + \sum_{D_i \in D_{xy}} d_{\mathrm{DSCJ}}(M', D_i)$$
$$+ |\overline{D_{xy}}| - (|D_{xy}| + 1)$$

We know that $|D_{xy}| + 1 \leq \lfloor \frac{k+1}{2} \rfloor$. If $k$ is even, $|\overline{D_{xy}}| > |D_{xy}| + 1$. Hence,

$$d_{\mathrm{DSCJ}}(A, M_a) + \sum_{D_i \in D_{xy}} d_{\mathrm{DSCJ}}(M, D_i) > d_{\mathrm{DSCJ}}(A, M_a') + \sum_{D_i \in D_{xy}} d_{\mathrm{DSCJ}}(M', D_i)$$

Thus, the cost of $M'$ is better than that of the optimal median $M$ and we have a contradiction. If $k$ is odd, then $|\overline{D_{xy}}| = |D_{xy}| + 1$ and hence both $M$ and $M'$ incur the same overall cost. In other words, the removal of a non-candidate adjacency does not increase the cost of the optimal median. Thus, iteratively removing all such adjacencies will yield an optimal median that consists solely of candidate adjacencies.

Case 2: $A$ does not contain $xy$. Then $|D_{xy}| \leq \lfloor \frac{k+1}{2} \rfloor$.

$$d_{\text{DSCJ}}(A, M_a) = d_{\text{DSCJ}}(A, M') + 1$$
$$d_{\text{DSCJ}}(M, D_i) = d_{\text{DSCJ}}(M', D_i) - 1 \qquad \forall D_i \in D_{xy}$$
$$d_{\text{DSCJ}}(M, D_i) = d_{\text{DSCJ}}(M', D_i) + 1 \qquad \forall D_i \notin D_{xy}$$

The analysis in this case is similar to Case 1. On adding all the equations and using $|D_{xy}| \leq \lfloor \frac{k+1}{2} \rfloor$, once again we reach a contradiction when $k$ is even. When $k$ is odd, both $M$ and $M'$ yield the same overall distance. Thus, we can still obtain the optimal median by iteratively removing non-candidate adjacencies.

Thus, when $k$ is odd, there exists at least one optimal median consisting only of candidate adjacencies. However, when $k$ is even, the optimal median must consist only of candidate adjacencies.                                            □

# References

1. Angibaud, S., Fertin, G., Rusu, I., Thévenin, A., Vialette, S.: On the approximability of comparing genomes with duplicates. J. Graph Algorithms Appl. **13**(1), 19–53 (2009)
2. Berman, P., Karpinski, M., Scott, A.D.: Approximation hardness of short symmetric instances of MAX-3SAT. Technical report TR03-049, Electronic Colloquium on Computational Complexity (ECCC) (2003)
3. Blanchette, M., Bourque, G., Sankoff, D.: Breakpoint phylogenies. Genome Inform. **8**, 25–34 (1997)
4. Blin, G., Chauve, C., Fertin, G., Rizzi, R., Vialette, S.: Comparing genomes with duplications: a computational complexity point of view. IEEE/ACM Trans. Comput. Biol. Bioinform. **4**(4), 523–534 (2007)
5. Boyd, S.C., Haghighi, M.: Mixed and circular multichromosomal genomic median problem. SIAM J. Discret. Math. **27**(1), 63–74 (2013)
6. Bryant, D.: The complexity of calculating exemplar distances. In: Sankoff, D., Nadeau, J.H. (eds.) Comparative Genomics: Empirical and Analytical Approaches to Gene Order Dynamics, Map Alignment and the Evolution of Gene Families, pp. 207–211. Springer, Dordrecht (2000). https://doi.org/10.1007/978-94-011-4309-7
7. Bryant, D.: A lower bound for the breakpoint phylogeny problem. J. Discret. Algorithms **2**(2), 229–255 (2004)
8. Davin, A.A., Tricou, T., Tannier, E., de Vienne, D.M., Szollosi, G.J.: Zombi: a simulator of species, genes and genomes that accounts for extinct lineages. bioRxiv (2018). https://doi.org/10.1101/339473
9. Doerr, D., Balaban, M., Feijão, P., Chauve, C.: The gene family-free median of three. Algorithms Mol. Biol. **12**(1), 14:1–14:14 (2017)

10. Feijão, P., Mane, A.C., Chauve, C.: A tractable variant of the single cut or join distance with duplicated genes. In: Meidanis, J., Nakhleh, L. (eds.) RECOMB CG 2017. LNCS, vol. 10562, pp. 14–30. Springer, Cham (2017). https://doi.org/10. 1007/978-3-319-67979-2_2

11. Feijão, P., Meidanis, J.: SCJ: a breakpoint-like distance that simplifies several rearrangement problems. IEEE/ACM Trans. Comput. Biol. Bioinform. **8**(5), 1318–1329 (2011)

12. Fertin, G., Labarre, A., Rusu, I., Tannier, E., Vialette, S.: Combinatorics of Genome Rearrangements. Computational Molecular Biology. MIT Press, Cambridge (2009)

13. Kondrashov, F.A.: Gene duplication as a mechanism of genomic adaptation to a changing environment. Proc. R. Soc. Lond. B Biol. Sci. **279**(1749), 5048–5057 (2012)

14. Kovác, J.: On the complexity of rearrangement problems under the breakpoint distance. J. Comput. Biol. **21**(1), 1–15 (2014)

15. Levasseur, A., Pontarotti, P.: The role of duplications in the evolution of genomes highlights the need for evolutionary-based approaches in comparative genomics. Biol. Direct **6**(1), 11 (2011)

16. Luhmann, N., Lafond, M., Thèvenin, A., Ouangraoua, A., Wittler, R., Chauve, C.: The SCJ small parsimony problem for weighted gene adjacencies. IEEE/ACM Trans. Comput. Biol. Bioinform. (2017). https://doi.org/10.1109/TCBB.2017. 2661761

17. Ming, R., VanBuren, R., Wai, C.M., et al.: The pineapple genome and the evolution of CAM photosynthesis. Nat. Genet. **47**(12), 1435–1442 (2015)

18. Moret, B.M.E., Wyman, S.K., Bader, D.A., Warnow, T.J., Yan, M.: A new implementation and detailed study of breakpoint analysis. In: Pacific Symposium on Biocomputing, pp. 583–594 (2001)

19. Neafsey, D., Waterhouse, R., Abai, M., et al.: Highly evolvable malaria vectors: the genomes of 16 Anopheles mosquitoes. Science **347**(6217), 1258522 (2015)

20. Pe'er, I., Shamir, R.: The median problems for breakpoints are np-complete. Technical report TR98-071, Electronic Colloquium on Computational Complexity (ECCC) (1998)

21. Sankoff, D., Sundaram, G., Kececioglu, J.D.: Steiner points in the space of genome rearrangements. Int. J. Found. Comput. Sci. **7**(1), 1–9 (1996)

22. Tannier, E., Zheng, C., Sankoff, D.: Multichromosomal median and halving problems under different genomic distances. BMC Bioinform. **10**, 120 (2009)

23. Zeira, R., Shamir, R.: Sorting by cuts, joins, and whole chromosome duplications. J. Comput. Biol. **24**(2), 127–137 (2017)

# A General Framework for Genome Rearrangement with Biological Constraints

Pijus Simonaitis[1], Annie Chateau[1,2], and Krister M. Swenson[1,2(✉)]

[1] LIRMM, CNRS – Université Montpellier, 161 rue Ada, 34392 Montpellier, France
swenson@lirmm.fr
[2] Institut de Biologie Computationnelle (IBC), Montpellier, France

**Abstract.** This paper generalizes previous studies on genome rearrangement under biological constraints, using double cut and join (DCJ). We propose a model for weighted DCJ, along with a family of optimization problems called $\varphi$-MCPS (MINIMUM COST PARSIMONIOUS SCENARIO), that are based on edge labeled graphs. After embedding known results in our framework, we show how to compute solutions to general instances of $\varphi$-MCPS, given an algorithm to compute $\varphi$-MCPS on a circular genome with exactly one occurrence of each gene. These general instances can have an arbitrary number of circular and linear chromosomes, and arbitrary gene content. The practicality of the framework is displayed by generalizing the results of Bulteau, Fertin, and Tannier on the SORTING BY wDCJs AND indels IN INTERGENES problem, and by generalizing previous results on the MINIMUM LOCAL PARSIMONIOUS SCENARIO problem.

**Keywords:** Double Cut and Join (DCJ)
Weighted genome rearrangement · Parsimonious scenario
Breakpoint graph · Maximum alternating cycle decomposition

## 1 Introduction

### 1.1 Context

The practical study of genome rearrangement between evolutionarily distant species has been limited by a lack of mathematical models capable of incorporating biological constraints. Without such constraints the number of *parsimonious* (shortest length) rearrangement scenarios between two gene orders grows exponentially with respect to the minimum number of rearrangements between genomes [11]. A natural way to mitigate this problem is to develop models that weight rearrangements according to their likelihood of occurring; a breakpoint may be more likely to occur in some intergenic regions than others.

To this end, the study of length-weighted reversals was started in the late nineties by Blanchette, Kunisawa, and Sankoff [9]. Baudet, Dias, and Dias

© Springer Nature Switzerland AG 2018
M. Blanchette and A. Ouangraoua (Eds.): RECOMB-CG 2018, LNBI 11183, pp. 49–71, 2018.
https://doi.org/10.1007/978-3-030-00834-5_3

present a summary of work done in this area, along with work on reversals centered around the origin of replication [2]. Recently, Tannier has published a series of papers focused on weighting intergenic regions by their length in nucleotides. In [7], Biller, Guéguen, Knibbe, and Tannier pointed out that, according to the Nadeau-Taylor model of uniform random breakage [17,18], a breakpoint is more likely to occur in a longer intergenic region. Subsequent papers by Fertin, Jean, and Tannier [15], and Bulteau, Fertin, and Tannier [12] present algorithmic results for models that take into account the length of intergenic regions. Using Hi-C data [16], Veron et *al.* along with our own study, have pointed out the importance of weighting pairs of breakpoints according to how close they tend to be in physical space [19,24]. In order to use this physical constraint, we partitioned intergenic regions into co-localized areas, and developed algorithms for computing distances that minimize the number of rearrangements that operate on breakpoints between different areas [22,23].

Much of this work is based on the mathematically clean model for genome rearrangement called *Double Cut and Join*, or *DCJ* [4,25]. Genomes are partitioned into $n$ orthologous syntenic blocks that we will simply call *genes*. Each gene is represented by two extremities, and each chromosome is represented by an ordering of these extremities. Those extremities that are adjacent in this ordering are paired, and transformations of these pairs occur by swapping extremities of two pairs. DCJ can naturally be interpreted as a graph edit model with the use of the *breakpoint graph*, where there is an edge between gene extremities $a$ and $b$ for each adjacent pair. A DCJ operation replaces an edge pair $\{\{a,b\},\{c,d\}\}$ of the graph by $\{\{a,c\},\{b,d\}\}$ or $\{\{a,d\},\{b,c\}\}$. This edge edit operation on a graph is called a *2-break*.

This paper establishes a general framework for weighting rearrangements. The results are based on the problem of transforming one edge-labeled graph into another through a scenario of operations, each weighted by an arbitrary function $\varphi$. The problem, called $\varphi$-MINIMUM COST PARSIMONIOUS SCENARIO (or $\varphi$-MCPS), asks for a scenario with a minimum number of 2-breaks, such that the sum of the costs for the operations is minimized.

## 1.2    Applications of Our Framework

While our framework is general, we use it to render two previous studies more practical. The first study is our work relating the likelihood of rearrangement breakpoints to the physical proximity in the nucleus [23]. This work is based on the hypothesis that two breakpoints could be confused when they are physically close. The model in this study labels the breakpoint graph edges (corresponding to intergenic regions) with fixed "colors", and the cost of a DCJ has a weight of one if the labels are different and a weight of zero if they are the same. Using that cost function, we colored intergenic regions by grouping them according to their physical proximity, as inferred by Hi-C data. Although this technique of grouping proved to make biological sense [19,22], it is far from ideal since much of the information given by the Hi-C data is lost in the labeling, and it is not immediately clear how to best compute the grouping. Our results here bypass the complexity of grouping by allowing each DCJ to be weighted by the values taken directly from the Hi-C contact maps. We give an algorithm for $\varphi$-MCPS on a breakpoint graph with an arbitrary $\varphi$ and fixed edge labels, that runs in $O(n^5)$ time in the worst case but has better parameterized complexity in practice (see Example 1). We give in Sect. 10.1 other reasons why the running times for this algorithm should remain practical.

The second study that we improve is that of Bulteau, Fertin, and Tannier [12]. Their biological constraint is based on the number of nucleotides in the intergenic regions containing breakpoints; they compute parsimonious scenarios that minimize the number of nucleotides inserted and deleted in intergenic regions. Their algorithm is restricted to instances where the breakpoint graph has only cycles (and no paths — sometimes referred to as *co-tailed* genomes). Using their $O(n \log n)$ algorithm, our framework gives an $O(n^3)$ algorithm on any breakpoint graph (see Example 3).

This is an example of how our framework simplifies algorithm design on weighted DCJs. For a weight function adhering to our general criteria of Sect. 4, future algorithm designers now need only to concentrate on developing an efficient algorithm that works on a single cycle of a breakpoint graph. Thanks to Theorem 4, they will get a polynomial time algorithm that works on a general instance for free. Section 8 shows that the same is true for approximation algorithms.

This paper is based on general results we obtain on weighted transformations of edge-labeled multi-graphs. The permitted transformations can change the connectivity of the graph through a 2-break, or change the edge labels, or both. This model not only proves to be powerful enough to subsume the previously mentioned results, but also offers other advantages. It is flexible enough so that DCJ costs can be based on the labels of edges in the breakpoint graph, or on the vertices, or a combination of both. Also, since single-gene insertions and deletions can be represented as "ghost" adjacencies [20], all of this paper applies to genomes where genes could be missing in one genome or the other. Most results can be applied to genomes with duplicate genes (as depicted in Fig. 1).

## 1.3   Our Model and General Results

The foundation of this paper is a renewed understanding of scenarios of 2-breaks on Eulerian graphs, a subject that has been studied not only in a restricted setting for genome rearrangement [1,4], but also in the more general settings of network design [5,6]. Although our results are about the transformation of one arbitrary Eulerian multi-graph $G$ into another one $H$ having the same vertex set, we find it convenient to reason in an equivalent but different setting. In the alternative setting we are given an Eulerian 2-edge-colored multi-graph with black and gray edges, the black edges being from $G$ and the gray from $H$. We transform the connectivity of the black edges into the connectivity of the gray edges. Therefore, whenever we use the word *graph*, *path* (resp. *cycle*), we are referring to an Eulerian 2-edge-colored multi-graph, a path (resp. cycle) that alternates between black and gray edges. Naturally, a *cycle decomposition* of a

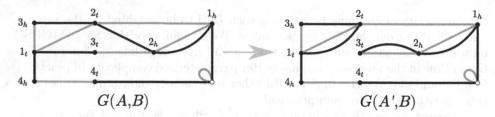

**Fig. 1.** Eulerian 2-edge-color multi-graphs for genomes $A = (\{3_t, 1_t\}, \{1_h, 2_h\}, \{2_t, 3_h\})$, $(\{4_t\}, \{4_h, 1_t\}, \{1_h\})$, $B = (\{1_h, 2_h\}, \{2_t, 1_t\})$, $(\{3_t, 2_h\}, \{2_t, 1_h\}, \{1_t, 3_h\})$, and $A' = (\{3_t, 2_h\}, \{2_t, 1_t\}, \{1_h, 2_h\}, \{2_t, 3_h\})$, $(\{4_t\}, \{4_h, 1_t\}, \{1_h\})$. Edges adjacent to a special vertex $\circ$ represent the endpoints of linear chromosomes (*e.g.* black edges $\{1_h, \circ\}$ and $\{4_t, \circ\}$). Extra edges are added for the missing genes (*e.g.* the black edge $\{2_t, 2_h\}$ and the gray edge $\{4_h, 4_t\}$), called *ghost adjacencies* in [20]. In the genomes $A$ and $A'$, gene 1 is repeated twice, and the operation transforming $A$ into $A'$ is an insertion of a gene 2, corresponding to the 2-break $G(A, B) \rightarrow G(A', B)$. A DCJ scenario transforming $A'$ into the linear genome $B$ includes a deletion of a gene 4.

graph is a partition of the edges of an Eulerian 2-edge-colored multi-graph into a set of alternating cycles. A *breakpoint graph* is a graph with a vertex for each gene extremity — each incident to exactly one gray and one black vertex — along with one chromosome endpoint vertex $\circ$ that could have degree as high as $2n$ (see Fig. 2). Section 2 introduces the breakpoint graph in detail, and defines the Double Cut and Join (DCJ) model.

Our model for weighting operations is primarily based on a labeling $\mathcal{L}$ of the edges, a set $\mathcal{O}$ of valid operations, and a weight function $\varphi : \mathcal{O} \rightarrow \mathbb{R}_+$. Roughly speaking, a labeled input graph can be transformed through a series of operations in $\mathcal{O}$, where an operation can change the connectivity of the black edges of the graph, and/or change the labels of the edges. Any weight function $\varphi$ defines an optimization problem $\varphi$-MCPS, which asks for a scenario that minimizes the total weight of the operations. This model subsumes many previously studied weighted DCJ models, as described in Sect. 4.1.

The spine of our results is built from successive theorems that speak to the decomposability into subproblems of a $\varphi$-MCPS instance. Theorem 1 shows that a parsimonious scenario of 2-breaks transforming the black edges into the gray implies a MAXIMUM ALTERNATING EDGE-DISJOINT CYCLE DECOMPOSITION (or MAECD) [13]. Theorem 2 says that an optimal solution to $\varphi$-MCPS can be found using solutions to the MAECD problem, so that if $\varphi$-MCPS can be solved on a simple alternating cycle, then it can be solved on any instance. Theorem 3 says that an optimal solution to $\varphi$-MCPS on a simple alternating cycle can be found using a solution to the $\varphi$-MCPS problem on what we call a *circle*, that is, an alternating cycle that does not visit the same vertex twice (see Fig. 4).

Under the common genome model, where each gene occurs exactly once in each genome, a relationship exists between parsimonious DCJ scenarios and solutions to MAECD on a breakpoint graph [4,10]. We exploit this link in Sect. 7. Theorem 4 ties everything together; an amortized analysis shows that, given an $O(r^t)$ algorithm for computing $\varphi$-MCPS on a circle with $r$ edges, $\varphi$-MCPS can be calculated on a breakpoint graph in $O(n^{t+1})$ time.

Under a more general genome model, that allows for changes in copy numbers of genes (*e.g.* insertions, deletions, and duplications), the spine of our results still holds due to the convenient representation of missing genes as *ghost adjacencies* in an Eulerian 2-edge-colored multi-graph [20] (See Fig. 1). All of our results hold for pairs of genomes with non-duplicated genes, but unequal gene content. Indeed, a breakpoint graph (*i.e.* graph with limited degree for most nodes) can still represent the pair of genomes in this case.

Caprara proved that MAECD is NP-Hard for Eulerian 2-edge-colored multi-graphs where each vertex is incident to at most two gray and two black edges (which is the case when there are two copies of each gene) [13]. We present a simple integer linear program (or ILP) that solves $\varphi$-MCPS for these types of graphs, given a method to solve $\varphi$-MCPS on a circle. This ILP is likely to be unwieldy in general, since the number of variables is exponential in the number of simple alternating cycles. In the case of breakpoint graphs on specific genomes, this may not always be intractable, as the number of duplicate genes may be limited. See Sect. 10.1 for a discussion of these practical matters.

## 2    DCJ Scenarios for Genomes and Breakpoint Graphs

A *genome* consists of *chromosomes* that are linear or circular orders of genes separated by potential *breakpoint* regions. In Fig. 2 the tail of an arrow represents the *tail extremity*, and the head of an arrow represents the *head extremity* of a gene. We can represent a genome by a set of *adjacencies* between the gene extremities. An *adjacency* is either *internal*: an unordered pair of the extremities that are adjacent on a chromosome, or *external*: a single extremity adjacent to one of the two ends of a linear chromosome. In what follows we will suppose that two genomes $A$ and $B$ are partitioned into $n$ genes each occurring exactly once in each genome, and our goal will be to transform $A$ into $B$ using a sequence of DCJs.

**Fig. 2.** Genomes $A$ and $B$ with their respective sets of adjacencies $\{\{1_t\}, \{1_h, 2_t\},$ $\{2_h, 3_h\}, \{3_t\}\}$ and $\{\{1_t\}, \{1_h, 2_h\}, \{2_t, 3_h\}, \{3_t\}\}$. A DCJ $\{1_h, 2_t\}, \{2_h, 3_h\}$ $\rightarrow$ $\{1_h, 2_h\}, \{2_t, 3_h\}$ transforms $A$ into $B$. The transformation $G(A, B) \rightarrow G(B, B)$ is a 2-break and $G(B, B)$ is a terminal graph.

**Definition 1 (double cut and join).** *A DCJ cuts one or two breakpoint regions and joins the resulting ends of the chromosomes back in one of the four following ways:* $\{a,b\}, \{c,d\} \rightarrow \{a,c\}, \{b,d\}$; $\{a,b\}, \{c\} \rightarrow \{a,c\}, \{b\}$; $\{a,b\} \rightarrow \{a\}, \{b\}$; *and* $\{a\}, \{b\} \rightarrow \{a,b\}$.

We represent the pairs of the genomes with a help of a breakpoint graph [1,25].

**Definition 2 (breakpoint graph).** $G(A, B)$ *for genomes* $A$ *and* $B$ *is a 2-edge-colored Eulerian undirected multi-graph.* $V$ *consists of* $2n$ *gene extremities and an additional vertex* $\circ$. *For every internal adjacency* $\{a,b\} \in A$ *(resp.* $\{a,b\} \in B$*) there is a black (resp. gray) edge* $\{a,b\}$ *in* $G(A,B)$ *and for every external adjacency* $\{a\} \in A$ *(resp.* $\{a\} \in B$*) there is a black (resp. gray) edge* $\{a,\circ\}$ *in* $G(A,B)$. *There is a number of black and gray loops* $\{\circ,\circ\}$ *ensuring that* $d^b(G(A,B),\circ) = d^g(G(A,B),\circ) = 2n$.

## 3   2-break Scenarios for 2-edge-colored Graphs

In this paper a *graph* is an Eulerian 2-edge-colored undirected multi-graph with edges colored black or gray as in Fig. 1. A graph with equal multi-sets of black and gray edges is called *terminal*, and our goal is to transform a given graph into a terminal one using 2-breaks.

**Definition 3 (2-break scenario).** *A 2-break replaces two black edges* $\{x_1, x_2\}$ *and* $\{x_3, x_4\}$ *by either* $\{x_1, x_3\}$ *and* $\{x_2, x_4\}$ *or* $\{x_1, x_4\}$ *and* $\{x_2, x_3\}$. *A 2-break scenario of length* $m$ *is a sequence of* $m$ *2-breaks transforming a graph into a terminal one.*

**Definition 4 (Eulerian graph and alternating cycle).** $G$ *is Eulerian if every vertex has equal black and gray degrees. A cycle is alternating if it is Eulerian. All use of the word cycle in this paper will be synonymous with alternating cycle.*

Define a MAXIMUM ALTERNATING EDGE-DISJOINT CYCLE DECOMPOSITION (MAECD) of a graph $G$ as a decomposition of $G$ into a maximum number of edge-disjoint alternating cycles. Denote the size of a MAECD of $G$ by $c(G)$ and the number of its black edges by $e(G)$. We make a distinction between simple cycles and circles (look at Fig. 4 to see a simple cycle that is not a circle).

**Definition 5 (simple cycle and circle).** *A graph* $G$ *is a simple cycle if the size of a MAECD,* $c(G) = 1$. *If in addition to that* $deg^b(G,v) = deg^g(G,v) = 1$ *for every vertex* $v$, *then* $G$ *is called a circle.*

## 3.1  Parsimonious 2-break Scenarios

The problem of finding a minimum length (or *parsimonious*) 2-break scenario was treated in several unrelated settings using different terminology. Lemma 1, proven in the appendix, was treated in [6] where the authors also showed that finding a minimum length 2-break scenario is NP-hard due to the NP-hardness of finding a MAECD of a graph. A variant of the problem for Eulerian digraphs where all the gray edges are loops was solved in [8].

**Lemma 1 (Bienstock et al. in [6]).** *The minimum length of a 2-break scenario on a graph $G$ is $d_{2b}(G) = e(G) - c(G)$.*

Since finding a MAECD for a breakpoint graph is easy, Lemma 1 leads to a linear time algorithm for finding a parsimonious DCJ scenario [25]. The algorithm is based on Lemma 2, proven in the appendix.

**Lemma 2 (Yancopoulos et al. in [25]).** *The minimum length of a DCJ scenario transforming genome $A$ into $B$ is equal to $d_{2b}(G(A,B)) = e(G(A,B)) - c(G(A,B))$.*

## 3.2  Decomposition of a 2-break Scenario

In this section we will show how a 2-break scenario $\rho$ of length $m$ can be partitioned into subscenarios $\rho^1, \ldots, \rho^k$ and $G$ can be decomposed into edge-disjoint Eulerian subgraphs $H^1, \ldots, H^k$ where $\rho^i$ is a scenario for $H^i$, and $k \geq e(G) - m$. We will use this decomposition in Sect. 5 to show that $\varphi$-MCPS on a graph can be solved by solving $\varphi$-MCPS on its simple cycles.

For a graph $G$ and a 2-break scenario $\rho$ we define a directed 1-edge-colored edge-labeled graph $\mathcal{D}(G, \rho)$, akin to the *trajectory graph* introduced by Shao, Lin, and Moret [21]. Denote the sequence of the first $l$ 2-breaks of $\rho$ by $\rho_l$ and the graph obtained from $G$ after these 2-breaks by $G_l$. Define $\mathcal{D}(G, \rho_0)$ in the following way: for each black edge $e$ of $G$ we have two new vertices connected by a directed edge labeled by $e$ (see Fig. 3). For the $l$-th 2-break of $\rho$, $\{x_1, x_2\}, \{x_3, x_4\} \rightarrow \{x_1, x_3\}, \{x_2, x_4\}$, merge the endpoints of the edges labeled $\{x_1, x_2\}$ and $\{x_3, x_4\}$ in $\mathcal{D}(G, \rho_{l-1})$. Proceed by adding two new vertices to $\mathcal{D}(G, \rho_{l-1})$ and two edges labeled $\{x_1, x_3\}$ and $\{x_2, x_4\}$ from the merged vertex to the newly added ones to obtain $\mathcal{D}(G, \rho_l)$. Continue until $\mathcal{D}(G, \rho_m)$ is obtained, where $m$ is the length of $\rho$, and denote it by $\mathcal{D}(G, \rho)$.

Shao, Lin, and Moret [21] characterize the connected components of a trajectory graph for a parsimonious scenario. Using similar techniques we prove the following theorem in the appendix.

**Theorem 1.** *If $\mathcal{D}(G, \rho)$ has $k$ connected components then $\rho$ can be partitioned into $k$ subscenarios $\rho^i$ and $G$ can be partitioned into $k$ edge-disjoint Eulerian subgraphs $H^i$ in such a way that $\rho^i$ is a scenario for $H^i$ for every $i \in \{1, \ldots, k\}$. If $\rho$ is parsimonious, then $k = c(G)$ and $C(\rho) = \{H^1, \ldots, H^k\}$ is a MAECD of $G$.*

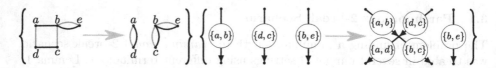

**Fig. 3.** A 2-break $\{a,b\}, \{d,c\} \rightarrow \{a,d\}, \{b,c\}$ transforming a graph $G$ into a terminal one is depicted on the left. A directed graph $\mathcal{D}(G, \rho)$ is obtained from $\mathcal{D}(G, \rho_0)$ on the right for this scenario $\rho$ of length 1. The endpoints of the edges labeled $\{a,b\}$ and $\{d,c\}$ are merged and two new edges labeled $\{a,d\}$ and $\{b,c\}$ are introduced. $\mathcal{D}(G,\rho)$ has 2 connected components that correspond to the 2 simple cycles of $G$.

## 4    Labeled 2-breaks and Their Costs

In this section we outline our model for assigning costs to 2-breaks on a graph $G$. We associate labels to edges, and then describe a set of valid operations $\mathcal{O}$ where each operation may transform the connectivity of $G$, the labeling of $G$, or both. Our cost function is defined on $\mathcal{O}$. This model is general enough to treat the edge labeled DCJ problems of [12] and [23].

For a set of vertices $V$ and a set of labels $\mathcal{L}$ a *labeled edge* is an unordered pair of vertices plus a label, denoted $(\{a,b\}, x)$ for $a, b \in V$ and $x \in \mathcal{L}$. A *label change* $(\{a,b\}, x) \rightarrow (\{a,b\}, y)$ changes the label of an edge. A *labeled 2-break* $(\{a,b\}, x), (\{c,d\}, y) \rightarrow (\{a,c\}, z), (\{b,d\}, t)$ is a 2-break that replaces two labeled edges. Take a set $\mathcal{O}$ containing labeled 2-breaks and label changes, and a graph $G$ with a labeling of its edges $\lambda : E \rightarrow \mathcal{L}$. An $\mathcal{O}$-*scenario* $\rho_{\mathcal{O}}$ for $(G, \lambda)$, is a sequence of operations in $\mathcal{O}$ transforming $(G, \lambda)$ into $(\bar{G}, \bar{\lambda})$ such that $\bar{G}$ is terminal, and the multi-sets of black and gray labeled edges of $\bar{G}$ are equal. The number of 2-breaks in $\rho_{\mathcal{O}}$ will be called the *2-break-length* of the scenario. If a $\rho_{\mathcal{O}}$ exists for $(G, \lambda)$, then $d_{2b\mathcal{O}}(G, \lambda)$ denotes the minimum 2-break-length of an $\mathcal{O}$-scenario.

An $\mathcal{O}$-scenario does not necessarily exist for a given $(G, \lambda)$, however if it exists, then the inequality $d_{2b\mathcal{O}}(G, \lambda) \geq d_{2b}(G)$ holds, where $d_{2b}(G)$ is the minimum length of a 2-break scenario on a graph $G$. In this paper we deal with the sets $\mathcal{O}$ that have the necessary operations to parsimoniously transform $(G, \lambda)$ into $(\bar{G}, \bar{\lambda})$.

**Definition 6 (p-sufficient $\mathcal{O}$ for $(G, \lambda)$).** *A set $\mathcal{O}$ is parsimonious-sufficient or p-sufficient for $(G, \lambda)$ if we have $d_{2b\mathcal{O}}(G, \lambda) = d_{2b}(G)$.*

The cost function that we consider is $\varphi : \mathcal{O} \rightarrow \mathbb{R}_+$. The cost of an $\mathcal{O}$-scenario is the sum of the costs of its constituent operations. If $\mathcal{O}$ is p-sufficient for $(G, \lambda)$, then $\mathrm{MCPS}_\varphi(G, \lambda)$ is the minimum cost of an $\mathcal{O}$-scenario of the 2-break-length equal to $d_{2b}(G)$, otherwise $\mathrm{MCPS}_\varphi(G, \lambda)$ is $\infty$. We consider the following problem:

*Problem 1 ($\varphi$-*Minimum Cost Parsimonious Scenario *or* $\varphi$-MCPS*).*

> INPUT : A graph $G$, and a labeling of its edges $\lambda$.
> OUTPUT : $\mathrm{MCPS}_\varphi(G, \lambda)$.

## 4.1   Examples of the Weighted DCJ Problems in the Literature

*Example 1 (*MINIMUM LOCAL PARSIMONIOUS SCENARIO*).* In [23] we suppose
the adjacencies of genome $A$ to be partitioned into spatial regions represented
by different colors. We then develop a polynomial time algorithm for finding
a parsimonious DCJ scenario minimizing the number of rearrangements whose
breakpoints appear in different regions. The problem, as it is stated in [23], differs
slightly from $\varphi$-MCPS as in that study we do not have colors for the adjacencies
of genome $B$. However, we can bridge this gap as follows.

Take a set of labels $\mathcal{L} = \mathcal{L}_c \cup \{\tau\}$ consisting of the colors $\mathcal{L}_c$ representing the
different spatial regions of a genome and an additional terminal label $\tau$. Define $\mathcal{O}$
as containing the labeled 2-breaks $(\{a,b\},x),(\{c,d\},y) \to (\{a,c\},x),(\{b,d\},y)$
for $a,b,c,d \in V$ and $x,y \in \mathcal{L}_c$, and a label change $(\{a,b\},x) \to (\{a,b\},\tau)$ for
$a,b \in V$ and $x \in \mathcal{L}_c$. The cost $\varphi_c$ of a labeled 2-break in $\mathcal{O}$ is 0 if the labels of
the edges being replaced are equal and 1 otherwise. The cost of a label change
is 0.

In [23] we presented an $O(n^4)$ time algorithm solving $\varphi_c$-MCPS for a labeled
breakpoint graph with the gray edges labeled by $\tau$. In [22] we demonstrated
that finding a minimum cost $\mathcal{O}$ scenario for such a breakpoint graph, when the
parsimonious criteria is disregarded, is NP-hard, and proposed an algorithm that
is exponential in the number of colors but not in the number of genes.

In Sect. 9 we use the same $\mathcal{O}$. We fix a symmetric function $\Phi : \mathcal{L}^2 \to \mathbb{R}_+$ and
define $\varphi_f((\{a,b\},x),(\{c,d\},y) \to (\{a,c\},x),(\{b,d\},y)) = \Phi(x,y)$. This drastically enriches the model introduced in [23]. In Sect. 7 we provide an $O(n^5)$ time
algorithm solving $\varphi_f$-MCPS for a labeled breakpoint graph.

*Example 2 (*DCJ WEIGHTED BY HI-C*).* In [19] we weight each DCJ by the
value taken directly from the Hi-C contact map. In this model every intergenic
region of genome $A$ gets assigned an interval corresponding to its genomic coordinates on a chromosome. The *weight* of a DCJ acting on two intergenic regions
is then equal to the average Hi-C value for their corresponding intervals. In [19]
we provide an algorithm greedily maximizing the weight of a parsimonious scenario and find that the obtained weight is significantly higher than the weight
of a random parsimonious scenario.

Take a set of labels consisting of the genomic intervals corresponding to the
intergenic regions of a genome $A$ plus an additional terminal label. Keep the
same $\mathcal{O}$ as in Example 1. Define $\Phi_{HiC}(x,y)$ on two genomic intervals to be
their average Hi-C value. The problem that maximizes Hi-C values can be easily
transformed into a minimization problem by setting the cost of a labeled 2-break
$(\{a,b\},x),(\{c,d\},y) \to (\{a,c\},x),(\{b,d\},y)$ to $\Phi_{\max} - \Phi_{HiC}(x,y)$, where $\Phi_{\max}$
is the maximum $\Phi_{HiC}(x,y)$ over all $x,y$.

In [19] the optimality of the proposed greedy algorithm is not discussed, but
our work presented in Sect. 9 of this paper provides us with a polynomial time
algorithm for solving this problem exactly.

*Example 3 (*SORTING BY WDCJS AND INDELS IN INTERGENES*).* Bulteau,
Fertin, and Tannier [12] introduce a problem where adjacencies of genomes are

labeled with their genetic length (number of nucleotides). A $wDCJ$ is a labeled DCJ that preserves the sum of the genetic lengths of the adjacencies and an *indel* $\delta$ is a label change that increases or decreases the genetic length of an adjacency by $\delta$. The cost of a wDCJ is 0 and the cost of an indel $\delta$ is $|\delta|$. A scenario of wDCJs and indels for $(G, \lambda)$ is said to be *valid* if its wDCJ-length is $d_{2b}(G)$. The paper presents an $O(n \log n)$ algorithm for finding a minimum cost scenario among the *valid* ones, for the genomes with circular chromosomes and $n$ genes.

Translating this into our formalism yields the following $\varphi$-MCPS problem. The labels $\mathcal{L}$ would be the natural numbers, while $\mathcal{O}$ contains labeled 2-breaks $(\{a, b\}, w_1), (\{c, d\}, w_2) \rightarrow (\{a, c\}, w_3), (\{b, d\}, w_4)$ for every $a, b, c, d \in V$, and $w_i \in \mathcal{L}$ satisfies $w_1 + w_2 = w_3 + w_4$. $\mathcal{O}$ also contains label changes $(\{a, b\}, w_1) \rightarrow (\{a, b\}, w_2)$ for every $a, b \in V$ and $w_i \in \mathcal{L}$. $\mathcal{O}$ is p-sufficient for any $(G, \lambda)$ since $G$ can be first transformed into a terminal graph using any parsimonious 2-break scenario and then its labels can be adjusted. The cost $\varphi_l$ of a labeled 2-break is 0 and the cost $\varphi_l$ of a label change $(\{a, b\}, w_1) \rightarrow (\{a, b\}, w_2)$ is $|w_1 - w_2|$.

In [12] the authors present an $O(r \log r)$ time algorithm for solving $\varphi_l$-MCPS on a circle with $r$ vertices. Combining this algorithm with our results from Sect. 7 gives an algorithm solving $\varphi_l$-MCPS in $O(n^3)$ time for a labeled breakpoint graph. The ILP defined in Sect. 5 solves $\varphi_l$-MCPS for any labeled graph.

*Example 4 (*wDCJ-DIST*).* Fertin, Jean, and Tannier [15] treat a problem wDCJ-DIST where wDCJs without indels are allowed, and the sums of the genetic lengths of the adjacencies of two genomes are equal.

In this case we keep the same $\mathcal{L}$ and $\mathcal{O}$ as in Example 3 except that the label changes are excluded from $\mathcal{O}$. A labeled graph is said to be *balanced* if the sums of the labels of black and gray edges are equal. wDCJ-DIST is the problem of finding $d_{2bo}$ for a balanced graph whose connected components are circles. The authors show that wDCJ-DIST is strongly NP-complete. However they also prove that $d_{2bo}(O, \lambda) = d_{2b}(O)$ for a balanced circle and that $\mathcal{O}$ is p-sufficient for a graph whose connected components are balanced circles.

*Example 5.* Although ignored in the previous examples, the weighting of operations based on the vertices is also possible under our framework. For example, take $\mathcal{L} = \{\tau\}$, $\mathcal{O}$ containing labeled 2-breaks $(\{a, b\}, \tau), (\{c, d\}, \tau) \rightarrow (\{a, c\}, \tau), (\{b, d\}, \tau)$ and any cost function $\varphi_v : \mathcal{O} \rightarrow \mathbb{R}_+$. The costs of the labeled 2-breaks in $\mathcal{O}$ could be a function of the genomic coordinates of the participating gene extremities.

Note that the set $\mathcal{O}$ is implicit, rather than explicit. In Example 3, $\mathcal{O}$ would be too large to represent explicitly since every pair of genetic lengths for every pair of edges would exist.

## 5   $\varphi$-MCPS for a Graph

**Theorem 2.** *Denote the $\varphi$-cost of a MAECD as the sum of the MCPS$_\varphi$ on its cycles. MCPS$_\varphi$ for a graph is equal to the minimum $\varphi$-cost of its MAECD.*

*Proof.* For a cycle $S$ of a labeled graph $(G, \lambda)$, $\lambda^S$ denotes the labeling of the edges of $S$ according to $\lambda$. We suppose that $min(\emptyset) = \infty$ and prove the following:

$$\text{MCPS}_\varphi(G, \lambda) = min\Big\{ \sum_{S \in C} \text{MCPS}_\varphi(S, \lambda^S) \mid C \text{ is a MAECD of } G\Big\}.$$

Suppose that there exists a MAECD $C$ of $G$ consisting of the simple cycles for which $\mathcal{O}$ is p-sufficient. For every $S \in C$ take an $\mathcal{O}$-scenario $\rho_\mathcal{O}^S$ of cost $\text{MCPS}_\varphi(S, \lambda^S)$ and 2-break-length $d_{2b}(S)$. By performing these scenarios one after another we obtain an $\mathcal{O}$-scenario $\rho_\mathcal{O}$ for $(G, \lambda)$ of 2-break-length $\sum_{S \in C} d_{2b}(S) = d_{2b}(G)$ and of cost $\sum_{S \in C} \text{MCPS}_\varphi(S, \lambda^S)$. This yields a scenario such that $\text{MCPS}_\varphi(G, \lambda) \leq \sum_{S \in C} \text{MCPS}_\varphi(S, \lambda^S)$.

On the other hand, suppose that $\mathcal{O}$ is p-sufficient for $(G, \lambda)$ and take an $\mathcal{O}$-scenario $\rho_\mathcal{O}$ for $(G, \lambda)$ of length $d_{2b}(G)$. For $\rho$, a 2-break scenario obtained from $\rho_\mathcal{O}$ when the labels of the edges are neglected, a decomposition $C(\rho)$ corresponding to $\rho$ is a MAECD of $G$ due to Theorem 1. A subsequence $\rho_\mathcal{O}^S$ of $\rho_\mathcal{O}$, consisting of the operations acting on the edges of a cycle $S \in C(\rho)$, is an $\mathcal{O}$-scenario for $(S, \lambda^S)$ of 2-break-length $d_{2b}(S)$. A sequence of operations $\hat{\rho}_\mathcal{O}$ obtained by performing the subsequences $\rho_\mathcal{O}^S$ one after another for each $S \in C(\rho)$ is an $\mathcal{O}$-scenario for $(G, \lambda)$. By construction the 2-break-length of $\hat{\rho}_\mathcal{O}$ is equal to the 2-break-length of $\rho_\mathcal{O}$. The costs of $\rho_\mathcal{O}$ and $\hat{\rho}_\mathcal{O}$ are also equal, as they consist of exactly the same operations that are performed in different orders, thus the cost of $\rho_\mathcal{O}$ is greater or equal to $\sum_{S \in C(\rho)} \text{MCPS}_\varphi(S, \lambda^S) \geq min\Big\{ \sum_{S \in C} \text{MCPS}_\varphi(S, \lambda^S) \mid C \text{ is a MAECD of } G\Big\}.$   □

Take the set $\mathcal{S}$ of simple labeled cycles of $(G, \lambda)$. If one can solve $\varphi$-MCPS for every $S \in \mathcal{S}$, then Theorem 2 provides a straightforward way to solve $\varphi$-MCPS for $(G, \lambda)$ as a set packing problem. First compute $c(G)$ by solving the ILP in the left column. Then proceed by solving the other ILP to compute $\text{MCPS}_\varphi(G, \lambda)$.

Maximize $\sum_{S \in \mathcal{S}} x_S$

Subject to $\sum_{S : e \in S} x_S \leq 1$ for each edge $e$ of $G$

and $x_S \in \{0, 1\}$ for simple cycle $S \in \mathcal{S}$.

Minimize $\sum_{S \in \mathcal{S}} x_S \text{MCPS}_\varphi(S, \lambda^S)$

Subject to $\sum_{S : e \in S} x_S \leq 1$ for each edge $e$ of $G$,

$\sum_{S \in \mathcal{S}} x_S = c(G)$

and $x_S \in \{0, 1\}$ for simple cycle $S \in \mathcal{S}$.

The size of $\mathcal{S}$ may be exponential in the size of $G$, which might make these ILPs intractable in general. For graphs representing genomes with duplicate genes, the number of simple cycles can grow exponentially as a function of the number of duplicate genes. For breakpoint graphs, the number grows quadratically.

## 6   $\varphi$-MCPS for a Simple Cycle

The decomposition theorem of Sect. 5 reduces the computation of $\varphi$-MCPS on a graph to the computation of $\varphi$-MCPS on a simple alternating cycle. In this

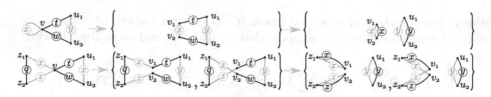

**Fig. 4.** Two simple cycles having a vertex $v$ of degree two are depicted in the first column. Their sets of the corresponding circles obtained by splitting $v$ into $v_1$ and $v_2$ are depicted in the second column. This set is of size 1 for the upper simple cycle containing the gray loop $\{v, v\}$, and of size 2 for the lower simple cycle. An $\mathcal{O}$-scenario for a simple cycle provides a scenario of the same cost and length transforming the graphs in the second column to the ones that become terminal once $v_1$ and $v_2$ are merged. One possible outcome of such a scenario is presented in the third column.

section we further decompose the problem into simpler versions of cycles, called circles, which are alternating cycles that contain a vertex only once.

Denote $deg_2(G)$ for a graph $G$ as the number of vertices with black and gray degree equal to two. It is easy to check that $deg^b(S, v) = deg^g(S, v) \leq 2$ for any vertex $v$ of a simple cycle $S$. If $deg_2(S) = 0$, then $S$ is a circle. See the first column of Fig. 4 for examples of simple cycles that are not circles.

Take a simple cycle $S$ on vertices $V$, a labeling of its edges $\lambda$ and denote $S_0$ as $\{(S, \lambda)\}$. Choose a vertex of degree two in $S$. If it is incident to a gray loop, then split it into two vertices, as depicted on the top row of Fig. 4, to obtain a set $S_1$ consisting of a single simple cycle. Otherwise split it into two vertices, as depicted on the bottom row of Fig. 4, to obtain a set $S_1$ consisting of two simple cycles. The simple cycles in $S_1$, by construction, share the same set of vertices, that we denote $\hat{V}$, and the same multi-set of labeled black edges. $\mathcal{O}$ and a cost function $\varphi$ defined for vertices $V$ can be extended in a natural way to $\hat{\mathcal{O}}$ and $\hat{\varphi}$ defined for vertices $\hat{V}$. For example if a vertex $v$ was split into $v_1$ and $v_2$, then $\hat{\varphi}((\{v_1, u\}, x) \to (\{v_1, u\}, y)) = \varphi((\{v, u\}, x) \to (\{v, u\}, y))$ for $u \in V \cap \hat{V}$ and labels $x, y$. In the appendix we prove the following lemma.

**Lemma 3.** $\mathrm{MCPS}_\varphi(S, \lambda) = min\{\mathrm{MCPS}_{\hat{\varphi}}(\hat{S}, \hat{\lambda})| \ (\hat{S}, \hat{\lambda}) \in \ S_1\}$

Simple cycles in $S_1$ share the same set of vertices of degree two. Choose such a vertex and split it simultaneously in all the cycles in $S_1$ as previously to obtain a set $S_2$ of at most 4 simple cycles sharing the same set of vertices and the same multi-set of labeled black edges. Continue this procedure until the set $circ(S, \lambda) = S_{deg_2(S)}$ of the labeled circles is obtained. We denote $\overline{V}$ as the set of vertices of these circles. $\mathcal{O}$ and a cost function $\varphi$ defined for vertices $V$ can be extended in a natural way to $\overline{\mathcal{O}}$ and $\overline{\varphi}$ defined for vertices $\overline{V}$.

**Theorem 3.** $\mathrm{MCPS}_\varphi$ for a simple cycle $(S, \lambda)$ is equal to the minimum of the $\mathrm{MCPS}_\varphi$ among the circles in $circ(S, \lambda)$.

*Proof.* We prove $\mathrm{MCPS}_\varphi(S, \lambda) = min\{\mathrm{MCPS}_{\overline{\varphi}}(O, \lambda^O)| \ (O, \lambda^O) \in \ circ(S, \lambda)\}$, which is clearly true for $deg_2(S) = 0$. We suppose it to be true for $deg_2(S) < t$

and prove it for $deg_2(S) = t$ by induction. By Lemma 3 we get $\text{MCPS}_\varphi(S, \lambda) = min\{\text{MCPS}_{\hat{\varphi}}(\hat{S}, \hat{\lambda})| \ (\hat{S}, \hat{\lambda}) \in \ S_1\}$. Since, for a simple cycle $(\hat{S}, \hat{\lambda}) \in S_1$ we have $deg_2(\hat{S}) = t - 1$, we use the inductive hypothesis to obtain $\text{MCPS}_{\hat{\varphi}}(\hat{S}, \hat{\lambda}) = min\{\text{MCPS}_{\overline{\varphi}}(O, \lambda^O)| \ (O, \lambda^O) \in \ circ(\hat{S}, \hat{\lambda})\}$. Further, we know that $circ(S, \lambda) = \cup_{(\hat{S},\hat{\lambda}) \in S_1} circ(\hat{S}, \hat{\lambda})$ by construction. Combining these results we obtain that the theorem is true for $deg_2(S) = t$.                                    □

# 7  $\varphi$-MCPS for a Breakpoint Graph

In this section we suppose that there exists an algorithm for computing $\text{MCPS}_\varphi$ on a labeled circle (e.g. the algorithm of Sect. 9). Using this algorithm as a subroutine we will construct an algorithm for finding $\text{MCPS}_\varphi$ for a labeled breakpoint graph. This is a generalization of the work first presented in [23].

Take genomes $A$ and $B$ partitioned into $n$ genes where each gene occurs exactly once in each genome, and a labeling $\lambda$ of the edges of $G(A, B)$. For all the vertices $v \neq \circ$ we have $deg^g(G(A, B), v) = deg^b(G(A, B), v) = 1$. Thus, if there is a circle in $G(A, B)$ containing an edge then this circle is the only simple cycle containing this edge. This means that every MAECD of $G(A, B)$ includes all of its circles. These set aside we are left with $G(A, B)'$, which is a union of alternating paths starting and ending at $\circ$ with end edges of the same color. If this color is black we call the path $AA$, and $BB$ otherwise.

We proceed by constructing a complete weighted bipartite graph $H$ having the $AA$ and $BB$ paths of $G(A, B)'$ as vertices. Every simple cycle of $G(A, B)'$ is a union of an $AA$ path and a $BB$ path. An edge joining these paths in $H$ will have the weight equal to $\text{MCPS}_\varphi$ for a union of these paths. A MAECD of $G(A, B)'$ provides us with a maximum matching of $H$ and every such matching provides a MAECD of $G(A, B)'$. Denote $\lambda'$ as the labeling of the edges of $G(A, B)'$ according to $\lambda$. Using Theorem 2 we obtain that $\text{MCPS}_\varphi(G(A, B)', \lambda')$ is equal to the minimum weight of a maximum matching of $H$. There is an equal number $p$ of $AA$ and $BB$ paths. Let $P$ denote the total number of edges in $G(A, B)'$. Using this notation we obtain the following lemma proven in the appendix.

**Lemma 4.** *For a function $f$ and an $O(f(r))$ time algorithm for $\varphi$-MCPS on a labeled circle on $r$ vertices, there exists an $O(p^2 f(P) + p^3 + f(n))$ time algorithm for $\varphi$-MCPS on a labeled breakpoint graph. If $f(r) = O(r^t)$ for some constant $t \geq 1$, then $\varphi$-MCPS on a labeled breakpoint graph can be solved in $O(pP^t + p^3 + n^t)$ time.*

Both $p$ and $P$ are $O(n)$, thus Lemma 4 leads to the following theorem.

**Theorem 4.** *Given a constant $t \geq 2$ and an $O(r^t)$ time algorithm for $\varphi$-MCPS on a labeled circle on $r$ vertices, $\varphi$-MCPS on a labeled breakpoint graph can be solved in $O(n^{t+1})$ time.*

**Corollary 1.** *Using the $O(r^4)$ algorithm from Sect. 9 we obtain an $O(n^5)$ algorithm for solving $\varphi_f$-MCPS on a labeled breakpoint graph with fixed labels.*

**Corollary 2.** *Using the $O(r \log r)$ algorithm from [12] for the* SORTING BY wDCJs AND INDELS IN INTERGENES *problem on a circle (see Example 3), we obtain an $O(n^3)$ algorithm for solving the problem on a breakpoint graph.*

# 8    $\alpha$-approximation for $\varphi$-MCPS

Theorems 2 and 3 demonstrate how $\varphi$-MCPS for any labeled graph can be solved if one is able to solve $\varphi$-MCPS for a labeled circle. This is exploited in Theorem 4 to solve $\varphi$-MCPS for a breakpoint graph. Analogous results hold if instead of an exact algorithm one has an $\alpha$-approximation for $\varphi$-MCPS for a labeled circle. This is illustrated with the following theorem proven in the appendix.

**Theorem 5.** *For a constant $t \geq 2$ and an $O(r^t)$ time $\alpha$-approximation algorithm for $\varphi$-MCPS on a labeled circle on $r$ vertices, there exists an $O(n^{t+1})$ time $\alpha$-approximation algorithm for $\varphi$-MCPS on a labeled breakpoint graph.*

# 9    $\varphi_f$-MCPS for a Circle with Fixed Labels

Here we define $\varphi_f$-MCPS, a particular instance of a $\varphi$-MCPS problem, and solve it for a circle. $\varphi_f$-MCPS generalizes our previous work presented in Example 1 and 2.

For a set $V$ of vertices and a set $\mathcal{L} \cup \{\tau\}$ of labels, define a set $\mathcal{O}$ consisting of labeled 2-breaks $(\{a,b\},x),(\{c,d\},y) \to (\{a,c\},x),(\{b,d\},y)$ for $a,b,c,d \in V$ and $x,y \in \mathcal{L}$, and label changes $(\{a,b\},x) \to (\{a,b\},\tau)$ for $a,b \in V$ and $x \in \mathcal{L}$. Fix a symmetric function $\Phi : \mathcal{L}^2 \to \mathbb{R}_+$ and define a cost function $\varphi_f((\{a,b\},x),(\{c,d\},y) \to (\{a,c\},x),(\{b,d\},y)) = \Phi(x,y)$ and $\varphi_f((\{a,b\},x) \to (\{a,b\},\tau)) = 0$.

We will provide a polynomial time algorithm for $\varphi_f$-MCPS on a labeled circle with the gray edges labeled by a terminal label $\tau$. Without loss of generality we can suppose that all of the black edges of a circle have different labels; if two edges are labeled with the same label $x$, then we simply replace one of these labels with a new label $\hat{x}$ and set $\hat{\Phi}(\hat{x},y) = \Phi(x,y)$ and $\hat{\Phi}(y,z) = \Phi(y,z)$ for $y,z \in \mathcal{L}$.

For a labeled circle having $r$ black edges, define a set $V_\mathcal{L}$ of $r$ vertices corresponding to their labels. For an $\mathcal{O}$-scenario $\rho_\mathcal{O}$ we define a 1-edge-colored undirected graph $\mathcal{T}(\rho_\mathcal{O})$ with vertices $V_\mathcal{L}$ and an edge $\{x,y\}$ for every labeled 2-break in $\rho_\mathcal{O}$ replacing the edges labeled with $x$ and $y$ (See Fig. 5). The *cost* of an edge $\{x,y\}$ is defined to be $\Phi(x,y)$ and the cost of a $\mathcal{T}(\rho_\mathcal{O})$ is the sum of the costs of its edges. The costs of $\rho_\mathcal{O}$ and $\mathcal{T}(\rho_\mathcal{O})$ are equal by construction.

Fix a circular embedding of $V_\mathcal{L}$ respecting the order of the black edges on the labeled circle (See Fig. 5). A graph with vertices $V_\mathcal{L}$ is said to be *planar on the circle* if none of its edges cross in this embedding. In the appendix we prove Lemma 5 linking planar trees and parsimonious scenarios.

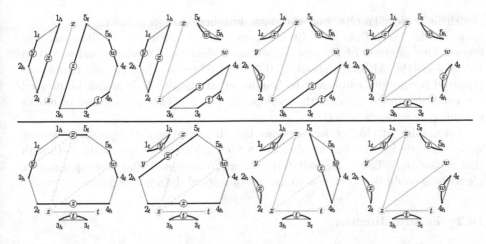

**Fig. 5. On the top:** 4 steps of a parsimonious $\mathcal{O}$-scenario for a circle are depicted together with each $\mathcal{T}$ corresponding to the scenario at that point colored in yellow. Vertices of $\mathcal{T}$ are superimposed on the corresponding edges of a circle providing their circular embedding. All of the $\mathcal{T}$ are planar trees. **On the bottom:** For a given planar tree $\mathcal{T}$ (dashed yellow) we provide a scenario $\rho_{\mathcal{O}}$ such that $\mathcal{T}(\rho_{\mathcal{O}}) = \mathcal{T}$.

**Lemma 5.** *If $\rho_{\mathcal{O}}$ is a minimum 2-break-length $\mathcal{O}$-scenario for a labeled circle $(O, \lambda)$, then $\mathcal{T}(\rho_{\mathcal{O}})$ is a planar tree on $(O, \lambda)$. In addition to that, for a planar tree $\mathcal{T}$ on $(O, \lambda)$ there exists an $\mathcal{O}$-scenario $\rho_{\mathcal{O}}$ such that $\mathcal{T}(\rho_{\mathcal{O}}) = \mathcal{T}$.*

Farnoud and Milenkovic in [14] pose the problem of sorting permutations by cost-constrained mathematical transpositions and provide a dynamic programming algorithm for finding a minimum cost planar tree on a circle. In the appendix we provide their proof for a following lemma which, together with Lemma 6, leads to Theorem 6.

**Lemma 6 (Farnoud et al. in [14]).** *A minimum cost planar tree on a circle can be found in $O(r^4)$ time, where $r$ is the number of vertices of a tree.*

**Theorem 6.** *$\varphi_f$-MCPS for a labeled circle on $r$ vertices can be solved in $O(r^4)$ time.*

## 10    Conclusions and Future Directions

### 10.1    Practical Matters

Our algorithm for $\varphi_f$-MCPS on a breakpoint graph with fixed labels has a running time of $O(n^5)$ in the worst case. Note that the running time is dominated, however, by the maximum bipartite matching step in Sect. 7. The size of this graph is determined by the number of $AA$ paths which is bounded by the number of chromosomes, so in practice it can be treated as a constant. Thus, the algorithm scales like $O(n^4)$ on real data. Further, since $n$ is the number of syntenic blocks — and not literally the genes as we call them — there are few blocks. Our analyses of *Drosophila* genomes yield no $AA$ paths, and less than 100 blocks [19]. Although about 13,000 blocks between human and mouse are reported in the files associated to Baudet et *al.*, many of them can be merged because they are co-linear in the two species [3]. The effective number of blocks for this pair is closer to 600.

For graphs with higher degree nodes, like those graphs that represent genomes with duplicated genes, the number of simple cycles can grow rapidly. Although this relationship is not immediately evident, we expect that fixed parameter algorithms could be developed to handle biological data in the future.

## 10.2    Future Direction

Our cost framework is liberal, and in our examples we have explored only a small portion of its capacities. Edges can be labeled by complex objects such as vectors or trees that encode the biological information of extant genomes and its modification throughout a scenario. Costs can be a function of a combination of the edge labels and vertices. We hope that a closer study of the graph $\mathcal{D}(G, \rho)$ from Sect. 3.2 will lead to polynomial time algorithms for $\varphi$-MCPS on circles for a large family of problems.

While all of our results apply to genomes with insertions or deletions of single genes, further study is required in order to increase efficiency on genomes with duplicate genes. Other improvements to our work could consider non-parsimonious 2-break scenarios.

**Acknowledgments.** This work is partially supported by the IBC (Institut de Biologie Computationnelle) (ANR-11-BINF-0002), by the Labex NUMEV flagship project GEM, and by the CNRS project Osez l'Interdisciplinarité.

# A    Proofs

## A.1    Lemma 1

**Lemma.** *The minimum length of a 2-break scenario on a graph $G$ is $d_{2b}(G) = e(G) - c(G)$.*

*Proof.* A 2-break can increase the size of a MAECD by at most 1 and the size of a MAECD of a terminal graph is $e(G)$. This leads to an inequality $d_{2b}(G) \geq e(G) - c(G)$.

In this paragraph the *length* of a cycle will be its number of black edges. For any cycle $c$ of length $l > 1$ there is a 2-break transforming $c$ into a union of length 1 and length $l-1$ cycles. This way we obtain a scenario of length $l-1$ for $c$, and can transform every cycle of a MAECD of $G$ independently, obtaining a 2-break scenario of length $e(G) - c(G)$. Thus, $d_{2b}(G) \leq e(G) - c(G)$.                                    □

## A.2    Lemma 2

**Lemma.** *The minimum length of a DCJ scenario transforming genome $A$ into $B$ is equal to $d_{2b}(G(A, B)) = e(G(A, B)) - c(G(A, B))$.*

*Proof.* $G(A, B)$ is constructed in such a way that for every DCJ $A \to A'$ the transformation $G(A, B) \to G(A', B)$ is a 2-break. Notably, a DCJ $\{a, b\} \to \{a\}, \{b\}$ results in a transformation $\{a, b\}, \{\circ, \circ\} \to \{a, \circ\}, \{b, \circ\}$, as the construction of a breakpoint graph guarantees that there are enough black loops $\{\circ, \circ\}$ to realize such a 2-break. For any 2-break $G(A, B) \to G'$ with $G' \neq G(A, B)$ there exists a DCJ $A \to A'$ such that $G(A', B) = G'$. Since $G(B, B)$ is terminal, it follows that the minimum length of a scenario transforming $A$ into $B$ is $d_{2b}(G(A, B))$ and we conclude using Lemma 1. □

## A.3    Theorem 1

**Theorem.** *If $\mathcal{D}(G, \rho)$ has $k$ connected components then $\rho$ can be partitioned into $k$ subscenarios $\rho^i$ and $G$ can be partitioned into $k$ edge-disjoint Eulerian subgraphs $H^i$ in such a way that $\rho^i$ is a scenario for $H^i$ for every $i \in \{1, \ldots, k\}$. If $\rho$ is parsimonious, then $k = c(G)$ and $C(\rho) = \{H^1, \ldots, H^k\}$ is a MAECD of $G$.*

*Proof.* Take a connected component $C$ of $\mathcal{D}(G, \rho)$. It has an equal number of vertices of indegree 0 and vertices of outdegree 0. Its edges incident to the vertices of indegree 0 are labeled with the black edges of $G$ and its edges incident to the vertices of outdegree 0 are labeled with the gray edges of $G$. Together these labels define a subgraph $H$ of $G$ that we will prove to be Eulerian.

Define $C_l$ to be a subgraph of $\mathcal{D}(G, \rho_l)$ consisting of its connected components containing the vertices of indegree 0 of $C$. This way $C_m = C$. Define $H_l$ to be a subgraph of $G_l$ containing the gray edges of $H$ and the black edges of $G_l$ labeling the edges of $C_l$ incident to the vertices of outdegree 0. This way $H_0 = H$ and $H_m$ is a terminal graph.

We prove that $H$ is Eulerian by induction. $H_m$ is Eulerian as it is terminal. Suppose that $H_l$ is Eulerian. By construction the two edges of $G_l$ replaced by the $l$-th 2-break of $\rho$ either both belong to $H_{l-1}$ or both are outside of $H_{l-1}$. In the first case, $H_l$ is obtained from $H_{l-1}$ via a 2-break and as $H_l$ is Eulerian this means that $H_{l-1}$ is also Eulerian. In the second case, $H_l = H_{l-1}$, thus the latter stays Eulerian. Thus $H = H_0$ is Eulerian and we obtain a subsequence of $\rho$ that is a scenario for $H$.

$\mathcal{D}(G, \rho_0)$ has $e(G)$ connected components. The $l$-th 2-break of $\rho$ merges two vertices of $\mathcal{D}(G, \rho_{l-1})$, thus reduces the number of the connected components by at most 1. This means that the number $k$ of the connected components of $\mathcal{D}(G, \rho)$ is greater or equal to $e(G) - m$.

If $\rho$ is parsimonious, then its length $m$ is $e(G) - c(G)$ using Lemma 1. This means that $k \geq c(G)$ and $G$ can be partitioned into $k$ edge-disjoint Eulerian subgraphs. Due to the maximality of $c(G)$, we have that $k = c(G)$ and all of the obtained edge-disjoint Eulerian subgraphs of $G$ are simple cycles. □

## A.4     Lemma 3

**Lemma.** $\text{MCPS}_\varphi(S, \lambda) = min\{\text{MCPS}_{\hat{\varphi}}(\hat{S}, \hat{\lambda}) | \ (\hat{S}, \hat{\lambda}) \in \ S_1\}$

*Proof.* For a labeled graph $(H, \mu)$ on vertices $\hat{V}$ we denote $r(H, \mu)$ as the labeled graph obtained from $(H, \mu)$ by merging the two vertices that were split in $S$. For $(\hat{S}, \hat{\lambda}) \in \ S_1$ we have $r(\hat{S}, \hat{\lambda}) = (S, \lambda)$ by construction. An operation in $\hat{\mathcal{O}}$ transforms $(\hat{S}, \hat{\lambda})$ into such $(\hat{S}', \hat{\lambda}')$ that there exists unique operation in $\mathcal{O}$ of the same cost transforming $(S, \lambda)$ into $r(\hat{S}', \hat{\lambda}')$. This leads to an observation that for an $\hat{\mathcal{O}}$-scenario for $(\hat{S}, \hat{\lambda})$ there exists an $\mathcal{O}$-scenario of the same cost and the same 2-break-length for $(S, \lambda)$.

On the other hand, for an operation in $\mathcal{O}$ transforming $(S, \lambda)$ into $(S', \lambda')$ there exists an operation in $\hat{\mathcal{O}}$ of the same cost transforming every $(\hat{S}, \hat{\lambda}) \in S_1$ into $(\hat{S}', \hat{\lambda}')$ such that $r(\hat{S}', \hat{\lambda}'_S) = (S', \lambda')$. This leads to an observation that an $\mathcal{O}$-scenario for $(S, \lambda)$ provides us with a sequence $\hat{\rho}_{\hat{\mathcal{O}}}$ of $\hat{\mathcal{O}}$ operations of the same cost and 2-break-length transforming every $(\hat{S}, \hat{\lambda}) \in S_1$ into such $(\overline{\hat{S}}, \overline{\hat{\lambda}})$ for which $r(\overline{\hat{S}}, \overline{\hat{\lambda}})$ is a terminal graph with equal multi-sets of labeled gray and black edges. As the later graph is obtained by merging two vertices of degree one of the former, we know that its structure is as well fairly simple. We can check all the possible cases by hand and show that there is $(\hat{S}, \hat{\lambda}) \in S_1$ such that $(\overline{\hat{S}}, \overline{\hat{\lambda}})$ is itself a terminal graph with equal multi-sets of labeled gray and black edges.

If $S_1$ is of size 1, then there is a single choice for $(\overline{\hat{S}}, \overline{\hat{\lambda}})$ such that $r(\hat{S}, \hat{\lambda})$ is a terminal graph with equal multi-sets of labeled gray and black edges (see the right upper corner of Fig. 4). If $S_1$ is of size 2, then there are more cases, but they are all easy to check and one of them is given in the right bottom corner of Fig. 4.                                                                        □

## A.5     Lemma 4

**Lemma.** *For a function $f$ and an $O(f(r))$ time algorithm for $\varphi$-MCPS on a labeled circle on $r$ vertices, there exists an $O(p^2 f(P) + p^3 + f(n))$ time algorithm for $\varphi$-MCPS on a labeled breakpoint graph. If $f(r) = O(r^t)$ for some constant $t \geq 1$, then $\varphi$-MCPS on a labeled breakpoint graph can be solved in $O(pP^t + p^3 + n^t)$ time.*

*Proof.* The $p^2$ edges of a bipartite graph $H$ can be weighted in $O(p^2 f(P))$ time due to Theorem 3 and the fact that the simple cycles of $G(A, B)$ have at most 1 vertex of degree 2. A minimum weight maximum matching of $H$ can be found in $O(p^3)$ time using the Hungarian algorithm. Finally, $\text{MCPS}_\varphi$ for the labeled circles in $G(A, B)$ can be computed in $O(f(n))$ time. Combining these results we obtain an $O(p^2 f(P) + p^3 + f(n))$ time algorithm for computing $\text{MCPS}_\varphi(G(A, B), \lambda)$.

Now suppose that $f(r) = O(r^t)$ for some constant $t \geq 1$. Let $a_1, \ldots, a_p$ and $b_1, \ldots, b_p$ denote the number of edges in $AA$ and $BB$ paths with $\sum_{i=0}^{p} a_i = P_A$, $\sum_{j=0}^{p} b_j = P_B$ and $P = P_A + P_B$.

MCPS$_\varphi$ for a union of an $AA$ path and a $BB$ path having $a$ and $b$ edges respectively can be obtained by computing MCPS$_\varphi$ for at most two circles on $a + b$ vertices due to Theorem 3. This can be done in less than $c(a+b)^t$ steps for some constant $c$ using the $O(r^t)$ time algorithm for computing MCPS$_\varphi$ for a circle. MCPS$_\varphi$ for every pair of $AA$ and $BB$ paths of $G(A, B)'$ can be computed in a number of steps bounded by:

$$\sum_{i=0}^{p}\sum_{j=0}^{p}c(a_i + b_j)^t = c\sum_{i=0}^{p}\sum_{j=0}^{p}\sum_{l=0}^{t}\binom{t}{l}a_i^l b_j^{t-l} = c\sum_{l=0}^{t}\binom{t}{l}\sum_{i=.For0}^{p}\sum_{j=0}^{p}a_i^l b_j^{t-l}$$

$$= c\sum_{j=0}^{p}\sum_{i=0}^{p}b_j^t + c\sum_{i=0}^{p}\sum_{j=0}^{p}a_i^t + c\sum_{l=1}^{t-1}\binom{t}{l}\sum_{i=0}^{p}a_i^l\sum_{j=0}^{p}b_j^{t-l}$$

$$= cp\sum_{j=0}^{p}b_j^t + cp\sum_{i=0}^{p}a_i^t + c\sum_{l=1}^{t-1}\binom{t}{l}\sum_{i=0}^{p}a_i^l\sum_{j=0}^{p}b_j^{t-l}$$

$$\leq cp(\sum_{j=0}^{p}b_j)^t + cp(\sum_{i=0}^{p}a_i)^t + c\sum_{l=1}^{t-1}\binom{t}{l}(\sum_{i=0}^{p}a_i)^l(\sum_{j=0}^{p}b_j)^{t-l}$$

$$\leq c(pP_B^t + pP_A^t) + pc\sum_{l=1}^{t-1}\binom{t}{l}P_B^{t-l}P_A^l = cp(P_B + P_A)^t = cpP^t$$

Thus, the weighting of $H$ can be performed in $O(pP^t)$ time. This provides us with an $O(pP^t + p^3 + n^t)$ time algorithm for computing MCPS$_\varphi(G(A, B), \lambda)$. $\square$

## A.6    Theorem 5

**Theorem.** *For a constant $t \geq 2$ and an $O(r^t)$ time $\alpha$-approximation algorithm for $\varphi$-MCPS on a labeled circle on $r$ vertices, there exists an $O(n^{t+1})$ time $\alpha$-approximation algorithm for $\varphi$-MCPS on a labeled breakpoint graph.*

*Proof.* In Theorem 3, MCPS$_\varphi$ on a simple cycle is expressed as the minimum of the MCPS$_\varphi$ for a set of corresponding circles. In Theorem 2, MCPS$_\varphi$ on a graph is expressed as the minimum of the sums of the MCPS$_\varphi$ for the simple cycles. We prove an auxiliary lemma establishing the following:

1. An $\alpha$-approximation for MCPS$_\varphi$ on a simple cycle can be obtained by taking the minimum of the $\alpha$-approximations for the corresponding circles.
2. An $\alpha$-approximation for MCPS$_\varphi$ on a graph can be obtained by taking the minimum of the sums of the $\alpha$-approximations for MCPS$_\varphi$ on the simple cycles.

**Lemma.** *Take $k \in \mathbb{N}$ and two sets of positive numbers $\{q_1^*, \ldots, q_k^*\}$ and $\{q_1, \ldots, q_k\}$ with $q_i \leq \alpha q_i^*$ for every $i$. The following inequalities hold:*

1. $min\{q_i | i \in \{1, \ldots, k\}\} \leq \alpha min\{q_i^* | i \in \{1, \ldots, k\}\}$
2. $\sum_{i=0}^{k} q_i \leq \alpha \sum_{i=0}^{k} q_i^*$

*Proof.* Take $u$ and $v$ such that $q_u^* = min\{q_i^*|i \in \{1,\ldots,k\}\}$ and $q_v = min\{q_i|i \in \{1,\ldots,k\}\}$. By construction $q_v \leq q_u \leq \alpha q_u^*$ which proves the first inequality. For the second inequality it suffice to observe that $\sum_{i=0}^{k} q_i \leq \sum_{i=0}^{k} \alpha q_i^* = \alpha \sum_{i=0}^{k} q_i^*$ □

A simple cycle of a breakpoint graph has at most one vertex of degree 2. This means that it has at most two corresponding circles (see Theorem 6). Taking the minimum of the $\alpha$-approximations for MCPS$_\varphi$ on these circles provides us with an $\alpha$-approximation for the simple cycle due to Theorem 6 and the first part of the lemma above. This way we obtain an $\alpha$-approximation algorithm for $\varphi$-MCPS on a simple cycle of a breakpoint graph that runs in $O(r^t)$ time where $r$ is the number of the vertices in the simple cycle.

We can reuse the structure of a bipartite graph $H$ presented in Sect. 7 with the weights of the edges now being the $\alpha$-approximations for the MCPS$_\varphi$ on the corresponding simple cycles. Following the same reasoning as in Sect. 7, we know that the minimum cost maximum matching of $H$ leads to a MAECD of a breakpoint graph minimizing the sum of the $\alpha$-approximations for the MCPS$_\varphi$ on its simple cycles. Combining Theorem 2, both parts of the lemma above, and the proof of Lemma 4, we obtain an $O(n^{t+1})$ time $\alpha$-approximation algorithm for $\varphi$-MCPS on a breakpoint graph. □

## A.7   Lemma 5

**Lemma.** *If $\rho_\mathcal{O}$ is a minimum 2-break-length $\mathcal{O}$-scenario for a labeled circle $(O, \lambda)$, then $T(\rho_\mathcal{O})$ is a planar tree on $(O, \lambda)$. In addition to that, for a planar tree $T$ on $(O, \lambda)$ there exists an $\mathcal{O}$-scenario $\rho_\mathcal{O}$ such that $T(\rho_\mathcal{O}) = T$.*

*Proof.* We prove the first statement by induction. It is trivially true if $O$ has 2 vertices. We suppose it to be true for all the circles having less than $2l$ vertices and prove it for a circle having $2l$ vertices. Fix a minimum 2-break-length scenario $\rho_\mathcal{O}$. Its length is $l-1$ due to Lemma 1. The first labeled 2-break of $\rho_\mathcal{O}$ transforms $(O, \lambda)$ into two vertex disjoint labeled circles $(O_1, \lambda_1)$ and $(O_2, \lambda_2)$ both having less vertices than $O$. The rest of the scenario $\rho_\mathcal{O}$ can be partitioned into $\rho_\mathcal{O}^1$ acting on the edges of $O_1$ and $\rho_\mathcal{O}^2$ acting on the edges of $O_2$. As $\rho_\mathcal{O}$ is a minimum 2-break-length scenario, $\rho_\mathcal{O}^1$ and $\rho_\mathcal{O}^2$ must also be minimum 2-break-length scenarios. By the inductive hypothesis, $T(\rho_\mathcal{O}^1)$ and $T(\rho_\mathcal{O}^2)$ are planar trees on $(O_1, \lambda_1)$ and $(O_2, \lambda_2)$ respectively. $T(\rho_\mathcal{O})$ can be easily obtained from $T(\rho_\mathcal{O}^1)$ and $T(\rho_\mathcal{O}^2)$ by taking the union of their edges and adding an edge corresponding to the first 2-break of $\rho_\mathcal{O}$. This way we obtain a planar tree $T(\rho_\mathcal{O})$ on $(O, \lambda)$ proving the first statement of the lemma.

Now define the *distance* of an edge $\{x, y\}$ in $T$ as the minimum number of vertices between $x$ and $y$ in the fixed circular embedding of $T$. For example, in the rightmost tree on the top of Fig. 5 the distance of the edge $\{w, z\}$ is one, because $t$ is in between $w$ and $z$, while the distance of the edge $\{x, y\}$ is 0. An edge is said to be *short* if its distance is 0. We prove an auxiliary lemma.

**Lemma.** *A planar tree $T$ on $(O, \lambda)$ has a short edge incident to a leaf.*

*Proof.* Choose a leaf $x$ in $\mathcal{T}$ incident to an edge of the minimum distance $d$. If $d \neq 0$, then in between the leaf and the vertex that it is adjacent to, there are $d$ other vertices. Since $\mathcal{T}$ is planar on $(O, \lambda)$, it is easy to see that there is at least one other leaf among these $d$ vertices, which contradicts the minimality of $x$. □

Now take a short edge $\{x, y\}$ incident to a leaf $x$ in $\mathcal{T}$. Take the black edges $\{a, b\}$ and $\{c, d\}$ in $(O, \lambda)$ labeled with $x$ and $y$ respectively and separated by a gray edge $\{b, c\}$. Perform a labeled 2-break $(\{b, a\}, x), (\{c, d\}, y) \to (\{b, c\}, x), (\{a, d\}, y)$. This 2-break results in two labeled circles. One of them is a terminal graph having two edges $\{b, c\}$ with the black one labeled with $x$. Remove the edge $\{x, y\}$ from $\mathcal{T}$. This way we have reduced the size of the problem. The number of the vertices in the circle was reduced by two and the number of the edges in the tree was reduced by 1. We iterate this procedure to construct a required scenario. See the bottom part of Fig. 5 for an example. □

## A.8   Lemma 6

**Lemma.** *A minimum cost planar tree on a circle can be found in $O(r^4)$ time, where $r$ is the number of vertices of a tree.*

*Proof.* Farnoud and Milenkovic pose the problem of sorting permutations by cost-constrained mathematical transpositions (a sorting scenario is called a *decomposition*) [14]. They define a cost function on the set of transpositions and treat the problem, called MIN-COST-MLD, of finding a minimum cost decomposition among the minimum length transposition decompositions of a permutation. They reduce this problem to finding a minimum cost planar tree on a circle, and propose the following $O(r^4)$ time dynamic programming algorithm for a tree having $r$ vertices.

Enumerate the vertices 1 to $r$ while respecting their order on the circle. Define $cost(i, j)$ as the minimum cost of a planar tree on the vertices $\{i, \ldots, j\}$ for $1 \le i < j \le r$ and set $cost(i, i) = 0$ for $1 \le i \le r$.

Take a planar tree $\mathcal{T}$ on the vertices $\{1, \ldots, r\}$. If $deg(1) = 1$ and 1 is on the edge $\{1, q\}$, then the cost of $\mathcal{T}$ is equal to $\Phi(1, q)$ plus the costs of the subgraphs of $\mathcal{T}$ induced by the vertices $\{2, \ldots, q\}$ and $\{q + 1, \ldots, r\}$. If $deg(1) > 1$, then take $q = \max(\{u | \{1, u\}$ belongs to $\mathcal{T}\})$ and $s = \max(\{u|$ there is a path in $\mathcal{T}$ joining 1 and $u$ but not visiting $q\})$. The cost of $\mathcal{T}$ is equal to $\Phi(1, q)$ plus the costs of the subgraphs of $\mathcal{T}$ induced by the vertices $\{1, \ldots, s\}$, $\{s + 1, \ldots, q\}$ and $\{q, \ldots, r\}$. This observation provides us with the following equality:

$$cost(i, j) = \max(cost(i, s) + cost(s + 1, q) + cost(q, j) + \Phi(i, q)| \ i \le s < q \le j)$$

for $1 \le i < j \le r$, that leads to an $O(r^4)$ time dynamic programming algorithm for finding $cost(1, r)$. □

# References

1. Bafna, V., Pevzner, P.A.: Genome rearrangements and sorting by reversals. SIAM J. Comput. **25**(2), 272–289 (1996)
2. Baudet, C., Dias, U., Dias, Z.: Sorting by weighted inversions considering length and symmetry. BMC Bioinform. **16**(19), S3 (2015)
3. Baudet, C., Lemaitre, C., Dias, Z., Gautier, C., Tannier, E., Sagot, M.-F.: Cassis: detection of genomic rearrangement breakpoints. Bioinformatics **26**(15), 1897–1898 (2010)
4. Bergeron, A., Mixtacki, J., Stoye, J.: A unifying view of genome rearrangements. In: Bücher, P., Moret, B.M.E. (eds.) WABI 2006. LNCS, vol. 4175, pp. 163–173. Springer, Heidelberg (2006). https://doi.org/10.1007/11851561_16
5. Bhuiyan, H., Chen, J., Khan, M., Marathe, M.: Fast parallel algorithms for edge-switching to achieve a target visit rate in heterogeneous graphs. In: 43rd International Conference on Parallel Processing (ICPP), pp. 60–69. IEEE (2014)
6. Bienstock, D., Günlük, O.: A degree sequence problem related to network design. Networks **24**(4), 195–205 (1994)
7. Biller, P., Knibbe, C., Guéguen, L., Tannier, E.: Breaking good: accounting for the diversity of fragile regions for estimating rearrangement distances. Genome Biol. Evol. **8**, 1427–39 (2016)
8. Bitner, J.R.: An asymptotically optimal algorithm for the dutch national flag problem. SIAM J. Comput. **11**(2), 243–262 (1982)
9. Blanchette, M., Kunisawa, T., Sankoff, D.: Parametric genome rearrangement. Gene **172**(1), 11–17 (1996)
10. Braga, M.D.V., Sagot, M.-F., Scornavacca, C., Tannier, E.: The solution space of sorting by reversals. In: Măndoiu, I., Zelikovsky, A. (eds.) ISBRA 2007. LNCS, vol. 4463, pp. 293–304. Springer, Heidelberg (2007). https://doi.org/10.1007/978-3-540-72031-7_27
11. Braga, M.D.V., Stoye, J.: The solution space of sorting by DCJ. J. Comput. Biol. **17**(9), 1145–1165 (2010)
12. Bulteau, L., Fertin, G., Tannier, E.: Genome rearrangements with indels in inter-genes restrict the scenario space. BMC Bioinform. **17**(14), 426 (2016)
13. Caprara, A.: Sorting by reversals is difficult. In Proceedings of the First Annual International Conference on Computational Molecular Biology, pp. 75–83. ACM (1997)
14. Farnoud, F., Milenkovic, O.: Sorting of permutations by cost-constrained transpositions. IEEE Trans. Inf. Theory **58**(1), 3–23 (2012)
15. Fertin, G., Jean, G., Tannier, E.: Algorithms for computing the double cut and join distance on both gene order and intergenic sizes. Algorithms Mol. Biol. **12**(1), 16 (2017)
16. Lieberman-Aiden, E., Van Berkum, N.L., Williams, L., Imakaev, M., Ragoczy, T., Telling, A., Amit, I., Lajoie, B.R., Sabo, P.J., Dorschner, M.O., et al.: Comprehensive mapping of long-range interactions reveals folding principles of the human genome. science **326**(5950), 289–293 (2009)
17. Nadeau, J.H., Taylor, B.A.: Lengths of chromosomal segments conserved since divergence of man and mouse. Proc. Natl. Acad. Sci. **81**(3), 814–818 (1984)
18. Ohno, S.: Evolution by Gene Duplication. Springer, Heidelberg (1970)
19. Pulicani, S., Simonaitis, P., Rivals, E., Swenson, K.M.: Rearrangement scenarios guided by chromatin structure. In: Meidanis, J., Nakhleh, L. (eds.) Comparative Genomics. RECOMB-CG 2017. LNCS, vol. 10562, pp. 141–155. Springer, Cham (2017)

20. Shao, M., Lin, Y.: Approximating the edit distance for genomes with duplicate genes under DCJ, insertion and deletion. In: BMC bioinformatics, vol. 13, p. S13. BioMed Central (2012)
21. Shao, M., Lin, Y., Moret, B.M.E.: Sorting genomes with rearrangements and segmental duplications through trajectory graphs. In: BMC bioinformatics, vol. 14, p. S9. BioMed Central (2013)
22. Simonaitis, P., Swenson, K.M.: Finding local genome rearrangements. Algorithms Mol. Biol. **13**(1), 9 (2018)
23. Swenson, K.M., Simonaitis, P., Blanchette, M.: Models and algorithms for genome rearrangement with positional constraints. Algorithms Mol. Biol. **11**(1), 13 (2016)
24. Veron, A., Lemaitre, C., Gautier, C., Lacroix, V., Sagot, M.-F.: Close 3D proximity of evolutionary breakpoints argues for the notion of spatial synteny. BMC Genomics **12**(1), 303 (2011)
25. Yancopoulos, S., Attie, O., Friedberg, R.: Efficient sorting of genomic permutations by translocation, inversion and block interchange. Bioinformatics **21**(16), 3340–3346 (2005)

# Estimation of the True Evolutionary Distance Under the INFER Model

Alexey Zabelkin and Nikita Alexeev[(✉)]

ITMO University, Saint Petersburg, Russia
nikita.v.alexeev@gmail.com

**Abstract.** Genome rearrangements are evolutionary events that shuffle genomic architectures. Usually the rearrangement distance between two genomes is estimated as the minimal number of rearrangements needed to transform one genome into another, which is usually referred to as the parsimony assumption.

Since in reality the parsimony assumption may or may not hold, the question arises of estimating the true evolutionary distance (i.e., the actual number of genome rearrangements between the genomes of two species). While several methods for solving this problem have been developed, all of them have their own disadvantages. In the current paper we consider a very general model and provide a flexible estimator as well as the limits of applicability for the most popular estimation methods, such as the maximum parsimony method.

## 1 Introduction

Genome rearrangements are evolutionary events that shuffle genomic architectures. Most frequent genome rearrangements are *reversals* (that flip segments of a chromosome), *translocations* (that exchange segments of two chromosomes), *fusions* (that merge two chromosomes into one), and *fissions* (that split a single chromosome into two). These four types of rearrangements can be modeled by Double-Cut-and-Join (DCJ) operations [22], which break the genome at two positions and glue the resulting fragments in a new order.

The ability to estimate the evolutionary distance between extant genomes plays a crucial role in many phylogenomic studies. Often such estimation is based on the parsimony assumption, implying that the distance between two genomes can be estimated as the *rearrangement distance* equal to the minimal number of genome rearrangements required to transform one genome into the other. However, in reality the parsimony assumption may not always hold, emphasizing the need for estimation that does not rely on the (minimal) rearrangement distance. The evolutionary distance that accounts for the actual (rather than the minimal) number of genome rearrangements between two genomes is often referred to as the *true evolutionary distance*.

The first method for estimating of the true evolutionary distance was introduced in [21]. This approach takes into account only reversals and transpositions

© Springer Nature Switzerland AG 2018
M. Blanchette and A. Ouangraoua (Eds.): RECOMB-CG 2018, LNBI 11183, pp. 72–87, 2018.
https://doi.org/10.1007/978-3-030-00834-5_4

on unichromosomal genomes. The method that considered general DCJs and an arbitrary number of chromosomes was introduced in [12]. This method implicitly assumes that all intergenic regions are prone to rearrangements. This assumption is usually referred to as the *random breakage model* (RBM) of chromosome evolution [14,15]. Since some intergenic regions may be under selection and so can not be involved in rearrangements, we proposed an estimation method [1] under the so-called *fragile breakage model* (FBM) [16] postulating that only certain "fragile" genomic regions are prone to rearrangements. However, the method is based on the assumption that all fragile regions are equally likely to be involved in rearrangements, which is not always true. Recently this model was generalized by Biller *et al.* [5], who proposed the *INFER* model taking into account the fact that each fragile region has its own "fragility" (the probability to be involved in a rearrangement).

It is important to note that all the mentioned models make the so-called unique gene content assumption, that is, they assume that two genomes contain the same set of genes and each gene is present in each genome in one copy. This assumption is very strong and does not usually hold for biological data. Indeed, since insertions and deletions (indels) of genes happen in the course of evolution, some genes could be present in one genome and not present in another. Moreover, duplications are the reason that some genes are present in the same genome in several copies. See [6,8–10,17–19,23] for more information. However, rearrangements themselves explain a great deal of genome diversity, and developing solid mathematical background for the rearrangement analysis is necessary for investigating more complex models.

While the INFER model is much more biologically relevant than other unique gene content based models, the computational method provided in [5] has a number of limitations. In particular, the method does not converge for distant genomes, and it is not very robust since it relies on just two parameters of the model which have high variance. It is also hard to implement and its running time is relatively high. In the current study, we propose a new method for estimating the true evolutionary distance between two genomes under the INFER model which does not have the limitations mentioned above. We estimated the parameters of the model for both the general case (when there are no assumptions about the distribution of the regions' fragilities) and the case when fragilities are distributed according to the Dirichlet distribution. We obtained surprisingly nice mathematical formulas, and showed that the estimation results are very accurate.

## 2   Background

### 2.1   Breakpoint Graphs and DCJs

Our analysis is essentially based on the models proposed in [1,5]. Below we remind the readers about the relevant definitions and notations.

In this paper we focus on the analysis of circular genomes (i.e., genomes with circular chromosomes) and address linear genomes later. Our preliminary simulations show that circularization of the chromosomes does not affect the estimations significantly. We represent a genome with $n$ blocks as a *genome graph* composed of $n$ directed edges encoding blocks and their strands and $n$ undirected edges encoding adjacencies between blocks.

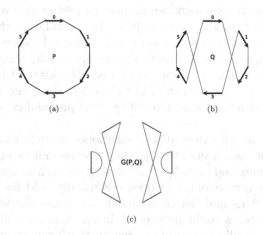

**Fig. 1.** (a) Genome graph of unichromosomal genome $P = (0, 1, 2, 3, 4, 5)$ with adjacency edges colored black. (b) Genome graph of unichromosomal genome $Q = (0, -2, -1, 3, -5, -4)$ with adjacency edges colored red. (c) The breakpoint graph $G(P, Q)$ of genomes $P$ and $Q$ represents a collection of black-red cycles. (Color figure online)

Let $P$ and $Q$ be genomes on the same set of blocks. We assume that in their genome graphs the adjacency edges of $P$ are colored black (Fig. 1a) and the adjacency edges of $Q$ are colored red (Fig. 1b). The *breakpoint graph* $G(P, Q)$ is the superposition of the genome graphs of $P$ and $Q$ with the block edges removed (Fig. 1c). The black and red edges in $G(P, Q)$ form a collection of alternating black-red cycles.

We say that a black-red cycle is an *m-cycle* if it contains $m$ black edges (and $m$ red edges) and let $c_m(P, Q)$ be the number of $m$-cycles in $G(P, Q)$. We refer to 1-cycles as *trivial*[1] and to the other cycles as *non-trivial*. The vertices of non-trivial cycles are called *breakpoints*.

A *DCJ* in genome $Q$ replaces any pair $\{x, y\}$, $\{u, v\}$ of red adjacency edges with either a pair of edges $\{x, u\}$, $\{y, v\}$ or a pair of edges $\{u, y\}$, $\{v, x\}$. We say that such a DCJ *operates* on the edges $\{x, y\}$, $\{u, v\}$ and their endpoints $x, y, u, v$. A DCJ in genome $Q$ transforming it into a genome $Q'$ corresponds to the transformation of the breakpoint graph $G(P, Q)$ into the breakpoint graph

---

[1] In the breakpoint graph constructed on synteny blocks of two genomes, there are no trivial cycles since no adjacency is shared by both genomes. However, the breakpoint graph constructed on orthologous genes or multi-genome synteny blocks may contain trivial cycles.

$G(P, Q')$ (Fig. 2). Each DCJ in the breakpoint graph can merge two black-red cycles into one (if edges $\{x, y\}$, $\{u, v\}$ belong to distinct cycles), split one cycle into two or keep the number of cycles intact (if edges $\{x, y\}$, $\{u, v\}$ belong to the same cycle). The *DCJ distance* between genomes $P$ and $Q$ is the minimum number of DCJs required to transform $Q$ into $P$. It can be evaluated as $d(P, Q) = b(P, Q) - c(P, Q)$, where $b(P, Q) = \sum_{m \geq 2} m \cdot c_m(P, Q)$ is half the number of breakpoints and $c(P, Q) = \sum_{m \geq 2} c_m(P, Q)$ is the number of non-trivial cycles in the breakpoint graph $G(P, Q)$ [22].

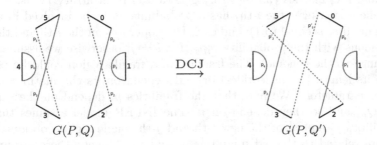

**Fig. 2.** A DCJ in genome $Q$ replaces a pair of red edges in the breakpoint graph $G(P, Q)$ with another pair of red edges forming a matching on the same 4 vertices. (Color figure online)

## 2.2 Evolutionary Model

To estimate the true evolutionary distance between genomes $P$ and $Q$ on the same set of blocks, we view the evolution between them as a discrete Markov process that transforms genome $P$ into genome $Q$ through a sequence of DCJs which occur independently. The process starts at genome $X_0 = P$ and after $k$ steps ends at $X_k = Q$, and corresponds to the transformation of the breakpoint graphs starting at $G(P, P)$ (formed by a collection of trivial cycles) and ending at $G(P, Q)$. We assume that $P$ and $Q$ are composed of a large unknown number $n$ of solid regions interspaced with the same number of fragile regions, some of which remain conserved by chance.

Let $P_n$ and $Q_n$ denote representations of $P$ and $Q$ as sequences of the solid regions. It is important to mention that while we do not know the number $n$ of solid regions, the breakpoint graphs $G(P, Q)$ and $G(P_n, Q_n)$ have the same cycle structure, except for trivial cycles. That is, we have $c_m(P_n, Q_n) = c_m(P, Q)$ for every $m \geq 2$, implying, in particular, that $b(P_n, Q_n) = b(P, Q)$, $c(P_n, Q_n) = c(P, Q)$, and $d(P_n, Q_n) = d(P, Q)$. Indeed, if genomes $P'$ and $Q'$ are obtained from $P$ and $Q$ by replacing a single block $a$ with two consecutive smaller blocks $a_1, a_2$, then $G(P', Q')$ can be obtained from $G(P, Q)$ by adding one trivial cycle (corresponding to the shared adjacency $a_1, a_2$). Since the genomes $P_n$ and $Q_n$ can be obtained from $P$ and $Q$ with a number of such operations, the breakpoint graphs $G(P, Q)$ and $G(P_n, Q_n)$ may differ only in the number of trivial cycles. We view genome $Q_n$ as obtained from $P_n$ with a sequence of $k$ DCJs, each operating on two randomly selected fragile regions. In contrast to $c_1(P_n, Q_n)$, the value of $c_1(P, Q)$ is rather arbitrary and thus is ignored in our model. We view genome $Q_n$ as obtained from $P_n$ with a sequence of $k$ DCJs each operating on two randomly selected fragile regions.

The INFER model introduced in [5] takes into account different probabilities to break for different fragile regions. While the original paper gave a detailed description of only one type of genome rearrangements – inversions, – the definitions and the analysis can be easily generalized to arbitrary DCJs. Namely, each fragile region (say, the $i$-th) has a probability $p_i$ to be involved in a DCJ (the sum of all $p_i$ is equal to 1); and each DCJ operates on the $i$-th and the $j$-th fragile regions with the probability $p_i p_j$ (for $i \neq j$; otherwise we assume there is no change to the genome). The fragility of a fragile region could be proportional to its length (in nucleotides) or/and depend on its chromatin structure and other parameters. We note that the fragilities $p_i$ depend on the genomes $P_n$ and $Q_n$, and so they depend on $n$. The INFER model assumes that the new fragilities $p'_i$ and $p'_j$ of the new $i$-th and $j$-th fragile regions obtained after a DCJ are related to the old fragilities $p_i$ and $p_j$. Namely, there are uniform random variables $r_1$ and $r_2$ ($r_1, r_2 \in [0, 1]$) such that $p'_i = r_1 p_i + r_2 p_j$ and $p'_j = (1 - r_1)p_i + (1 - r_2)p_j$. Note that they satisfy the property $p'_i + p'_j = p_i + p_j$. Since the fragilities are updated during the process, it is natural to assume that they follow the equilibrium distribution of the updating process. This distribution is the flat Dirichlet distribution, i.e. the uniform distribution on a standard simplex $\{(x_1, x_2, \ldots, x_n) : x_i \geq 0, \sum x_i = 1\}$.

In the current paper we consider two cases:

- The general case, when the only assumption on $p_i$'s is that $p_i = \Theta(\frac{1}{n})$ for each $i$. In this case we introduce random variables $\alpha_i = np_i$, and we assume that these variables $\alpha_i$ are identically distributed (since there is no information about the positions of fragile regions).
- The case when fragilities $p_i$ are distributed according to the Dirichlet distribution.

We note that it is important to consider the general case, since, while the flat Dirichlet distribution is the equilibrium distribution of $p_i$'s under the INFER model, their actual distribution is not known because INFER does not take into account such factors as duplications, indels, and dispersion of transposable elements (see [5]). Thus, as soon as new insights into the fragilities distribution appear, one can easily modify Theorem 1 below to apply their modified method to the data.

Our evolutionary model has the following observable parameters:

- $c_m = c_m(P_n, Q_n) = c_m(P, Q)$ for any $m \geq 2$, i.e., the number of $m$-cycles in $G(P, Q)$;
- $b = b(P_n, Q_n) = b(P, Q) = \sum_{m \geq 2} m \cdot c_m$, the number of broken fragile regions between $P$ and $Q$, which is also the number of synteny blocks between $P$ and $Q$, or half of the total length of all non-trivial cycles in $G(P, Q)$;

- $d = d(P_n, Q_n) = d(P, Q) = b - \sum_{m \geq 2} c_m$, the DCJ distance between $P$ and $Q$;

while the following parameters are hidden:

- $c_1 = c_1(P_n, Q_n)$, the number of trivial cycles in $G(P_n, Q_n)$ (under the FBM, $c_1(P_n, Q_n) \neq c_1(P, Q)$);
- $n = n(P) = n(Q)$, the number of fragile regions in each of genomes $P$ and $Q$, half the total length of all cycles in $G(P_n, Q_n)$;
- $k = k(P, Q)$, the number of DCJs in the Markov process, the true evolutionary distance between $P$ and $Q$;
- $p_i = \frac{\alpha_i}{n}$, the fragility of the $i$-th fragile region.

## 3  Methods

First, we will analytically estimate the number of $m$-cycles $c_m$ ($m \geq 2$), considering only relatively small $m$ and assuming that $n$ and $k$ are sufficiently large (see Theorem 2). Then, based on this analysis, we propose a method to estimate the true evolutionary distance $k$.

### 3.1  Theoretical Analysis

**Theorem 1.** *Let genome $P_n$ be a genome with $n$ fragile regions and genome $Q_n$ be obtained from $P_n$ with $k = \lfloor \gamma n/2 \rfloor$ random DCJs for some $\gamma > 0$, and $\alpha_i = np_i$ are identically distributed. Then, for any fixed $m$, the proportion of edges that belong to $m$-cycles in $G(P_n, Q_n)$ is*

$$\lim_{n \to \infty} \mathbb{E} \frac{mc_m}{n} = \frac{\gamma^{m-1}}{(m-1)!} \mathbb{E} \, \alpha_1 \dots \alpha_m (\alpha_1 + \dots + \alpha_m)^{m-2} e^{-\gamma \sum_{i=1}^{m} \alpha_i}. \quad (1)$$

**Theorem 2.** *Let genome $P_n$ be a genome with $n$ fragile regions and genome $Q_n$ be obtained from $P_n$ with $k = \lfloor \gamma n/2 \rfloor$ random DCJs for some $\gamma > 0$, and the fragilities $p_i$ are distributed according to the Dirichlet distribution. Then, for any fixed $m$, the proportion of edges that belong to $m$-cycles in $G(P_n, Q_n)$ is*

$$\lim_{n \to \infty} \mathbb{E} \frac{mc_m}{n} = \frac{(3m-3)! \gamma^{m-1}}{(m-1)!(2m-1)!(\gamma+1)^{3m-2}}. \quad (2)$$

Let us prove Theorem 1 first.

*Proof.* Let us note that the majority of $m$-cycles are the result of merging smaller cycles, and the number of $m$-cycles obtained as a result of splitting larger cycles is negligible. Indeed, the probability that some $\ell$-cycle (with $\ell \geq m$) is split into an $m$-cycle and an $(\ell - m)$-cycle during an individual DCJ has the order $\frac{1}{n}$, and thus after $k$ DCJs the expected number of "split" $m$-cycles is finite. At the same time, the number of"merged" $m$-cycles after $k$ DCJs has the order $n$ (See Lemma 3 in [1] for the rigorous proof).

Consider a set of $m$ black edges $A_m$. Without loss of generality, we assume that the corresponding red edges have the labels $1, 2, \ldots, m$ and the fragilities $p_1, p_2, \ldots, p_m$. Let us find the probability that after $k$ DCJs the black edges of $A_m$ would form an $m$-cycle, obtained as a result of merging smaller cycles. Since an $m$-cycle is formed within $m - 1$ DCJs, there are $\binom{k}{m-1}$ ways to choose the corresponding steps. Let us call the $m - 1$ steps forming an $m$-cycle from $A_m$ a *merging scenario* (see Fig. 3 for an example). To each merging scenario $S$ we assign a labeled tree $T_S$ on the vertices labeled $1, 2, \ldots, m$ (see Fig. 4) in the following way: there is an edge between the vertices $i$ and $j$ iff there is a DCJ operating on the $i$-th and $j$-th red edges in the scenario. We note that such a map is $2^{m-1}(m-1)!$ to 1, since the tree does not reflect the order of DCJs (which gives the factor $(m-1)!$) and the way of merging the cycles in each DCJ (which gives the factor $2^{m-1}$).

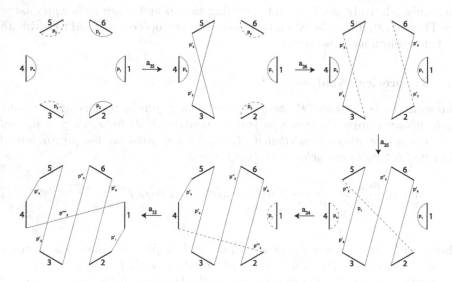

**Fig. 3.** An example of a merging scenario with 5 DCJs. Its probability is equal to $p_1 \cdot p_2 \cdot p_2' \cdot p_2'' \cdot p_2''' \cdot p_3 \cdot p_4 \cdot p_5 \cdot p_5' \cdot p_6$. (Color figure online)

If the fragility of the $i$-th region was $p_i$ during the whole process, then the probability of the scenario $S$ is equal to $\prod_{i=1}^{m} p_i^{d_i}$, where $d_i$ is a degree of the vertex $i$ in $T_S$. By Cayley's theorem [7], the sum of all such probabilities is $p_1 \ldots p_m (p_1 + \cdots + p_m)^{m-2}$. We note that this sum stays the same even if the fragilities are updated after each DCJ as long as the sum of the fragilities of edges in each cycle does not change. The last property holds since the INFER model updating rule for fragilities requires $p_i' + p_j' = p_i + p_j$.

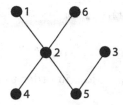

**Fig. 4.** The tree corresponding to the merging scenario from Fig. 3. (Color figure online)

Thus, the probability that the edges of $A_m$ form an $m$-cycle is

$$\binom{k}{m-1} 2^{m-1}(m-1)! p_1 \ldots p_m (p_1 + \cdots + p_m)^{m-2} \left(1 - \sum_{i=1}^{m} p_i\right)^{2(k-m+1)}, \quad (3)$$

where the last factor $(1 - \sum_{i=1}^{m} p_i)^{2(k-m+1)}$ corresponds to the fact that the red edges $1, 2, \ldots, m$ are not involved in the $k - m + 1$ DCJs.

There are $\binom{n}{m}$ ways to choose the set $A_m$. Thus, the expected normalized number of $m$-cycles converges to

$$\frac{m}{n}\binom{n}{m}\binom{k}{m-1} 2^{m-1}(m-1)! p_1 \ldots p_m \left(\sum_{i=1}^{m} p_i\right)^{m-2} \left(1 - \sum_{i=1}^{m} p_i\right)^{2(k-m+1)} \sim$$

$$\frac{m}{n}\frac{n^m}{m!}(2k)^{m-1}\frac{1}{n^{2m-2}}\alpha_1 \ldots \alpha_m \left(\sum_{i=1}^{m} \alpha_i\right)^{m-2} \left(1 - \frac{\sum_{i=1}^{m} \alpha_i}{n}\right)^{2(k-m+1)} \sim$$

$$\frac{1}{(m-1)!}\frac{2k^{m-1}}{n}\alpha_1 \ldots \alpha_m \left(\sum_{i=1}^{m} \alpha_i\right)^{m-2} \left(1 - \frac{\sum_{i=1}^{m} \alpha_i}{n}\right)^{2(k-m+1)} \sim$$

$$\frac{\gamma^{m-1}}{(m-1)!}\alpha_1 \ldots \alpha_m (\alpha_1 + \cdots + \alpha_m)^{m-2} \exp\left(-\gamma \sum_{i=1}^{m} \alpha_i\right), \quad (4)$$

which finishes the proof.

Note that Theorem 1 allows to estimate $\frac{mc_m}{n}$ for various assumptions. In the case of degenerate distribution (all $\alpha_i = 1$) we obtain the main theorem of [1] as a particular case (except for the fact that Theorem 1 does not provide any bound for the rate of convergence). Theorem 2 can also be considered as a particular case of Theorem 1.

*Proof.* Let us recall that a vector $(p_1, \ldots, p_n)$ from the flat Dirichlet distribution can be generated in the following way. Let $\beta_i$, $(i = 1, 2, \ldots, n)$ be exponential random variables with parameter 1. Then $p_i = \frac{\beta_i}{\beta_1 + \beta_2 + \cdots + \beta_n}$ are distributed according to the Dirichlet distribution. Note that by the law of large numbers $\beta_1 + \beta_2 + \cdots + \beta_n$ strongly converges to $n$. So we can consider the values $\alpha_i = \lim n p_i$ as exponential random variables, moreover, for small $m$ all the dependencies between $\alpha_1, \alpha_2, \ldots, \alpha_m$ are negligible. Thus, we only need to find the expected value of

$$\alpha_1 \ldots \alpha_m (\alpha_1 + \ldots + \alpha_m)^{m-2} e^{-\gamma \sum_{i=1}^{m} \alpha_i} \tag{5}$$

with the measure $e^{-\sum_{i=1}^{m} \alpha_i} d\alpha_1 \ldots d\alpha_m$ in order to prove Theorem 2.

Let us prove that

$$\int \cdots \int_{\mathbb{R}_+^m} \alpha_1 \cdot \ldots \cdot \alpha_m (\alpha_1 + \ldots + \alpha_m)^{m-2} e^{-\sum_{i=1}^{m} ((\gamma+1)\alpha_i)} d\alpha_1 \ldots d\alpha_m$$

$$= \frac{(3m-3)!}{(2m-1)!(\gamma+1)^{3m-2}}. \tag{6}$$

First, change the variable $t_i = \alpha_i(\gamma+1)$:

$$\int \cdots \int_{\mathbb{R}_+^m} \frac{t_1}{\gamma+1} \cdot \ldots \cdot \frac{t_m}{\gamma+1} \left( \frac{t_1 + \ldots + t_m}{\gamma+1} \right)^{m-2} e^{-\sum_{i=1}^{m} t_i} \frac{dt_1}{\gamma+1} \ldots \frac{dt_m}{\gamma+1} =$$

$$\frac{1}{(\gamma+1)^{3m-2}} \int \cdots \int_{\mathbb{R}_+^m} t_1 \cdot \ldots \cdot t_m (t_1 + \ldots + t_m)^{m-2} e^{-\sum_{i=1}^{m} t_i} dt_1 \ldots dt_m. \tag{7}$$

Then introduce another change $u = t_1 + \ldots + t_m$:

$$\frac{1}{(\gamma+1)^{3m-2}} \int_0^\infty \int \cdots \int_{\sum_{i=1}^{m-1} t_i \leq u} t_1 \ldots t_{n-1} \left( u - \sum_{i=1}^{m-1} t_i \right) u^{m-2} e^{-u} dt_1 \ldots dt_{m-1} du$$

$$= \frac{1}{(\gamma+1)^{3m-2}} \left( \int_0^\infty \int \cdots \int_{\sum_{i=1}^{m-1} t_i \leq u} t_1 \cdot \ldots \cdot t_{m-1} u^{m-1} e^{-u} dt_1 \ldots dt_{m-1} du \right.$$

$$- (m-1) \int_0^\infty \int \cdots \int_{\sum_{i=1}^{m-1} t_i \leq u} t_1^2 \cdot t_2 \cdot \ldots \cdot t_{m-1} u^{m-2} e^{-u} dt_1 \ldots dt_{m-1} du \right)$$

$$= \frac{1}{(\gamma+1)^{3m-2}} \left( \int_0^\infty \left( \int \cdots \int_{\sum_{i=1}^{m-1} t_i \leq u} t_1 \cdot \ldots \cdot t_{m-1} dt_1 \ldots dt_{m-1} \right) u^{m-1} e^{-u} du \right.$$

$$- (m-1) \int_0^\infty \left( \int \cdots \int_{\sum_{i=1}^{m-1} t_i \leq u} t_1^2 \cdot t_2 \cdot \ldots \cdot t_{m-1} dt_1 \ldots dt_{m-1} \right) u^{m-2} e^{-u} du \right). \tag{8}$$

The latter two integrals can be computed recursively:

$$\int \cdots \int_{\sum_{i=1}^{m-1} t_i \leq u} t_1 \cdot \ldots \cdot t_{m-1} dt_1 \ldots dt_{m-1} = \frac{u^{2m-2}}{(2m-2)!} \tag{9}$$

and

$$\int \cdots \int_{\sum_{i=1}^{m-1} t_i \leq u} t_1^2 \cdot \ldots \cdot t_{m-1} dt_1 \ldots dt_{m-1} = \frac{u^{2m-1}}{(2m-1)!}. \tag{10}$$

After simplification we obtain (6) and so prove the theorem.

*Remark 1.* We note that the sequence $\frac{(3m-3)!}{(m-1)!(2m-1)!}$ $(1, 1, 3, 12, 55, \dots)$ is well-known as *Fuss–Catalan numbers* and appears in the On-Line Encyclopedia of Integer Sequences (OEIS) [20] as the sequence A001764. This sequence appears in combinatorics and probability theory in many different contexts, but such an interpretation is new (to the best of our knowledge).

## 3.2 Estimation Method

Since Biller *et al.* [5] showed the biological relevance of the assumption that the fragilities are distributed according to the Dirichlet distribution, we develop an estimation method for this case based on our analysis. We run the simulations under the INFER model, and they show that the estimations of the number of $m$-cycles

$$\hat{c}_m := \frac{(3m-3)!\gamma^{m-1}}{m!(2m-1)!(\gamma+1)^{3m-2}}n \qquad (11)$$

are very accurate (see Fig. 5). We performed 200 runs of the Markov process, for each we chose the parameter $n$ randomly from $[500, 3{,}000]$ and changed $k$ from 0 to $n$.

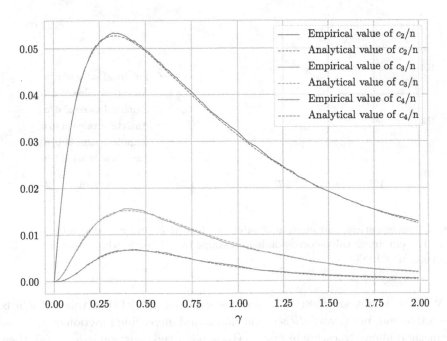

**Fig. 5.** The empirical and analytical values of $c_2/n$, $c_3/n$, and $c_4/n$ depending on $\gamma = 2k/n$. The empirical value is averaged over 200 runs of the Markov process with the parameters $n$ from $[500, 3{,}000]$ and $k$ from $[0, n]$.

Based on this, we can estimate the normalized number of breakpoints and the normalized minimal distance:

$$\frac{\hat{b}}{n} = 1 - \frac{\hat{c}_1}{n} = 1 - \frac{1}{1+\gamma} = \frac{\gamma}{1+\gamma}, \tag{12}$$

$$\frac{\hat{d}}{n} = \sum_{m=2}^{\infty} \frac{\hat{c}_m}{n}(m-1) = 1 - \frac{(1+\gamma)^2 \left( {}_2F_1\left(-\frac{2}{3}, -\frac{1}{3}, \frac{1}{2}, \frac{27\gamma}{4(1+\gamma)^3}\right) - 1\right)}{3\gamma}, \tag{13}$$

where $_2F_1$ stands for the hypergeometric function. These estimations are very accurate (see Fig. 6) even for a single run (!) of the Markov process.

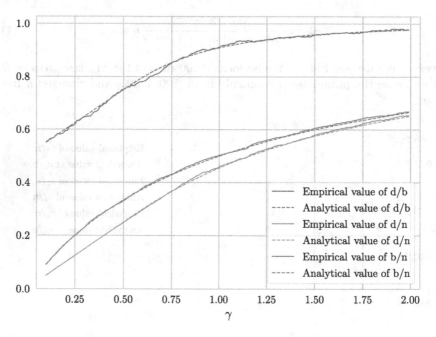

**Fig. 6.** The empirical and analytical values of $b/n$, $d/b$, and $b/n$ depending on $\gamma = 2k/n$. The empirical value corresponds to a single run of the Markov process with the parameter $n = 1,000$.

While $n$ is not an observable parameter of our model, the value of $d/b$ is observable and, moreover, $\hat{d}/\hat{b}$ is a continuous and increasing function of $\gamma$. Thus, our method allows to estimate first $\gamma_e$ (by solving the equation $\hat{d}/\hat{b} = d/b$), then find $n_e$ as $b\frac{1+\gamma_e}{\gamma_e}$ and $k_e$ as $\frac{n_e \gamma_e}{2}$. This estimation method is very accurate (see the boxplots for the relative error $\frac{k-k_e}{k}$ in Fig. 7).

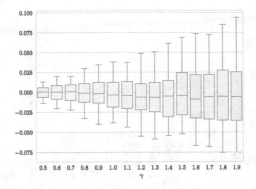

**Fig. 7.** The relative error $(k - k_e)/k$ of our method depending on $\gamma = 2k/n$.

## 4 Discussion

In this paper we propose a method to estimate the true evolutionary distance under the INFER model introduced in [5]. While the original paper [5] proposed not only the biologically relevant model, but a method to estimate the distance as well, the method had a number of computational limitations. In particular, the method does not converge if $2k \geq n$, and it relies on the estimations of $b$ and $c_2$ only, which makes it very sensitive to random outliers of these two parameters. We simulate a dataset according to the INFER model with the parameter $n$ randomly chosen from $[500, 3000]$ and the parameter $k$ from $[0, n]$. We run both methods (ours and the one from [5]) on a laptop and compare them (see Table 1). As one can see, our method is easier to implement, its relative error $(\frac{k - k_e}{k})$ is smaller and the method itself is faster. Despite the fact that our method relies on asymptotic analysis and assumes that $n$ is a large number, the method performs well for relatively small values of $n$ as well (see Fig. 8).

**Table 1.** Comparison of the methods' performance

| – | The method from [5] | Our method |
|---|---|---|
| Average running time | 3.02 s | 0.00017 s |
| Average relative error | 1.99% | 0.68% |
| Maximal relative error | 8% | 5% |
| Formula for $\hat{c}_1$ | $\sum_{l=0}^{\infty} \dfrac{(-2k)^l}{\prod_{u=0}^{l-1}(n+u)}$ | $\dfrac{n^2}{2k+n}$ |
| Formula for $\hat{c}_2$ | $kn^2 \sum_{l=0}^{\infty} \sum_{m=0}^{\infty} \dfrac{(-2(k-1))^{l+m}(l+1)(m+1)}{\prod_{u=0}^{l+m+1}(n+u)}$ | $\dfrac{kn^4}{(2k+n)^4}$ |
| Does it work if $2k \geq n$ | No | Yes |
| Search approach | Modified gradient descent method in $\mathbb{R}^2$ | Binary search |

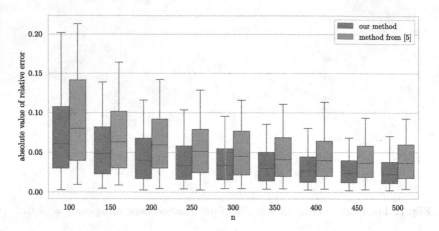

**Fig. 8.** The absolute value of the relative error $(k - k_e)/k$ for small values of $n$.

Moreover, our method allows to find the limits of the parsimony approach under the INFER model. One can easily see (Fig. 9) and even prove (we will do that in further publications) that the minimal distance $d$ is very close to $k$ as long as $k < \frac{n}{4}$ (or $\gamma < 0.5$). At the same time, if $k$ is close to $n$, the relative error provided by parsimony approach reaches 30%.

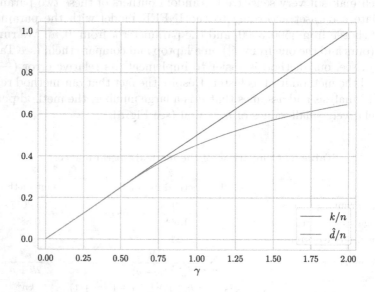

**Fig. 9.** The values of $\hat{d}/n$ and $k/n$ depending on $\gamma = 2k/n$.

We analyzed the real genomes of three subfamilies from the Rosaceae and Vitaceae families, based on the data provided in [11], and six Mammalian species (human, chimpanzee, macaque, mouse, rat, dog), based on the data provided in [13]. Since the genomes have linear chromosomes and we analyze genomes with circular chromosomes only, we add "virtual" adjacencies in order to circularize the genomes in such a way as to minimize the pairwise minimal distance between the genomes. This operation does not introduce much noise into the data, since the number of such virtual adjacencies is about 10, which is an order of magnitude lower than the number of real adjacencies. For each pair of species we build a breakpoint graph and find the values of $b$, $d$, and $c_m$ for $m \geq 2$. Then we estimate the parameters $n$ and $k$ of our model, and run simulations with similar parameters. The results show that the values of $c_m$ for the real data have similar behavior to the simulated ones (see Fig. 10). At the same time the behavior of the estimated value of the number of fragile regions $n$ is not very stable. Namely, for three genomes $P$, $Q_1$ and $Q_2$ (say, *Prunus*, *Fragaria*, and *Vitis*), the value of $n_1$ estimated from $G(P, Q_1)$ can be different from the value of $n_2$ estimated from $G(P, Q_2)$. We will address the question of more accurate estimation of the parameter $n$ for several genomes simultaneously in future research.

Moreover, while the mammal species seem to be in the parsimony phase, the results for plant species in Table 2 show that the parsimony method could underestimate the true evolutionary distance by 10%.

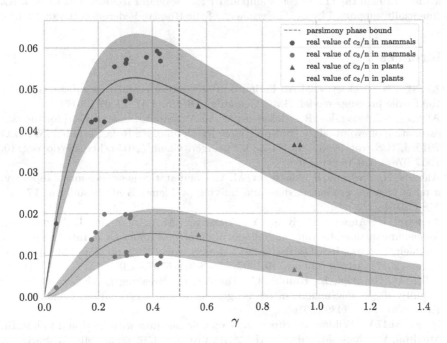

**Fig. 10.** The values of $c_2/n$ and $c_3/n$ for pairs of mammal species (dots) and plant species (triangles). The parameters $n$, $k$ and $\gamma = 2k/n$ are estimated with our method, the curves represent the theoretical expected values of $c_2/n$ and $c_3/n$ depending on $\gamma$, and the shaded regions represent 95% confidence intervals (based on simulations).

**Table 2.** Estimations of the true distance between three plant genomes.

| Genome pair | Minimal distance | Our method | The method from [5] | The method from [1] |
|---|---|---|---|---|
| *Prunus — Fragaria* | 265 | 291 | 281 | 269 |
| *Prunus — Vitis* | 242 | 248 | 251 | 241 |
| *Fragaria — Vitis* | 396 | 437 | 420 | 401 |

As we mention in Introduction, the INFER model makes the unique gene content assumption and does not take into account indels and duplication. In future research we are going to address this issue and combine the methods from the current paper together with the topological and ILP approaches implemented in [3,4]. Moreover, we are going to take into account more complex types of rearrangements, such as transpositions, and properly update the techniques from [2].

**Acknowledgments.** The authors thank Artem Vasilyev and Pavel Avdeyev for fruitful discussions and the anonymous reviewers for valuable comments.

The work of NA was financially supported by the Government of the Russian Federation through the ITMO Fellowship and Professorship Program. The work of AZ was financially supported by the Government of the Russian Federation (Grant 08-08).

# References

1. Alexeev, N., Alekseyev, M.A.: Estimation of the true evolutionary distance under the fragile breakage model. BMC Genomics **18**(Suppl 4), 19–27 (2017)
2. Alexeev, N., Aidagulov, R., Alekseyev, M.A.: A computational method for the rate estimation of evolutionary transpositions. In: Ortuño, F., Rojas, I. (eds.) IWBBIO 2015. LNCS, vol. 9043, pp. 471–480. Springer, Cham (2015). https://doi.org/10. 1007/978-3-319-16483-0_46
3. Alexeev, N., Avdeyev, P., Alekseyev, M.A.: Comparative genomics meets topology: a novel view on genome median and halving problems. BMC Bioinform. **17**(14), 418 (2016)
4. Avdeyev, P., Alexeev, N., Rong, Y., Alekseyev, M.A.: A unified ILP framework for genome median, halving, and aliquoting problems under DCJ. In: Meidanis, J., Nakhleh, L. (eds.) RECOMB-CG 2017. LNCS, vol. 10562, pp. 156–178. Springer, Cham (2017). https://doi.org/10.1007/978-3-319-67979-2_9
5. Biller, P., Gueguen, L., Knibbe, C., Tannier, E.: Breaking good: accounting for fragility of genomic regions in rearrangement distance estimation. Genome Biol. Evol. **8**(5), 1427–1439 (2016)
6. Braga, M.D.V., Willing, E., Stoye, J.: Genomic distance with DCJ and indels. In: Moulton, V., Singh, M. (eds.) WABI 2010. LNCS, vol. 6293, pp. 90–101. Springer, Heidelberg (2010). https://doi.org/10.1007/978-3-642-15294-8_8
7. Cayley, A.: A theorem on trees. Q. J. Math. **23**, 376–378 (1889)
8. Compeau, P.E.C.: A simplified view of DCJ-indel distance. In: Raphael, B., Tang, J. (eds.) WABI 2012. LNCS, vol. 7534, pp. 365–377. Springer, Heidelberg (2012). https://doi.org/10.1007/978-3-642-33122-0_29
9. Compeau, P.E.C.: DCJ-Indel sorting revisited. Algorithms Mol. Biol. **8**(1), 6 (2013)

10. El-Mabrouk, N.: Genome rearrangement by reversals and insertions/deletions of contiguous segments. In: Giancarlo, R., Sankoff, D. (eds.) CPM 2000. LNCS, vol. 1848, pp. 222–234. Springer, Heidelberg (2000). https://doi.org/10.1007/3-540-45123-4_20

11. Jung, S., Cestaro, A., Troggio, M., Main, D.: Whole genome comparisons of fragaria, prunus and malus reveal different modes of evolution between rosaceous subfamilies. BMC Genomics 13(129), 1–12 (2012)

12. Lin, Y., Moret, B.M.: Estimating true evolutionary distances under the DCJ model. Bioinformatics 24(13), i114–i122 (2008). https://doi.org/10.1093/bioinformatics/btn148

13. Ma, J., et al.: Reconstructing contiguous regions of an ancestral genome. Genome Res. 16(12), 1557–1565 (2006)

14. Nadeau, J.H., Taylor, B.A.: Lengths of chromosomal segments conserved since divergence of man and mouse. Proc. Natl. Acad. Sci. 81(3), 814–818 (1984). https://doi.org/10.1073/pnas.81.3.814

15. Ohno, S.: Evolution by Gene Duplication. Springer, Berlin (1970). https://doi.org/10.1007/978-3-642-86659-3

16. Pevzner, P.A., Tesler, G.: Human and mouse genomic sequences reveal extensive breakpoint reuse in mammalian evolution. Proc. Natl. Acad. Sci. 100, 7672–7677 (2003)

17. Shao, M., Lin, Y.: Approximating the edit distance for genomes with duplicate genes under DCJ, insertion and deletion. BMC Bioinform. 13, S13 (2012). BioMed Central

18. Shao, M., Lin, Y., Moret, B.: An exact algorithm to compute the DCJ distance for genomes with duplicate genes. In: Sharan, R. (ed.) RECOMB 2014. LNCS, vol. 8394, pp. 280–292. Springer, Cham (2014). https://doi.org/10.1007/978-3-319-05269-4_22

19. Swenson, K.M., Marron, M., Earnest-DeYoung, J.V., Moret, B.M.E.: Approximating the true evolutionary distance between two genomes. J. Exp. Algorithmics 12, 3–5 (2008)

20. The OEIS Foundation: The On-Line Encyclopedia of Integer Sequences (2018). http://oeis.org

21. Wang, L.S., Warnow, T.: Estimating true evolutionary distances between genomes. In: Proceedings of the Thirty-Third Annual ACM Symposium on Theory of Computing, pp. 637–646. ACM (2001)

22. Yancopoulos, S., Attie, O., Friedberg, R.: Efficient sorting of genomic permutations by translocation, inversion and block interchange. Bioinformatics 21(16), 3340–3346 (2005)

23. Yancopoulos, S., Friedberg, R.: Sorting genomes with insertions, deletions and duplications by DCJ. In: Nelson, C.E., Vialette, S. (eds.) RECOMB-CG 2008. LNCS, vol. 5267, pp. 170–183. Springer, Heidelberg (2008). https://doi.org/10.1007/978-3-540-87989-3_13

# Genome Sequencing

# On the Hardness of Approximating Linearization of Scaffolds Sharing Repeated Contigs

Tom Davot[1]([⊠]), Annie Chateau[1,2], Rodolphe Giroudeau[1], and Mathias Weller[3]

[1] LIRMM - CNRS UMR 5506, Montpellier, France
{davot,chateau,rgirou}@lirmm.fr
[2] IBC, Montpellier, France
[3] CNRS, LIGM, Université Paris Est, Marne-la-Vallée, France
mathias.weller@u-pem.fr

**Abstract.** Solutions to genome scaffolding problems can be represented as paths and cycles in a "solution graph". However, when working with repetitions, such solution graph may contain branchings and they may not be uniquely convertible into sequences. Having introduced, in a previous work, various ways of extracting the unique parts of such solutions, we extend previously known NP-hardness results to the case that the solution graph is planar, bipartite, and subcubic, and show the APX-completeness in this case. We also provide some practical tests.

## 1 Introduction

**Motivation.** The process of generating proper biological genomes, from Next-Generation Sequencing (NGS) data to a full sequence of nucleotides, is a path strewn with pitfalls [6]. NGS data are going to evolve towards longer and longer sequences, but most of the available sequencing data in public databases are huge collections of billions of *short reads* (*i.e.* words of between fifteen and hundreds of characters) [17] which have to be assembled into longer sequences called *contigs*. Those contigs represent fragments of the final genome but they usually do not reach the size of chromosomes and the thusly obtained *draft genomes* may therefore be highly fragmented, especially due to repeats in the genomes [18]. Though some emerging methods aim to use partially assembled genomes to infer global information on genomes [3], reducing this fragmentation is of great interest when it comes to consider whole-genome rearrangements. This fragmentation can be reduced by an additional operation, the *scaffolding*, that aims at providing an order and relative orientation of contigs that is consistent with most of the original NGS data [14]. Especially when reads are paired, it is possible to construct a *scaffold graph* summarizing the putative hypotheses concerning ordering and orientation of contigs [4]. Herein, a scaffold graph is a weighted, undirected graph $G$ consisting of 1. a perfect matching $M^*$ that corresponds to the contigs and 2. non-contig edges $uv$ whose weights indicate the confidence that the contig-extremity $u$ is adjacent to the contig-extremity $v$ in the target genome.

© Springer Nature Switzerland AG 2018
M. Blanchette and A. Ouangraoua (Eds.): RECOMB-CG 2018, LNBI 11183, pp. 91–107, 2018.
https://doi.org/10.1007/978-3-030-00834-5_5

This paper focuses on the following problem: suppose that (a) we know for each contig how often it occurs in the genome (its *multiplicity* – which can be inferred using various possibilities), and (b) an optimal subgraph (the *solution graph*) has been extracted from the input scaffold graph (see [4,20,22] for methods to infer such solution graphs), which we consider as a given input. Then, the task is to infer sequences from the solution graph (see Fig. 1). We consider several score functions, examine several special cases and performed tests on a dataset of various species.

**Repeats.** If each contig occurs exactly once in the target genome, then all vertices of the solution graph will have degree at most two and the problem becomes easy. However, in numerous organisms, a significant part of the genome *is* repeated. Such repeats may be of various sizes and present variable copy numbers, according to the species and individuals [2]. Due to the conservatism of some assembly methods, a repeat may cover an entire contig which is separated from the other genomic side fragments [18]. It turns out that, in presence of repeated contigs, a solution graph implies a unique set of sequences if and only if it does not contain so called *ambiguous paths* [21]. Thus, the task above can be achieved by destroying all ambiguous paths in the solution graph. A brutal way to do this is to cut the non-contig edges incident to both extremities of each ambiguous path. However, this solution may erase potentially important information. Indeed, to destroy an ambiguous path, it is sufficient to remove the non-contig edges incident to one of its extremities. The problem of finding a most parsimonious (with respect to some cost function ) set $X$ of edges such that removing $X$ from the given solution graph destroys all ambiguous paths is called SEMI-BRUTAL CUT.

**Definitions and Problems.** We denote by $E(G)$ and $V(G)$ the set of edges and vertices, respectively, of a graph $G$ (or $E$ and $V$ if no ambiguity occurs). A scaffold graph $(G, M^*, \omega)$ consists of a simple loopless multigraph $G$ associated with a perfect matching $M^*$, a weight function $\omega : E \setminus M^* \to \mathbb{N}$. The matching $M^*$ represents the contigs and $\omega$ represents the confidence that two contigs occur consecutively (respecting relative orientation implied by the edge) in the target sequence. The maximum degree of a graph $G$ is denoted by $\Delta(G)$. For a vertex $v$, we define $M^*(v)$ as the unique vertex $u$ with $uv \in M^*$. A *path* (resp. a *cycle*) is a sequence $(u_1, u_2, \ldots, u_\ell)$ of distinct vertices (resp. distinct vertices except the first and the last) such that, for each two consecutive vertices $u_i$ and $u_{i+1}$, we have $u_i u_{i+1} \in E$. A path (or a cycle) $p$ is called *alternating* with respect to $M^*$ if, for all vertices $u$ of $p$, also $M^*(u)$ is a vertex of $p$. The SCAFFOLDING problem is defined as follows:

> SCAFFOLDING (SCA)
> **Input:** a scaffold graph $(G, M^*, \omega)$ and integers $\sigma_p, \sigma_c, k \in \mathbb{N}$
> **Question:** Is there some $S \subseteq E \setminus M^*$ such that $S \cup M^*$ is a collection of $\leq \sigma_p$ alternating paths and $\leq \sigma_c$ alternating cycles and the weight-score $\sum_{e \in S} \omega(e) \geq k$?

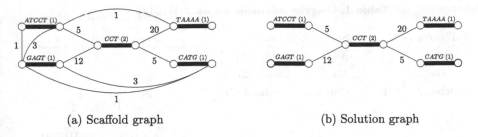

(a) Scaffold graph                              (b) Solution graph

**Fig. 1.** A scaffold graph, its solution graph, and sequences that can be inferred. Contigs are represented by bold edges, labeled by the corresponding sequence and their multiplicity (in parentheses). Inter-contig edges are labeled by their weight. The solution graph is obtained as a solution for the MSCA instance asking for two walks with total weight $\geq 42$. In the solution graph, the contig of multiplicity two labeled CCT constitutes an ambiguous path, yielding two possible sets of sequences {ATCCT..CCT..TAAAA, GAGT..CCT..CATG} and {ATCCT..CCT..CATG, GAGT..CCT..TAAAA}. Brutal cut would provide a set of six independent sequences of total weight zero (the initial set of contigs), whereas SEMI-BRUTAL CUT with weight-score provides a unique set of four sequences {ATCCT, GAGT, CCT..TAAAA, CCT..CATG}, and weight 25 (minimal weight-score 17).

SCAFFOLDING has been studied in the framework of complexity and approximation [4,20,22]. If contigs may appear repeatedly in the genome, we add a multiplicity function $m : E \to \mathbb{N}$ to the scaffold graph. For contig edges, the multiplicity equals the number of times the contig occurs in the genome and this can be estimated from the data [9]. For each non-contig edge $uv$, its multiplicity $m(uv)$ equals the smaller of the multiplicities of the contig edges incident to $u$ and $v$. A *walk* $W$ is a sequence $(u_1, u_2, \ldots, u_\ell)$ of vertices such that, for each two consecutive vertices $u_i$ and $u_{i+1}$, we have $u_i u_{i+1} \in E$. Then, $W$ is called *closed* if $u_1 = u_\ell$ and $W$ is called *alternating* with respect to $M^*$ if $\ell$ is even and, for each odd $i$, we have $u_i u_{i+1} \in M^*$. The difference between path and walk (resp. cycle and closed walk) is that the vertices do not need to be distinct. The SCAFFOLDING WITH MULTIPLICITIES problem is the following:

> SCAFFOLDING WITH MULTIPLICITIES (MSCA)
> **Input:** a scaffold graph $(G, M^*, \omega, m)$ and $\sigma_p, \sigma_c, k \in \mathbb{N}$
> **Question:** Is there a multiset $S$ of $\leq \sigma_c$ closed and $\leq \sigma_p$ non-closed alternating walks in $G$ such that each $e \in M^*$ occurs at most $m(e)$ times in across all walks of $S$ and $\sum_{e \in E(S) \setminus M^*} \omega(e) \geq k$?

In this setting, a scaffold graph $(G^*, M^*, \omega^*, m^*)$ is called *solution graph* for $(G, M^*, \omega, m)$ if (a) $G^*$ is a subgraph of $G$, (b) $\omega^*$ is the restriction of $\omega$ to $G^*$, (c) $m^*(uv) \leq m(uv)$ for all $uv \in E$, (d) $G^*$ can be decomposed into $\leq \sigma_c$ closed and $\leq \sigma_p$ non-closed walks. Such a decomposition into walks is called a *linearization* of the solution graph and, in general, it is not necessarily unique (see Fig. 1).

**Table 1.** Overview of results for SEMI-BRUTAL CUT.

| Topologies | Score | Complexity | Lower bound |
|---|---|---|---|
| General | All | NP-hard [21] | |
| Trees | All | Linear [21] | |
| Planar, $\Delta \leq 4$ | Cut-score | NP-hard [21] | Approx: 1.37 $(P \neq NP)$ [21], $2 - \epsilon$ (UGC) [21], exact: $2^{o(n)}$(ETH) [21] |
| General, $\Delta \leq 2$ | All | Linear (Proposition 1) | |
| Complete bipartite | Cut-score | Linear (Proposition 2) | |
| Bip. plan., $\Delta \leq 3$ | Cut-score | NP-hard (Theorem 1) | APX-Hard (Theorem 2) exact: $2^{o(\sqrt{n+m})}n^{O(1)}$ (ETH) (Corollary 1) |

**Observation 1.** *For each vertex u of a solution graph, the sum of multiplicities of its incident non-matching edges is at most the multiplicity of its incident matching edge.*

In earlier work [21], we showed that a largest uniquely linearizable subgraph can be obtained by destroying all *ambiguous paths*, that is, all alternating paths $p$ such that all edges of the path have the same multiplicity $m_p$ and both extremities of $p$ are incident to a non-contig edge with multiplicity strictly less than $m_p$. Thus, the main problem considered in this work is the following.

SEMI-BRUTAL CUT (SBC)
**Input:** a solution graph $(G^*, M^*, \omega, m)$ and some $k \in \mathbb{N}$
**Question:** Is there a set $X$ of non-contig edges of $G$ such that $G - X$ does not contain ambiguous paths and the score of $X$ is at most $k$?

We choose to separate MSCA and SEMI-BRUTAL CUT, which is justified by the danger of producing chimeric sequencing when combining optimisation of weight on the scaffold graph and the linearisability constraint (see Fig. 2). Moreover, the solution graph is by itself an interesting object to study. It embeds all *possibilities*, and may be a reasonable representation of a genome under our current knowledge. We expect that additional information may disambiguate a solution graph, such as finer study of the nature of involved repeats, dynamic of transposed elements, etc. Thus, SEMI-BRUTAL CUT was raised as a problem aiming to propose a standard output (*e.g.* fasta files) from the solution graph. Several cost-functions $\omega'$ make sense in this setting.

**Definition 1.** *A weight function $\omega' : 2^E \to \mathbb{N}$ is called*

1. cut-score, *if $\omega'$ counts one per cut vertex (that is, $\omega'(X)$ is the size of a smallest vertex cover of $X$),*
2. path-score, *if $\omega'$ counts one per removed edge (that is, $\omega'(X) = |X|$), and*

*3.* weight-score, *if* $\omega'$ *counts the total weight of the removed edges (that is,* $\omega'(X) = \sum_{e \in X} \omega(e))$.

Note that, from the perspective of computational complexity, the path-score is a special case of the weight score, since we can just set $\omega(e) = 1$ for all edges $e$. Thus, when saying "both scores" we refer to cut- and weight-score. Further, when talking about cut-score, we sometimes say "to cut a vertex $v$", by which we mean cutting all non-contig edges incident with $v$. In context of approximation, SEMI-BRUTAL CUT refers to its optimization variant, minimizing the score of $X$.

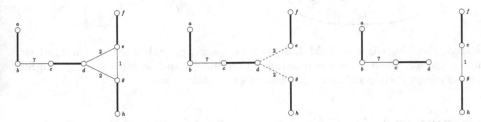

**Fig. 2. (Left)** Edge $\{c, d\}$ has multiplicity two. Other multiplicities are equal to one. The labels on the edges correspond to their weight. In the input scaffold graph, the real sequences are both paths $(a, b, c, d, e, f)$ and $(c, d, g, h)$. **(Middle)** After resolving successively MSCA (with $\sigma_p = 2$ and $\sigma_c = 0$) and SBC (dashed edges are cut), the solution is compatible with the initial hypothesis. The only ambiguous path is the matching edge $\{c, d\}$ and the cut vertex is $d$. **(Right)** Directly searching two maximum weighted alternating paths such that the solution graph does not contain ambiguity yields a chimeric sequence $(f, e, g, h)$.

**Related Works.** In previous work [21], we proposed the first results concerning the complexity of SEMI-BRUTAL CUT according to the scoring functions mentioned in Definition 1. In that article, two main results are proved: the NP-completeness for general graphs and a polynomial-time algorithm for trees based on dynamic programming. Here, we push this hardness result to bipartite, planar, subcubic graphs and give polynomial-time algorithms for more special cases (especially for $\Delta \leq 2$) marking the boundary between the NP-completeness and the polynomiality. Table 1 summarizes the complexity results.

## 2    Computational Hardness

While SEMI-BRUTAL CUT is known to be NP-hard for both cut- and weight-score [21], we extend the cut-score hardness to planar, bipartite, subcubic graphs. To this end, we reduce the classic NP-complete 3-SAT [7] problem to SBC.

   3-SATISFIABILITY (3-SAT)
   **Input:** A boolean formula $\varphi$ in conjunctive normal form where each clause
          contains exactly three literals.
   **Question:** Is there a satisfying assignment $\beta$ for $\varphi$?

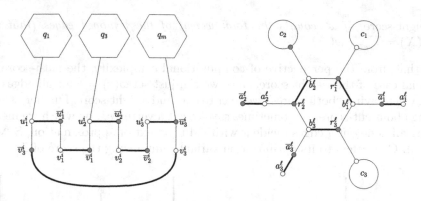

**Fig. 3.** Matching edges are bold. **Left:** variable gadget $c_{x_i}$ linked to the clause gadgets $q_1$, $q_3$ and $q_m$, where $x_i$ occurs positively in $C_1$ and $C_3$ and negatively in $C_m$. **Right:** clause gadget corresponding to the clause $C_\ell = (x_1 \vee \overline{x_2} \vee x_3)$.

**Construction 1.** *Let $\varphi$ be an instance of 3-SAT with $n$ variables $x_1, \ldots, x_n$ and $m$ clauses $C_1, \ldots, C_m$. For each variable $x_i$, let $\psi_i$ be the list of indices $\ell$ such that $C_\ell$ contains $x_i$ and $|\psi_i|$ is the number of occurrences of $x_i$ in $\varphi$. We construct the following solution graph $(G^*, M^*, \omega, m)$ with a 2-coloring of $G^*$ (see Fig. 3).*

- *For each $x_i$, we construct a cycle $c_i$ on the vertex set $\bigcup_{j \leq |\psi_i|} \{u_j^i, \overline{u}_j^i, v_j^i, \overline{v}_j^i\}$ such that, for all $j \leq |\psi_i|$,*
  - *$\{u_j^i, \overline{u}_j^i\}, \{v_j^i, \overline{v}_j^i\} \in M^*$, and*
  - *the vertices $u_j^i$ and $v_j^i$ are blue and the vertices $\overline{u}_j^i$ and $\overline{v}_j^i$ are red.*
- *For each $C_\ell$, we construct an alternating 6-cycle $q_\ell$ on the vertex set $\bigcup_{j \leq 3} \{r_j^\ell, b_j^\ell\}$ such that, for all $j \leq 3$, $\{r_j^\ell, b_j^\ell\} \in M^*$, and $r_j^\ell$ is red and $b_j^\ell$ is blue.*
- *For each clause $C_\ell$ and each $j \leq 3$, let $x_i$ be the $j^{th}$ literal of $C_\ell$ and let $t$ be such that $C_\ell$ is the $t^{th}$ clause in which $x_i$ occurs. Then,*
  - *create a single matching edge $\{a_j^\ell, \overline{a}_j^\ell\}$, where $a_j^\ell$ is blue and $\overline{a}_j^\ell$ is red,*
  - *if $x_i$ is a positive literal, introduce the edges $\{r_j^\ell, u_t^i\}$ and $\{b_j^\ell, \overline{a}_j^\ell\}$, and*
  - *if $x_i$ is a negative literal, introduce the edges $\{b_j^\ell, \overline{u}_t^i\}$ and $\{r_j^\ell, a_j^\ell\}$.*
- *Each non matching edge has multiplicity 1 and all matching edges have multiplicity 2 (thus, each matching edge except the $\{a_i^\ell, \overline{a}_i^\ell\}$ is an ambiguous path).*

Clearly, Construction 1 can be carried out in polynomial time. Further, the resulting graph $G^*$ is bipartite and $\Delta(G^*) = 3$. In the following, we call a matching edge *clean* if one of its endpoints has degree one. Note that a scaffold graph whose every matching edge is clean does not contain ambiguous paths.

**Theorem 1.** SEMI-BRUTAL CUT *is NP-complete for the cut-score, even if the graph is planar, bipartite, subcubic and all multiplicities are one or two.*

In order to prove Theorem 1, we use the following properties of Construction 1, yielding a "canonical" set of cuts.

**Lemma 1.** *Let $S \subseteq V(G^*)$ be a set of vertex-cuts destroying all ambiguous paths in $(G^*, M^*, \omega, m)$, let $c_i$ be a variable gadget and let $q_\ell$ be a clause gadget. There is a set $S'$ of cuts with $|S'| \leq |S|$ that also destroys all ambiguous paths and*

*(a)* $|S \cap V(c_i)| = |S' \cap V(c_i)| \geq |\psi_i|$ *and* $|S \cap V(q_\ell)| = |S' \cap V(q_\ell)| \geq 2$ *(S and $S'$ have the same cut partition in variable gadgets and clause gadgets),*

*(b)* *if* $|S' \cap V(c_i)| = |\psi_i|$, *then* $S' \cap V(c_i)$ *is either* $\bigcup_{j \leq |\psi_i|}\{u_j^i\}$ *or* $\bigcup_{j \leq |\psi_i|}\{\overline{u}_j^i\}$ *(if $S'$ is optimal on a variable gadget, cuts are only on positive sides or only on negative sides),*

*(c)* $|S' \cap V(q_\ell)| = 2$ *if and only if $S'$ contains a vertex adjacent to $q_\ell$ (only two cuts are needed in a clause gadget iff it has been isolated by a cut in an adjacent variable gadget, meaning that the variable satisfies the clause).*

*Proof.* (a): Since, in a cycle with $y$ ambiguous paths, each cut can destroy at most two of them, we need at least $\lceil y/2 \rceil$ cuts to linearize this cycle. A clause gadget and a variable gadget contain a cycle of three ambiguous paths and a cycle of $2|\psi_i|$ ambiguous paths, respectively. Thus, we need at least two cuts in a clause gadget and $|\psi_i|$ cuts in a variable gadget to linearize $G^*$.

(b): If $v_j^i \in S$ for some $j$, then we can swap it for $\overline{u}_j^i$ in $S$ (and analogously for $\overline{v}_j^i$). This operation does not increases the cardinality of $S$. Thus, we suppose that $S$ contains neither $v_j^i$ nor $\overline{v}_j^i$ for any $j$, implying that $S$ contains either $u_j^i$ or $\overline{u}_j^i$ for each $j$. Since $|S \cap V(c_i)| = |\psi_i|$ we know that exactly one of $u_j^i$ and $\overline{u}_j^i$ is in $S$ for each $j$. Now, if $S$ contains some $u_j^i$ and some $\overline{u}_{j'}^i$, then it also contains such vertices for consecutive $j$ and $j'$. Hence, we can suppose that $j' = (j + 1)$ mod $|\psi_i|$, implying that $\{v_j^i, \overline{v}_j^i\}$ is an ambiguous path.

(c): Note that both $r_1^\ell$ and $b_1^\ell$ are incident with a non-matching edge leaving $q_\ell$. To destroy the ambiguous path $\{r_1^\ell, b_1^\ell\}$, one of these edges has to be removed. By symmetry the same holds for $\{r_2^\ell, b_2^\ell\}$ and $\{r_3^\ell, b_3^\ell\}$. Since at least three edges leaving $q_\ell$ have to be removed, we need a total of three cuts inside of $q_\ell$ unless a vertex adjacent to $q_\ell$ is cut. Conversely, suppose by symmetry that the edge incident to the vertex $r_2^\ell$ is cut. Then, $q_\ell$ can be linearized by cutting $b_1^\ell$ and $b_3^\ell$ (see Fig. 4). □

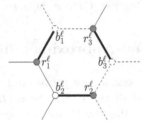

**Fig. 4.** A cut of size 2 in $q_\ell$ when 1 incident edge to $q_\ell$ is cut. Dashed edges and vertices are part of the cut.

*Proof of Theorem 1.*

Recall that 3-SAT remains NP-complete if the input formula is planar [12] and, in this case, since each gadget is planar and the edges between the clause gadget and the variable gadget can be placed in any order on the gadgets, the graph produced by Construction 1 can also be assumed to be planar. Since, clearly, SEMI-BRUTAL CUT $\in$ NP, it remains to show that Construction 1 is correct, that is $\varphi$ is satisfiable if and only if the scaffold graph $(G^*, M^*, \omega, m)$ resulting from Construction 1 can be linearized with $5m$ cuts.

**Fig. 5.** Matching edges are bold. Example of variable gadget $r_{x_i}$ linked to the clause gadgets $q_p$, $q_{p'}$, $q_n$ and $q_{n'}$, where $x_i$ occurs positively in $C_p$ and $C_{p'}$ and negatively in $C_n$ and $C_{n'}$.

"$\Rightarrow$": Let $\beta$ be a satisfying assignment for $\varphi$. Then, for each variable $x_i$ and for all $j \leq |\psi_i|$, we cut the vertices $u_j^i$ if $\beta(x_i) = 1$ and the vertices $\overline{u}_j^i$ otherwise. As $\beta$ is satisfying, this removes at least one edge adjacent to each clause gadget. Thus, according to Lemma 1(c), we can cut two vertices in each clause gadget $q_j$ to turn every matching edge in $q_j$ clean. Since we also cut either the vertices $u_j^i$ or the vertices $\overline{u}_j^i$ for each vertex gadget, we conclude that all matching edges of the result are clean and we cut exactly $2m + \sum_i |\psi_i| = 5m$ vertices.

"$\Leftarrow$": Let $S \subseteq V$ be the set of vertices such that cutting each vertex of $S$ destroys all ambiguous paths in $(G^*, M^*, \omega, m)$ and $|S| = 5m$. According to Lemma 1(a), each variable gadget contains $|\psi_i|$ cuts and each clause gadget contains two cuts. Moreover, by Lemma 1(b), for each variable gadget $c_i$, we can suppose that $S \cap V(c_i)$ equals $\bigcup_{j \leq |\psi_i|} \{u_j^i\}$ or $\bigcup_{j \leq |\psi_i|} \{\overline{u}_j^i\}$. In the former case, we set $\beta(x_i) = 1$ and, in the latter, we set $\beta(x_i) = 0$. To show that $\beta$ satisfies $\varphi$, assume that there is a clause $C_\ell$ that is not satisfied by $\beta$. Then, none of the edges incident to $q_\ell$ is cut which, by Lemma 1(c), contradicts the fact that there are two cuts in $q_\ell$. $\square$

Since Construction 1 is linear in the number of vertices and planar 3-SAT does not admit a $2^{o(\sqrt{n+m})}n^{O(1)}$-time algorithm [13], there is also no $2^{o(\sqrt{n+m})}n^{O(1)}$-time algorithm for SEMI-BRUTAL CUT.

**Corollary 1.** *Assuming ETH, there is no $2^{o(\sqrt{n+m})}n^{O(1)}$-time algorithm for* SEMI-BRUTAL CUT *in bipartite planar subcubic graphs for the cut-score.*

## 3   Non-Approximability

In this section, we prove approximation lower bounds for SEMI-BRUTAL CUT. First recall the definition of *L-reduction* between two hard problems $\Pi$ and $\Pi'$, described by Papadimitriou and Yannakakis [16]. This reduction consists of polynomial-time computable functions $f$ and $g$ such that, for each instance $x$ of $\Pi$, $f(x)$ is an instance of $\Pi'$ and for each feasible solution $y'$ for $f(x)$, $g(y')$ is a feasible solution for $x$. Moreover, let $\Pi'' \in \{\Pi, \Pi'\}$, we denote by $OPT_{\Pi''}$ the value of an optimal solution of $\Pi''$ and by $val_{\Pi''}(y'')$ the value of a solution $y''$ of an instance of $\Pi''$. There are constants $\alpha, \beta > 0$ such that:

1. $OPT_{\Pi'}(f(x)) \leq \alpha OPT_\Pi(x)$ and
2. $|val_\Pi(g(y')) - OPT_\Pi(x)| \leq \beta |val_{\Pi'}(y') - OPT_{\Pi'}(f(x))|$.

In the following, we present an *L-reduction* from the classical problem MAX 3-SAT(4) to SEMI-BRUTAL CUT.

Max 3-SAT(4)

**Input:** A boolean formula $\varphi$ in exact 3-CNF where every variable occurs in 4 clauses

**Task:** Find an assignment that satisfies a maximum number of clauses.

**Construction 2.** *We reuse Construction 1 and change some variable gadgets. Let $x_i$ be a variable which occurs positively in the clauses $C_p$ and $C_{p'}$ and negatively in the clauses $C_n$ and $C_{n'}$. We replace the variable gadget associated to $x_i$ by the following gadget $r_i$:*

- *Construct a cycle $c_i$ on the vertex set $\bigcup_{j \leq 2}\{u_j^i, \overline{u}_j^i, v_j^i, \overline{v}_j^i\}$ such that, for all $j \leq 2$, $\{u_j^i, \overline{u}_j^i\}, \{v_j^i, \overline{v}_j^i\} \in M^*$, the vertices $u_j^i$ and $v_j^i$ are blue and $\overline{u}_j^i$ and $\overline{v}_j^i$ are red.*
- *Give multiplicity 1 to all non-matching edges and multiplicity 2 to all matching edges.*
- *Link the clause gadgets $q_p, q_{p'}, q_n$ and $q_{n'}$ to vertices $u_1^i, u_2^i, \overline{u}_1^i$ and $\overline{u}_2^i$ respectively in the same way as in Construction 1.*

*Note that all matching edges are ambiguous paths in the variable gadget. The clause gadgets and the other variable gadgets remain unchanged.*

The resulting graph $G^*$ is bipartite and $\Delta(G^*) = 3$. In the following, when we want to differentiate the variable gadgets, we designate by *rectangle* variable gadget those defined in Construction 2 and by *cycle* variable gadget those defined in Construction 1. An example of a rectangle variable gadget is given in Fig. 5. Notice that the properties (a) and (c) of Lemma 1 hold. We can add the following property:

**Lemma 2.** *Let $S \subseteq V(G^*)$ be an optimal set of vertex-cuts destroying all ambiguous paths in $(G^*, M^*, \omega, m)$, let $c_i$ be a cycle variable gadget and $r_{i'}$ be a rectangle variable gadget. There is a set $S'$ of cuts with $|S'| = |S|$ that also destroys all ambiguous paths, and*

*(a) $S' \cap V(c_i)$ is either $\bigcup_{j \leq |\psi_i|}\{u_j^i\}$ or $\bigcup_{j \leq |\psi_i|}\{\overline{u}_j^i\}$, and*

*(b) $S' \cap V(r_{i'})$ is either $\{u_1^{i'}, u_2^{i'}\}$ or $\{\overline{u}_1^{i'}, \overline{u}_2^{i'}\}$.*

*Proof.* Recall that $S$ covers the edges of $M^*$ and, by Lemma 1(a), $|S \cap V(c_i)| \geq |\psi_i|$.

"(a)": By symmetry, suppose that $x_i$ occurs mostly positively in $\varphi$. If $x_i$ occurs four times positively, then replacing $S \cap V(c_i)$ by $\bigcup_{j \leq |\psi_i|}\{u_j^i\}$ in $S$ yields a solution $S'$ as sought. Thus, suppose that $x_i$ occurs three times positively. Let $C_\ell$ be the clause where $x_i$ occurs negatively and let $z$ denote the neighbor of $\overline{u}_j^i$ in $c_\ell$. If $|S \cap V(c_i)| > |\psi_i|$, then replacing $S \cap c_i$ by $\bigcup_{j \leq |\psi_i|}\{u_j^i\}$ plus $z$ yields a solution $S'$ as sought. Finally, if $|S \cap V(c_i)| = |\psi_i|$, then $S$ already corresponds to (a) as, otherwise, some ambiguous path $\{v_j^i, \overline{v}_j^i\}$ is not destroyed.

"(b)": Note that one cut in $r_{i'}$ is not enough to destroy all ambiguous paths and cutting either the vertices $\{u_1^{i'}, u_2^{i'}\}$ or the vertices $\{\overline{u}_1^{i'}, \overline{u}_2^{i'}\}$ destroys all ambiguous paths in the rectangle variable gadget. Further if $S$ cuts $\{v_1^{i'}, v_2^{i'}\}$ or $\{\overline{v}_1^{i'}, \overline{v}_2^{i'}\}$, then we can instead cut $\{u_1^{i'}, u_2^{i'}\}$ or $\{\overline{u}_1^{i'}, \overline{u}_2^{i'}\}$, respectively, without creating ambiguous paths. Suppose without loss of generality that $\{u_1^{i'}, u_2^{i'}\} \subseteq S$. Suppose further that there is some $\overline{u} \in S \cap V(r_{i'}) \setminus \{u_1^{i'}, u_2^{i'}\}$. Then, there is some clause gadget $q_n$ linked to $\overline{u}$ since, otherwise, $S - \overline{u}$ is also a solution, contradicting optimality of $S$. Since all matching edges of $r_{i'}$ are already clean, the cut can only remove the edge between $\overline{u}$ and $q_n$. Thus we can replace $\overline{u}$ by its neighbor in $q_n$ without changing the cardinality of $S$. By swapping the one or two cuts in $S \cap V(r_{i'}) \setminus \{u_1^{i'}, u_2^{i'}\}$, we obtain $S' \cap V(r_j) = \{u_1^{i'}, u_2^{i'}\}$. $\qquad\square$

**Theorem 2.** *There is a constant $\epsilon_4' > 0$ (the value $\epsilon_4' > 0$ is defined in [1]) for which* SEMI-BRUTAL CUT *cannot be approximated to any factor better than $(1 + 7\epsilon_4'/41)$, even on bipartite graphs of maximum degree three, unless P=NP.*

*Proof.* Recall that, unless P=NP, MAX 3-SAT(4) cannot be approximated to a factor better than $\epsilon_4' = 1,00052$ [1] and that, in an optimal solution of MAX 3-SAT(4), at least $7/8$ of the clauses are satisfied [8], yielding

$$OPT(\varphi) \geq 7m/8. \tag{1}$$

To show that Construction 2 constitutes an $L$-reduction, let $f$ be a function transforming any instance $\varphi$ of MAX 3-SAT(4) into an instance $I$ of SEMI-BRUTAL CUT as above, let $S$ be a feasible solution for $I$ corresponding to the properties of Lemmas 1(a), 1(c) and 2, and let $g$ be the function that transforms $S$ into an assignment $\beta$ as constructed in the proof of Theorem 1: each variable $x_i$ is set to true if $S$ cuts $u_j^i$ for all $j$, and false, otherwise. By Lemma 2, for each clause gadget $q_\ell$ without an adjacent vertex in $S$, the "extra" cut occurs in $q_\ell$. Hence, for each of the at most $m/8$ unsatisfied clauses in $\varphi$, we have to spend another cut to linearize $I$. Thus,

$$OPT(I) \leq 5m + m/8 \overset{(1)}{\leq} 41/7 OPT(\varphi) \tag{2}$$

An important obstacle to overcome (and reason why Construction 1 is not enough for Theorem 2) is that an approximate solution to SBC might spend extra cuts in variable gadgets in order to "change the assignment" of a variable $x_i$ mid-way. However, since each variable occurs at most four times, this only happens for variables that occur two times positively and two times negatively. Now, with our modification to Construction 1, we can observe that each extra cut in any of the variable gadgets allows such a misuse only for a single clause gadget. Thus, the number of satisfied clauses of $\varphi$ and the clause gadgets in which we have to spend extra cuts adds up to $m$. Hence,

$$6m = val(g(S)) + val(S) = OPT(I) + OPT(\varphi) \tag{3}$$

Thus, we constructed an $L$-reduction with $\alpha = {}^{41}/_7$, $\beta = 1$ and, since $val(g(S)) < (1 - \epsilon'_4) \cdot OPT(\varphi)$, we conclude

$$val(S) \overset{(3)}{=} OPT(I) + OPT(\varphi) - val(g(S))$$
$$> OPT(I) + \epsilon'_4 OPT(\varphi)$$
$$\overset{(2)}{\geq} (1 + {}^{7\epsilon'_4}/_{41}) \cdot OPT(I)$$

□

Relaxing on planarity et or maximum degree, we obtain better lower bounds:

**Theorem 3.** *There is a constant $\epsilon'_4 > 0$ for which* SEMI-BRUTAL CUT *cannot be approximated to any factor better than $1 + {}^{\epsilon'_4}/_{10}$, even on bipartite, subcubic graphs with multiplicities in the set $\{1, 2\}$ unless P=NP.*

The value $\epsilon'_4 > 0$ is defined in [1].

*Proof.* We use a gap preserving reduction from MAX 3-SAT(4) (assume w.l.o.g. that each clause of $\phi$ has exactly three literals (this can be easily done by repeating the literals within a clause, if necessary)) to SEMI-BRUTAL CUT even on bipartite, subcubic graphs with multiplicities in the set $\{1, 2\}$ that transforms a Boolean formula $\phi$ to a graph using Construction 1 such that:

1. if $OPT(\varphi) = m$ then $OPT(G) = 5m$ (see Theorem 1), and
2. if $OPT(\varphi) < (1 - \epsilon'_4)m$ then $OPT(G) \geq 5m + \epsilon'_4 m/2$.

First item is obtained by Theorem 1. Second item comes from the observation that if the optimal solution of MAX 3-SAT(4) does not satisfy $m$ clauses, but $k \leq m - 1$, it means that an extra cut was necessary in the transformed instance to linearize the graph $G$. If this extra cut is placed in a variable gadget, it can linearize between one and two clauses (the extra cut can not linearize more than two clauses since otherwise changing the value of the variable increases the number of satisfied clauses which is a contradiction). Thus, for $k$ between $m - 1$ and $m - 2$, we know that at least one extra cut is needed. We generalize this argument to any number of satisfied clauses in the optimal solution, to finally get:

$$OPT(G) \geq 5m + \left\lceil \frac{m - OPT(\varphi)}{2} \right\rceil.$$

By hypothesis we have $OPT(\varphi) < (1 - \epsilon'_4)m$, thus $OPT(G) \geq 5m + \lceil \epsilon'_4 m/2 \rceil \geq 5m + \epsilon'_4 m/2$. The Theorem follows. □

Hereafter, we consider MAX 3-SAT(B) which is the restricted special case of MAX 3-SAT where every variable occurs in at most $B$ clauses. Recall that for MAX 3-SAT(B) the best possible approximation is at least $^7/_8 + \Omega(^1/_B)$ (as a function of $B$) and at most $^7/_8 + O(^1/\sqrt{B})$ unless NP=RP[19]. Based on the same arguments as given in Theorem 2, we have $val(S) \geq (^{335}/_{328} - O(^1/\sqrt{B}))OPT(\varphi)$, this leads us to following result:

**Theorem 4.** SEMI-BRUTAL CUT *cannot be approximated to any factor better than ${}^{335}/_{328} - O(^1/\sqrt{B})$, even on bipartite graphs of maximum degree three, unless RP=NP.*

## 4     Polynomial Cases

In this section, we consider the SEMI-BRUTAL CUT problem in graphs with maximum degree two and in complete bipartite graphs. We show a linear-time algorithm for both cut-score and weight-score. Recall that SEMI-BRUTAL CUT can be solved in linear time in trees [21]. In the following, we suppose that the input solution graph contains at least one ambiguous path and that $G$ is connected as otherwise, we can treat each connected component individually.

**Proposition 1.** SEMI-BRUTAL CUT *can be solved in linear time on a collection of paths and cycles ($\Delta(G) = 2$) for the weight function given by Definition 1 .*

*Proof.* First, since $\Delta(G) = 2$, the cut-score is a special case of the weight-score with $\omega(e) = 1$ for all non-matching edges $e$. Thus, suppose that the weight-score is used. Second, since SBC can be solved in linear time on paths [21], suppose that $G$ is a cycle. Let $p$ be any ambiguous path in $G$, let $e_1$ and $e_2$ be the two unique non-matching edges incident to the extremities of $p$. Since $G_1 = G - e_1$ and $G_2 = G - e_2$ are trees, we can find and compare optimal solutions $X_1$ and $X_2$ for $G_1$ and $G_2$, respectively, in linear time. Further, as $p$ is ambiguous, all optimal solutions delete $e_1$ or $e_2$ (or both) and, thus, one of $X_1 \cup \{e_1\}$ and $X_2 \cup \{e_2\}$ is optimal for $G$.                                                    □

**Proposition 2.** SEMI-BRUTAL CUT *can be solved in linear time for cut-score on complete bipartite graphs.*

*Proof.* Let $K_{n,n}$ be a complete bipartite graph and note that $n > 2$ as, otherwise, there is no ambiguous path in $K_{n,n}$. Also note that, by Observation 1, all matching edges are ambiguous paths. Then, it is sufficient to cut all but one vertex of any of the two cells of the bipartition to turn all matching edges clean. To show that $n - 1$ cuts are also necessary, assume that there is a solution $X$ with cut-score $n - 2$ and let $u$ and $v$ be the vertices that are not cut. Then, $u$ and $v$ are in the same cell of the bipartition since, otherwise, there is a matching edge $xy$ with $ux, vy \notin X$ and, thus, $xy$ is an ambiguous path in $G - X$. But then, $\{u, M^*(v)\}$ and $\{M^*(u), v\}$ are not in $X$, implying that $\{u, M^*(u)\}$ is an ambiguous path in $G - X$.                                                    □

## 5     Approximable Cases

We propose a greedy strategy (see Algorithm 1) for the SEMI-BRUTAL CUT problem under the weight-score function. Let $(G^*, M^*, \omega, m)$ be a solution graph and let $X \subseteq E \setminus M^*$ be a set of non-matching edges. For a vertex $x$, we let $\omega_X(x)$ denote the sum of the weights of all non-matching edges incident with $x$ that are not in $X$. More formally, we define $\omega_X(x) := \sum_{e \in E \setminus (M^* \cup X)} \omega(e) \cdot \chi_e(x)$, where $\chi_e(x) := |e \cap \{x\}|$ is the characteristic function of $e$. The principle of our algorithm is to successively visit each ambiguous path and cut the edges incident to the extremity with the lowest value of $w_S$, where $S$ contains all previously cut edges.

---

**Algorithm 1:** Greedy Algorithm

---

**Data**: A solution graph $(G^*, M^*, \omega, m)$.

**Result**: A set $X \subseteq E \setminus M^*$ whose removal makes $G^*$ uniquely linearizable.

1 $X \leftarrow \emptyset$ ;

2 $A \leftarrow$ list of extremities of ambiguous paths;

3 **while** $A \neq \emptyset$ **do**

4     $u \leftarrow \text{argmin}_{x \in A} \omega_X(x)$;

5     remove the two extremities of the ambiguous path containing $u$ from $A$;

6     add all non-matching edges incident with $u$ to $X$;

7 **end**

8 **return** $X$;

---

**Fig. 6.** Tightness of the approximation ratio. Edges are bold ($\in M^*$), solid ($\in X_{opt}$) or dashed ($\in X$) and all edges have weight one. The multiplicities of the matching edges are equal to two and the multiplicities of the non-matching edges are equal to one. Thus, $\omega(X) = 2$ and $\omega(X_{opt}) = 1$.

**Proposition 3.** *In $\mathcal{O}((|V| + |E|) \log |V|)$ time, Algorithm 1 computes a solution for* SEMI-BRUTAL CUT *under the weight-score with an approximation ratio of 2 and this ratio is tight.*

*Proof.* Since each time some extremities are removed from $A$, the ambiguous path they belonged to has been destroyed, there are no more ambiguous paths remaining when $A = \emptyset$. Thus, the set $X$ that is returned is indeed a solution. Let $X_{opt}$ be an optimal solution. Let $p_i$ denote the ambiguous path of $G^*$ considered in step $i$ of Algorithm 1, let $u$ and $v$ be its extremities, and let $X_i$ be the set of edges added to $X$ in step $i$. If $X_{opt}$ contains all non-matching edges incident to $u$, then let $Q_i$ contain them. Otherwise, $X_{opt}$ contains all non-matching edges incident to $v$, and we let $Q_i$ contain those. Then, $\omega'(X_i) \leq \omega'(Q_i)$ for all $i$ and, thus, $\omega'(X) \leq \sum_i \omega'(Q_i)$. Further, $\bigcup_i Q_i = X_{opt}$ and, since each edge of $G^*$ occurs in at most two sets $Q_i$, we conclude $\sum_i \omega'(Q_i) \leq 2\omega'(X_{opt})$. The claimed approximation factor of two follows and, by Fig. 6, it is tight.

Concerning the running time, the list of ambiguous paths is build in $\mathcal{O}(|E| + |V|)$ with a depth-first search algorithm. The sorting of this list can be done in $\mathcal{O}(|V| \log |V|)$. The maintain of the sorting of the list at each cut yields a $\mathcal{O}((|V| + |E|) \log |V|)$. $\square$

## 6 Experiments

In order to observe the behavior of our algorithm on real instances, we tested it on datasets described below and we compare the obtained solutions of three different algorithms.

**Table 2.** Sequences selected for experiments.

| Species | Tax | Alias | Size (bp) | Type | Acc. number |
|---|---|---|---|---|---|
| *Bacillus anthracis str. Sterne* | Bacteria | Anthrax | 5228663 | Chrom.[a] | NC_005945.1 |
| *Gloeobacter violaceus* PCC 7421 | Bacteria | Gloeobacter | 4659019 | Chrom.[a] | NC_005125.1 |
| *Lactobacillus acidophilus* NCFM | Bacteria | Lactobacillus | 1993560 | Chrom.[a] | NC_006814.3 |
| *Pandoravirus salinus* | Virus | Pandora | 2473870 | Comp.[b] | NC_022098.a |
| *Pseudomonas aeruginosa* PAO1 | Bacteria | Pseudomonas | 6264404 | Chrom.[a] | NC_002516.2 |
| *Oryza sativa Japonica* | Plant | Rice | 134525 | Chlor.[c] | X15901.1 |
| *Saccharomyces cerevisiae* | Yeast | Sacchr3 | 316613 | Chrom.[a] 3 | X59720.2 |
| *Saccharomyces cerevisiae* | Yeast | Sacchr12 | 1078177 | Chrom.[a] 12 | NC_001144.5 |

[a] chromosome
[b] complete genome
[c] chloroplast

*Description of the datasets.* They were generated in the following way:

1. A set of reference genomes in the Nucleotide NCBI database (see Table 2) have been chosen for their diversity in genome sizes, and types of organisms.
2. Paired-end reads, have been simulated using `wgsim` [11]. then assembled using the De Bruijn Graph based *de novo* assembly tool `minia` [5].
3. Reads have been mapped to the contigs, using `bwa` [10] and contigs on the reference genome, using `megablast` [15], in order to find their multiplicities and generate scaffold graphs. Table 3 presents some statistics about produced scaffolding graphs. Notice that those graphs may be large, however their sparsity and mean degree explain why we consider very constrained classes of graphs (degree bounded by three, planar, bipartite, etc.). They do not fit these constraints, but they are quite close to.
4. We generated the solution graphs from the scaffold graphs using our ILP formulation [23] for SCAFFOLDING WITH MULTIPLICITIES and using the *cplex* solver. Statistics on solution graphs are available on Table 4.

*Results.* We ran Algorithm 1 on the datasets and we compared it with two other algorithms:

1. an exact algorithm obtained by an ILP formulation of SEMI-BRUTAL CUT
2. a naive algorithm cutting arbitrary extremities of ambiguous path as long as such paths exist.

**Table 3.** Scaffold graphs.

| Data | $|V|$ | $|E|$ | Min/Max/Avg degree |
|---|---|---|---|
| Anthrax | 8110 | 11013 | 1 / 7 / 2.72 |
| Gloeobacter | 9034 | 12402 | 1 / 12 / 2.75 |
| Lactobacillus | 3796 | 5233 | 1 / 12 / 2.76 |
| Pandora | 4902 | 6722 | 1 / 7 / 2.74 |
| Pseudomonas | 10496 | 14334 | 1 / 9 / 2.73 |
| Rice | 168 | 223 | 1 / 6 / 2.65 |
| Sacchr3 | 592 | 823 | 1 / 7 / 2.78 |
| Sacchr12 | 1778 | 2411 | 1 / 10 / 2.12 |

**Table 4.** Sequences selected for experiments.

| Data | #AP[a] | #NAP[b] | Total weight | Avg. deg.[c] | Max/min deg |
|---|---|---|---|---|---|
| Anthrax | 13 | 260 | 329 | 5.31 | 4 / 2 |
| Gloeobacter | 44 | 432 | 694 | 5.68 | 6 / 2 |
| Lactobacillus | 15 | 135 | 225 | 5.27 | 5 / 2 |
| Pandora | 5 | 183 | 210 | 5.00 | 4 / 2 |
| Pseudomonas | 47 | 413 | 650 | 5.20 | 5 / 2 |
| Rice | 6 | 9 | 29 | 4.17 | 3 / 2 |
| Sacchr3 | 5 | 25 | 54 | 5.40 | 4 / 2 |
| Sacchr12 | 23 | 74 | 190 | 4.87 | 4 / 2 |

[a]ambiguous paths
[b]non-ambiguous paths
[c]average degree of extremities of amb. paths

The idea of implementing the naive algorithm is that it provides an upper bound of SEMI-BRUTAL CUT. We wanted answer to the following question: is the ratio of Algorithm 1 closer than 1 to those provided by the naive algorithm on real-world instances? Statistics on produced solutions are presented in Table 5. We notice that Algorithm 1 finds an optimal solution in most of the cases, even if the number of cuts is of the order of several dozens (i.e. gloeobacter, pseudomonas). The algorithm does not find an optimal solution for two instances: rice and sacchr12 and the ratio of the computed solutions are 1.33 and 1.11, respectively. The high ratio of the rice can be explained by the low score in the optimal solution. Thus, the answer to our question seems to be that the ratio of Algorithm 1 is close to 1. However, the tested instances are relatively small and it is interesting to run tests on bigger instances.

**Table 5.** Results statistics.

| Data | Exact | | Naive algorithm | | | Algorithm 1 | | |
|---|---|---|---|---|---|---|---|---|
| | Score | #cuts | Score | Ratio | #cuts | Score | Ratio | #cuts |
| Anthrax | 17 | 13 | 20 | 1.17 | 13 | 17 | 1.00 | 12 |
| Gloeobacter | 68 | 45 | 80 | 1.17 | 43 | 68 | 1.00 | 41 |
| Lactobacillus | 19 | 15 | 21 | 1.10 | 14 | 19 | 1.00 | 14 |
| Pandora | 6 | 5 | 7 | 1.16 | 5 | 6 | 1.00 | 5 |
| Pseudomonas | 51 | 50 | 65 | 1.27 | 41 | 51 | 1.00 | 40 |
| Rice | 3 | 6 | 5 | 1.66 | 4 | 4 | 1.33 | 4 |
| Sacchr3 | 6 | 5 | 6 | 1.00 | 4 | 6 | 1.00 | 5 |
| Sacchr12 | 18 | 23 | 24 | 1.33 | 16 | 20 | 1.11 | 17 |

## 7 Conclusion

In this article, we develop results concerning the complexity, lower bounds and approximability of the linearization problem for genome scaffolds sharing repeated contigs with two possible scoring functions. We managed to strengthen previously known NP-hardness to the very restricted class of planar bipartite subcubic graphs with only two multiplicities for the cut-score. Natural perspectives of this work are to extend this result to the weight-score, explore the possibility of FPT algorithms and approximations in the difficult cases, and examine the practical performance of the presented greedy algorithm on larger real-world instances.

**Acknowledgments.** This work was supported by the Institut de Biologie Computationnelle (http://www.ibc-montpellier.fr/) (ANR Projet Investissements d'Avenir en bioinformatique IBC).

## References

1. Berman, P., Karpinski, M., Scott, A.D.: Approximation hardness and satisfiability of bounded occurrence instances of SAT. Electronic Colloquium on Computational Complexity (ECCC), 10(022) (2003)
2. Biscotti, M.A., Olmo, E., Heslop-Harrison, J.S.: Repetitive DNA in eukaryotic genomes. Chromosome Res. **23**(3), 415–420 (2015)
3. Cameron, D.L., et al.: GRIDSS: sensitive and specific genomic rearrangement detection using positional de Bruijn graph assembly. Genome Res. **27**(12), 2050–2060 (2017)
4. Chateau, A., Giroudeau, R.: A complexity and approximation framework for the maximization scaffolding problem. Theor. Comput. Sci. **595**, 92–106 (2015). https://doi.org/10.1016/j.tcs.2015.06.023

5. Chikhi, R., Rizk, G.: Space-efficient and exact de Bruijn graph representation based on a bloom filter. In: Raphael, B., Tang, J. (eds.) WABI 2012. LNCS, vol. 7534, pp. 236–248. Springer, Heidelberg (2012). https://doi.org/10.1007/978-3-642-33122-0_19

6. Ekblom, R., Wolf, J.B.: A field guide to whole-genome sequencing, assembly and annotation. Evol. Appl. **7**(9), 1026–1042 (2014)

7. Garey, M.R., Johnson, D.S.: Computers and Intractability: A Guide to the Theory of NP-Completeness. W. H. Freeman & Co., New York (1979)

8. Håstad, J.: Some optimal inapproximability results. J. ACM **48**(4), 798–859 (2001)

9. Koch, P., Platzer, M., Downie, B.R.: RepARK-de novo creation of repeat libraries from whole-genome NGS reads. Nucleic Acids Res. **42**(9), e80 (2014)

10. Li, H., Durbin, R.: Fast and accurate long-read alignment with Burrows-Wheeler transform. Bioinformatics **26**(5), 589–595 (2010)

11. Li, H., et al.: The sequence alignment/map format and samtools. Bioinformatics **25**(16), 2078–2079 (2009)

12. Lichtenstein, D.: Planar formulae and their uses. SIAM J. Comput. **11**(2), 329–343 (1982)

13. Lokshtanov, D., Marx, D., Saurabh, S.: Lower bounds based on the exponential time hypothesis. Bull. EATCS **105**, 41–72 (2011)

14. Mandric, I., Lindsay, J., Măndoiu, I.I., Zelikovsky, A.: Scaffolding algorithms. In: Măndoiu, I., Zelikovsky, A. (eds.) Computational Methods for Next Generation Sequencing Data Analysis, pp. 107–132. Wiley (2016). Chapter 5

15. Morgulis, A., Coulouris, G., Raytselis, Y., Madden, T.L., Agarwala, R., Schäffer, A.A.: Database indexing for production megablast searches. Bioinformatics **24**(16), 1757–1764 (2008). https://doi.org/10.1093/bioinformatics/btn322

16. Papadimitriou, C.H., Yannakakis, M.: Optimization, approximation, and complexity classes. J. Comput. Syst. Sci. **43**(3), 425–440 (1991)

17. Quail, M.A.: A tale of three next generation sequencing platforms: comparison of ion torrent, pacific biosciences and illumina miseq sequencers. BMC Genomics **13**(1), 341 (2012)

18. Tang, H.: Genome assembly, rearrangement, and repeats. Chem. Rev. **107**(8), 3391–3406 (2007)

19. Trevisan, L.: Non-approximability results for optimization problems on bounded degree instances. In: Proceedings on 33rd Annual ACM Symposium on Theory of Computing, 6–8 July 2001, Heraklion, Crete, Greece, pp. 453–461 (2001)

20. Weller, M., Chateau, A., Giroudeau, R.: Exact approaches for scaffolding. BMC Bioinf. **16**(Suppl 14), S2 (2015)

21. Weller, M., Chateau, A., Giroudeau, R.: On the linearization of scaffolds sharing repeated contigs. In: Gao, X., Du, H., Han, M. (eds.) COCOA 2017. LNCS, vol. 10628, pp. 509–517. Springer, Cham (2017). https://doi.org/10.1007/978-3-319-71147-8_38

22. Weller, M., Chateau, A., Dallard, C., Giroudeau, R.: Scaffolding problems revisited: complexity, approximation and fixed parameter tractable algorithms, and some special cases. Algorithmica **80**(6), 1771–1803 (2018)

23. Weller, M., Chateau, A., Giroudeau, R., Poss, M.: Scaffolding with repeated contigs using flow formulations (2018)

# Detecting Large Indels Using Optical Map Data

Xian Fan[1,2](✉), Jie Xu[3], and Luay Nakhleh[1](✉)

[1] Rice University, Houston, TX 77005, USA
{xian.fan,nakhleh}@rice.edu
[2] MD Anderson Cancer Center, Houston, TX 77030, USA
[3] The Pennsylvania State University, Hershey, PA 17033, USA

**Abstract.** Optical Maps (OM) provide reads that are very long, and thus can be used to detect large indels not detectable by the shorter reads provided by sequence-based technologies such as Illumina and PacBio. Two existing tools for detecting large indels from OM data are Bio-Nano Solve and OMSV. However, these two tools may miss indels with weak signals. We propose a local-assembly based approach, OMIndel, to detect large indels with OM data. The results of applying OMIndel to empirical data demonstrate that it is able to detect indels with weak signal. Furthermore, compared with the other two OM-based methods, OMIndel has a lower false discovery rate. We also investigated the indels that can only be detected by OM but not Illumina, PacBio or 10X, and we found that they mostly fall into two categories: complex events or indels on repetitive regions. This implies that adding the OM data to sequence-based technologies can provide significant progress towards a more complete characterization of structural variants (SVs). The algorithm has been implemented in Perl and is publicly available on https://bitbucket.org/xianfan/optmethod.

## 1 Introduction

Structural variant (SV) detection is essential in understanding human genetic diseases such as cancer [11,23,37]. Detecting SVs is very challenging due to several factors, including the simple sequence context of the SV breakpoints [1], the multiple SVs aggregated to form a complex SV [2,31,39], and the repetitive nature of the human genome [13,34]. Advances in sequencing technology make it possible to detect SVs through computational tools [25]. Several SV detection

This work was supported in part by the National Cancer Institute (NCI) grant R01-CA172652 and National Human Genome Research Institute (NHGRI) grant U41-HG007497-01 to Ken Chen at MD Anderson Cancer Center, and the National Cancer Institute Cancer Center Support Grant P30-CA016672 to the MD Anderson cancer center.

© Springer Nature Switzerland AG 2018
M. Blanchette and A. Ouangraoua (Eds.): RECOMB-CG 2018, LNBI 11183, pp. 108–127, 2018.
https://doi.org/10.1007/978-3-030-00834-5_6

methods using Illumina paired-end reads have been devised [6,8,16,18,32,41]. However, due to their small length (typically, 300 bp read length), the focus was mainly on small indel detection and medium-sized simple SVs such as deletion and translocation [1,40]. Large SVs whose breakpoints fall at repetitive regions were not fully resolved by Illumina reads. PacBio single molecule reads [7,30], on the other hand, tackle those SVs in larger repetitive regions, and detectable SV types naturally generalize to insertion and inversion, due primarily to PacBio's larger read length (typically 12 kbp for RS II) [4]. Nevertheless, the read length is still not enough for spanning large repeats, leading to missing SVs.

Optical Maps [21,33] produce one of the longest read lengths among all. It utilizes restriction enzymes to make fluorescent labels on the molecule wherever there is a 6 or 7 bp sequence motif [3,17]. The molecule is then linearized and imaged. The subsequent image processing step measures the distance of the two neighboring fluorescent labels and outputs an array of integers, indicating the position (in bp) of each fluorescent label on the read. When the DNA has a structural variant with respect to the reference, the read has discordant patterns of integers with that of the *in silico* digested reference sequence. Read length is typically >150 kbp [24], which is one order of magnitude longer than PacBio reads and two orders of magnitude longer than Illumina reads. With such large length, Optical Maps data enables the detection of SVs that are missed by other technologies, and they have been applied to both normal and cancer patient samples [15,24]. Despite the large read length, computational methods are required for it to be widely used for SV detection by accounting carefully for OM data shortcomings, which include the small number of fluorescent labels in each read, and the various errors of additional labels (17%), missing labels (10%), and sizing difference [3].

The use of OM reads data for SV detection started from correcting [28], assessing [14] and scaffolding *de novo* whole genome assembly (WGA) from other sequencing technologies such as PacBio [30], Illumina [38], 10X [27] or a combination of multiple technologies [36]. Recently, there have been efforts for using OM alone for SV detection. There are two existing approaches to SV detection using OM data alone: assembly-based and alignment-based. In assembly-based methods, OM reads are assembled *de novo* into contigs, which are then compared with the *in silico* digested reference sequence [3,24]. Such *de novo* WGA strategy takes advantage of the randomness of errors in a cohort of reads for obtaining accurate and long contigs. However, due to the repetitive regions in the genome and the low resolution of the label coordinates, *de novo* WGA requires typically 70x read coverage in a diploid healthy human genome [12]. This makes it impossible to tackle those SVs with low coverage of reads. BioNano Solve [12] is an assembly-based approach and its recall is limited in low-coverage loci. Alignment-based methods, on the other hand, align the OM reads to the reference, and cluster the reads on focal regions where discordant patterns occur. OMSV [22] is an alignment-based method which uses the reads that can span the indels to infer insertions and deletions. It is computationally efficient as compared with BioNano Solve (at least one order

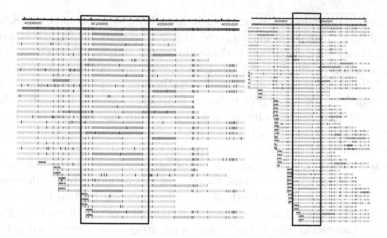

**Fig. 1.** OMView [19] illustrations of a deletion (left), as validated by parents' signal, and insertion (right), as validated by both parents' signal and orthogonal sequence-based methods (i.e., the methods that are applied to sequencing technologies other than OM), both of which are missed by OMSV and BionanoSolve on NA12878. Shown is the alignment of the OM reads to the reference. Reference is the top bar in red. OM reads are the bars below. On the reference and OM reads, the vertical lines indicate the presence of a restriction enzyme fluorescent label. On a read, we call the part in between two neighboring restriction enzymes a fragment. Fragment length is the distance between these two restriction enzymes. A read is composed of $N$ fragments if it has $N + 1$ restriction enzymes. Whenever such distance is consistent with that of the reference, the color of that fragment is set to yellow. When the distance on the read is smaller than that on the reference, the fragment is in green. When the distance on the read is larger than that on the reference, the fragment is in red. The intensity of green and red represents the intensity of contraction and stretch, respectively. It is possible that the two neighboring fragments are taken as one block in the alignment, in which case their colors are the same. Left: The fragments in green in the middle of the reads (highlighted by the black box) indicate a deletion (19:40101872-40147822). But due to their not having the same boundary (some green fragments in green protrude to the right and some to the left due to two or more fragments that are aligned as one block) and the same intensity (green colors vary), the signal is weak, leading to the missing of the call by the two existing methods. Right: The fragments in orange in the middle of the reads (highlighted by the black box) indicate an insertion (1:236385430-236394679). (Color figure online)

of magnitude less time and much smaller required memory) and is applicable to loci with lower coverage of reads. However, in inferring indels, it can only cluster the reads having the same indel boundary. The design of OMSV limits the detection power only to indels in which a significant number of reads are confidently well aligned, but cannot deal with the indels when the aligners render different boundaries for different reads because of data noise (illustrated in Fig. 1). Mak *et al.* [24] combined the assembly-based and alignment-based approaches and used both WGA and an alignment-based approach for

SV detection. However, their SV calling process involves heavy manual curation based on Illumina reads. Furthermore, no accompanying tool was released with the paper, making it hard to compare to other methods in a performance study.

In this paper, we propose OMIndel, an alignment-based method combined with local assembly-like approach for indel detection. It is sensitive on calls with weak signals, an improvement over both BioNano Solve and OMSV. A test on NA12878, a healthy diploid genome, and the whole CEU trio demonstrate that OMIndel is able to detect those indels not detectable by either of the two existing methods, while simultaneously maintaining a lower or comparable false discovery rate. Furthermore, we looked into the indels that are only detectable by OM but not by sequencing-based technologies such as Illumina or PacBio, and categorized them into complex events or indels falling on repetitive regions. The method is implemented in Perl and is publicly available for download.

## 2 Methods

### 2.1 General Overview of OMIndel

For aligning the reads to the reference genome, OMIndel uses the same strategy as OMSV, which integrates the results from two aligners RefAligner [24] and OMBlast [20]. From the alignment, we extract all reads that do not have high concordance with the reference (i.e., at least one of the fragment correspondences between read and reference has sizing difference larger than 2,000 bp). We detect indels > 2,000 bp as this size range is the strength of OM [22], and smaller indels can be covered by other sequence technologies. The information of the discordance is recorded, including the coordinates on the reference, sizing difference, etc. The subsequent read clustering involves two steps, coarse and fine, for achieving both fast and accurate clustering. First, the coarse clustering builds a graph for the discordant records and a graph-based union-find algorithm [35] is used to find all connected components of this graph. Fine clustering is then performed on each connected component. As the coarse clustering step may have multiple indels clustered together due to false edges, for reads in one connected component, we further apply a hierarchical clustering algorithm for breaking reads into groups that are truly corresponding to the same indel. The scoring system in the hierarchical clustering is a distance ranging from 0 to 1 between each pair of the reads (0 means the two reads are exactly the same on the focal indel region, and 1 means the two are completely different). Such distance is calculated by aligning the focal region of one read to another (a dynamic programming algorithm for alignment is described below). The alignment score is normalized and subsequently taken to calculate the distance. We then classify the putative indel calls from each group into homozygous reference, homozygous variant and heterozygous variant with a variant score, followed by filtering.

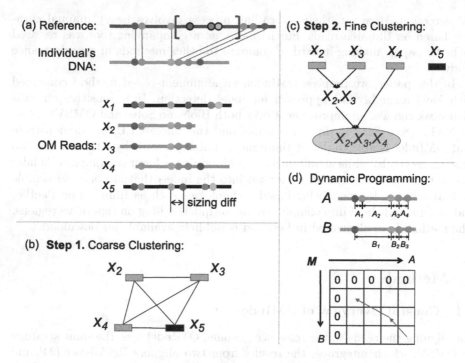

**Fig. 2.** Illustration of the OMIndel method. (a) Deletion of a segment (in purple brackets) with respect to the reference genome (in black horizontal line) is shown. The three labels with black outer circles are deleted in individual's DNA. Each label is shown in a different color for visualization. Correspondence between labels on the reference and the individual's DNA is shown in dashed lines. Five OM reads aligned to this locus are shown: the ones in black lines ($X_1$ and $X_5$) come from the reference allele, and the ones in dark blue lines ($X_2$, $X_3$, and $X_4$) come from a variant allele. Note that due to a sizing difference error on $X_5$, it is selected as a variant read along with $X_2$, $X_3$ and $X_4$ in coarse clustering (step 1), yet to be filtered in fine clustering (step 2). (b) Step 1, a coarse clustering is performed. $X_2$, $X_3$, $X_4$, and $X_5$ are all selected as variant reads. They represent nodes in a graph. Since they overlap with each other, they are all connected. A union-find algorithm is applied to the graph to cluster connected components, and the four reads are grouped in one cluster. (c) Step 2, a fine/hierarchical clustering is performed on individual connected components. The alignment score of each pair of two reads is calculated by a dynamic programming algorithm (described in (d)). The clustering starts from the two reads that have the smallest distance score, and stops when the distance score between two groups of reads is above a threshold. $X_5$, due to its large distance with $X_2$, $X_3$, and $X_4$, does not successfully make it into the cluster, as expected. (d) The process of dynamic programming algorithm to calculate the alignment score. $A$ and $B$ are two OM reads (notice that $B$ does not have the orange label as that in $A$ due to a missing label error). The topmost row and leftmost column of matrix $M$ was initialized with zeros. The entries are filled from top left to bottom right and the largest value is selected as the alignment score. The traceback path (shown in red arrows) retrieves the optimal alignment. In this example, $A_1$ and $A_2$ are joined as a group to be compared with $B_1$, with a gap penalty applied. (Color figure online)

An outline with the cartoons is shown in Fig. 2. We now turn to describing the various steps of OMIndel in detail.

## 2.2   Union-Find for Coarse Clustering

Before the first round of clustering, we align the OM reads to the reference (GRCh38 is used in the Results section below) in the same fashion as that of OMSV [22]. That is, the method uses integrated results from two aligners, RefAligner [24] and OMBlast [20]. We then extract the variant reads that have at least one abnormal sizing difference for all fragments in the read compared with the reference. As the two end labels on the reads have a higher error rate, sizing differences on these labels are skipped. Also, in local alignment, in case more than 5 consecutive fragments have to be aligned as one block and cannot be aligned separately, they are omitted as they probably contain large errors.

We then build an undirected graph, in which each read represents a node, and an edge links two nodes if their corresponding reads have their indicated indel coordinates overlapping with each other on the reference. A union-find algorithm [35] is applied to find all connected components in the graph, producing the clusters of reads. Following this step, each connected component is refined via fine clustering as we describe in the next section.

## 2.3   Local Assembly-Like Approach for Fine Clustering

To overcome random errors in OM reads and achieve high accuracy in indel calling, one more round of clustering is performed for each connected component obtained by the previous step. Within a connected component, for a pair of reads $A$ and $B$, a distance $D_{AB}$ is calculated, which is used in the subsequent hierarchical clustering step. The merging of clusters (in the hierarchical clustering) stops when the two clusters have their distance larger than a threshold. The distance between two groups of reads is calculated as the average distance of all pairs of reads between two groups. For a pair of reads $A$ and $B$, $D_{AB}$ is symmetric, composed of the scores from a dynamic programming algorithm for pairwise read alignment. Specifically,

$$D_{AB} = 1 - \frac{S_{AB} + S_{BA}}{2(\max(S_{AA}, S_{BB}))}, \tag{1}$$

where $S_{AB}$ is the score of aligning read $A$ to read $B$ using a dynamic programming algorithm described next. When $A$ and $B$ are exactly the same, $D_{AB}$ equals zero. The maximum value of $D$ is 1.

We now describe a dynamic programming algorithm for OM reads. In sequence-based pairwise alignment, dynamic programming algorithms such as Smith-Waterman [9] look for best matches between subsequences of the two sequences. A scoring system is used as a way to penalize gaps and mismatches but reward matches. In OM, the dynamic programming is designed in a similar fashion except that instead of penalizing mismatches of the nucleotide bases, we penalize the sizing difference between the two fragments. Also penalizing the indel is turned into penalizing the additional and missing labels. To allow errors that occur near each other, we take the matching of two merged sets of fragments into consideration, with a penalty to the number of fragments that are being merged. More formally, suppose OM read $A$ has fragments $A_1, \ldots, A_x$ and OM read $B$ has fragments $B_1, \ldots, B_y$. For example, if OM read $A$ consists of four coordinates $(5, 10, 12, 18)$, then the fragments are $A_1 = 5$ ($=10 - 5$), $A_2 = 2$ ($=12 - 10$), and $A_3 = 6$ ($=18 - 12$); i.e., $A_i$ is the number of positions that separate the $i$-th and $(i + 1)$-th coordinates in read $A$. The following is a dynamic programming algorithm for calculating the score of optimally aligning $A_{1 \ldots i}$ to $B_{1 \ldots j}$ ($1 \le i \le x$, $1 \le j \le y$), which is stored as entry $M(i, j)$.

- Initialization: $M(0, j) = 0$ for $1 \le j \le y$, and $M(i, 0) = 0$ for $1 \le i \le x$.
- Recursion:

$$M(i, j) = \theta + \max\Bigg( 0,$$

$$\max_{a=1,\ldots,\omega; b=1,\ldots,\omega} \left( M(i - a, j - b) - \left| \sum_{u=0}^{a-1} A_{i-u} - \sum_{v=0}^{b-1} B_{j-v} \right| - G(a, b) \right) \Bigg),$$

where $G(a, b) = \sigma \cdot (a + b - 2)$ is the gap penalty ($\sigma$ is a normalizing factor that makes sizing difference penalty and gap penalty comparable), $\theta$ is a reward for extending the alignment to make the matrix entries positive when there is a good alignment, and $\omega$ ($\le \min(x, y)$) is a user-specific threshold on the maximum number of the fragments to be counted as one block for alignment. In the equation, the summations are taken over $u$'s and $v$'s that satisfy $i - u \ge 1$ and $j - v \ge 1$.
- Termination: $i = x$ and $j = y$.

If two reads are on two different genomic loci, they are unlikely to have overlapping coordinates (with respect to the reference genome) on their respective OM reads and, consequently, are unlikely to belong to the same connected component as identified by the step given in Sect. 2.2 above. This is why the formula for $M(i, j)$ does not account for the actual coordinates, but only for the "spacings" between coordinates (fragments). Finally,

$$S_{AB} = \max_{1 \le i \le x, 1 \le j \le y} M(i, j). \tag{2}$$

## 2.4   Genotyping

We classify each call into homozygous reference, homozygous variant and heterozygous variant by a maximum likelihood approach. The likelihood of each genotype takes both supporting read number and concordance of their indicated indel size into account. Specifically, we model the supporting read number as a Gaussian distribution (the number of reads aligned to a focal region varies and the farther that number from the mean, the smaller its frequency, hence the choice of the Gaussian distribution). We model the sizing difference of each read in a cluster as a Cauchy distribution (the sizing differences from noise have a Gaussian distribution with long tails, hence the choice of the Cauchy distribution, which is also discussed in [22]; see Fig. 3 below). The likelihoods can be expressed as follows:

$$L(D|g = 0) = f_{gaus}(N; \mu, \sigma) \prod_{i=1}^{N} f_{cauchy}(d_i; x_0, \gamma),$$

$$L(D|g = 1) = \sum_{k=1}^{N-1} \left( f_{gaus}(k; \frac{\mu}{2}, \frac{\sigma}{2}) \prod_{i=1}^{k} f_{cauchy}(d_i; (x_0 + d_s), \gamma) f_{gaus}(N - k; \frac{\mu}{2}, \frac{\sigma}{2}) \right.$$
$$\left. \prod_{i=k+1}^{N} f_{cauchy}(d_i; x_0, \gamma) \right),$$

$$L(D|g = 2) = f_{gaus}(N; \mu, \sigma) \prod_{i=1}^{N} f_{cauchy}(d_i; (x_0 + d_s), \gamma).$$

In these expressions:

- $D$ is the OM data (all reads aligned to the local region of interest, given in terms of their fragment length and alignment);
- $g$ is the number of variant allele in the site ($g = 0$ for homozygous reference, $g = 1$ for heterozygous and $g = 2$ for homozygous variant);
- $N$ is the total number of reads on the site;
- $\mu$ and $\sigma$ are the parameters learned from the whole genome, representing the mean and standard deviation of the number of reads covering a site;
- $d_i$ is the inferred indel size from the $i^{th}$ OM read;
- $x_0$ and $\gamma$ are the location and scale parameters of the Cauchy distribution learned from the whole genome where the assumption is there is no indel; and,
- $d_s$ is the estimated indel size given from the previous local assembly-like step, which is the mean of the inferred indels from the reads that cluster.

For homozygous reference, there is no indel, and the location parameter of the sizing difference between read and reference is simply $x_0$, the one learnt from the whole genome. For homozygous variant, the sizing difference of every read on the site corresponds to the same Cauchy distribution learnt from the whole genome, except that the distribution shifts to the left by $d_s$. For heterozygous variant, suppose $k$ reads support the variant and the rest of $N - k$ reads support the reference. The variant and reference supporting read number should both be corresponding to a modified Gaussian distribution (i.e., the mean and variant are half of the $\mu$ and $\sigma$), with their sizing difference to the reference and variant Cauchy distribution, respectively.

To improve computation time, we approximate the heterozygous variant's likelihood as follows:

$$L(D|g=1) = f_{gaus}(k; \frac{\mu}{2}, \frac{\sigma}{2}) \prod_{i=1}^{k} f_{cauchy}(d_i; (x_0 + d_s), \gamma)$$

$$f_{gaus}(N-k; \frac{\mu}{2}, \frac{\sigma}{2}) \prod_{i=k+1}^{N} f_{cauchy}(d_i; x_0, \gamma),$$

where $k$ represents the number of variant-supporting reads that are clustered in the previous step, and $N - k$ is the number of remaining reads aligned to the site. As the previous step assembles all the reads supporting the same allele, this approximation is valid as the other terms in the summation (the first equation for $L(D|g=1)$ above) are close to zero and thus can be omitted. When a read supporting the variant is wrongly clustered as a reference read, as long as its sizing difference is >1,000 bp (some weak signal exists), such omission makes a difference of only less than 6.12e–05 (through the calculation of Cauchy distribution by setting $\gamma = 200$). Finally, a maximum likelihood estimate of the genotype is given by

$$g^* = \text{argmax}_g L(D|g). \tag{3}$$

The variant score can be calculated as

$$S_v = -10\log \frac{L(D|g=0)P_v(g=0)}{\sum_{l=0}^{2} L(D|g=l)P_v(g=l)}, \tag{4}$$

where $P_v(g)$ are the prior probabilities for the three genotypes, and

$$P_v(g=l) = \frac{1}{3} \tag{5}$$

for $l = 0, 1$ and 2.

## 3   Results

### 3.1   Simulated Data

Our simulation process involves two steps: simulating variant alleles and simulating OM reads. In simulating variant alleles, on the in silico digested human reference chromosome 20, we simulate 50 deletions and 50 insertions. For each indel, we uniformly sample its starting label. The indel size is sampled from a Cauchy distribution (locality = 0, scale = 300) and is at least 2,000 bp. We use a Cauchy distribution to simulate the real situation where medium-sized indels outnumber large indels. Labels that are covered in the deleted area are also deleted. For insertions, we simulate the inserted labels such that the distance between the current and the next inserted label is drawn from a Poisson distribution, where the mean is the average distance between two labels in the real case (10 kbp). This process of simulating the inserted label is repeated until no more labels can be sampled from the simulated insertion size. To avoid sampling overlapping indels, we constrain the distance between each pair of neighboring indels to be >100 kbp.

In simulating OM reads, we learned the statistics, including read length, error rates and sizing difference from the real data (CEU trio) and approximated with distributions described below. We simulated three total coverages: 80x, 100x, and 120x and four variant allele fractions (VAFs): 0.2, 0.3, 0.4 and 0.5, resulting in 12 genomes, each having one variant allele and one reference allele. In simulating a read from a given allele (reference or variant), we uniformly sample the starting point. From the real data, we estimated the median of read's length to be about 200 kbp. Since the minimum read length starting to contribute to SV detection is 150 kbp [3], we set the length of the read to be $l_0 + l_r$, where $l_0$ is 150 kbp and $l_r$ is sampled from a Poisson distribution with mean at 50 kbp. Next, we learned the error profiles from the high-confidence alignments (alignments whose reads have $\geq 12$ labels and clipped end is $\leq 4$ labels). The following items are the statistics learned for the three error types.

- Missing label error rate's median is 0.05 (similar to that reported in [22]);
- Additional label error rate is one per 200 kbp;
- Sizing difference's distribution is Cauchy-like as it has long tails (Fig. 3). The Cauchy distribution's parameters are approximated to be locality = 0 and scale = 200 (this is similar to [22]).

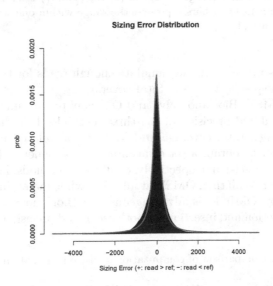

**Fig. 3.** Sizing difference distribution of NA12878 as approximated by a Cauchy distribution (red curve). (Color figure online)

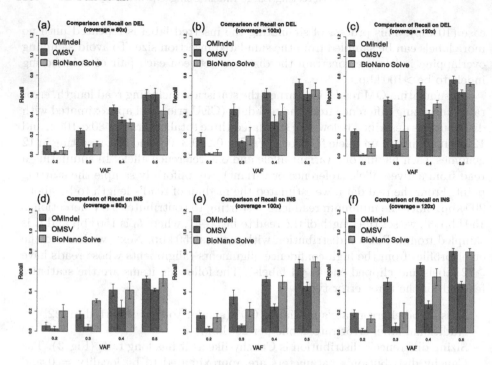

**Fig. 4.** Comparison of recall on simulation data on VAF = 0.2, 0.3, 0.4, 0.5 among OMIndel, OMSV and BioNano Solve, for deletion with coverage at (a) 80x (b) 100x (c) 120x and insertion with coverage at (d) 80x (e) 100x and (f) 120x. Height of the bars represents the mean; the error bars represent the range within one standard deviation whenever it is within $[0, 1]$.

We iterate this process until we simulate enough reads for the desired depth at this allele for a specific VAF and total coverage.

We applied OMSV, BioNano Solve and OMIndel to the simulated data, and measured the recall and precision of the three methods. To reduce the effect of randomness, for each total coverage and VAF, we simulated five data sets of OM reads, in order to obtain a set of accurate measurements. Figures 4 and 5 show the recall and precision, respectively, of all three methods. For all coverages, OMIndel has higher recall than OMSV at all VAFs while maintaining comparable precision. Similarly, OMIndel is advantageous over BioNano Solve on almost all VAFs for both deletion and insertion on both recall and precision. BioNano Solve

**Table 1.** Comparison of computational cost on simulation data.

|           | OMIndel | OMSV  | BioNano solve |
|-----------|---------|-------|---------------|
| CPU hours | 5 h     | 0.5 h | 140 h         |
| Memory    | 4G      | 1G    | 20G           |

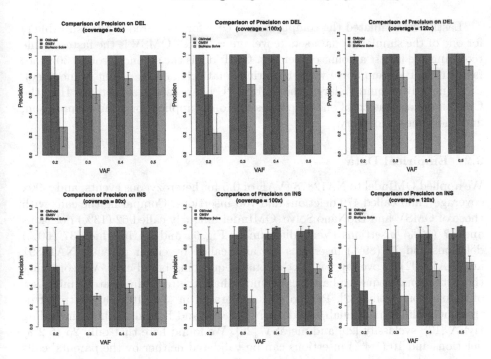

**Fig. 5.** Comparison of precision on simulation data on VAF = 0.2, 0.3, 0.4, 0.5 among OMIndel, OMSV and BioNano Solve, for deletion with coverage at (a) 80x (b) 100x (c) 120x and insertion with coverage at (d) 80x (e) 100x and (f) 120x. Height of the bars represents the mean; the error bars represent the range within one standard deviation whenever it is within $[0, 1]$.

has a slight advantage over OMIndel for insertion when VAF is at 0.5 for 100x, or when VAF is at 0.2 or 0.3 for 80x, at the cost of much lower precision. We observe that at low VAFs, with the increasing of the coverage, BioNano Solve's recall decreases. This shows the instability of BioNano Solve when reference reads greatly outnumber variant reads. Overall, our algorithm has the advantages on recall at small VAFs, with comparable precision with the other two methods for higher VAFs.

It is important to note here that BioNano Solve does not report the indel sizes along with the predictions, which is the reason why we cannot report the recall and precision of the methods broken down by indel sizes (as the indel size could often be a factor in a method's performance).

We investigated whether the low recall of OMSV is due to false alignments. We generated a "ground truth" alignment file given our knowledge of the indel and read errors for one of the twelve cases (VAF = 0.5, total coverage = 120x). We then applied OMSV to the true alignments and found that while maintaining a high precision (1 for both deletion and insertion), the recall for deletion and insertion are respectively 1 and 0.98, compared with 0.64 and 0.44 from the alignment that has errors. This shows that OMSV's recall was greatly affected by erroneous alignments.

Lastly, we evaluated the computational cost for the three methods (Table 1) for one of the simulation data sets (coverage = 100x). OMSV is the fastest while requiring the lowest amount of memory. OMIndel is the second (around 30 times faster than BioNano Solve while requiring relatively small amount of memory). BioNano Solve requires a large amount of CPU hours as well as memory. Here the CPU hours are the total ones if parallelization is applied for a fair comparison, and the same to memory.

## 3.2    Empirical Data

We applied OMIndel to NA12878 (VAF of 0.5 for heterozygous events, and ~90x coverage), and called 479 deletions and 700 insertions. Comparing the calls with those of OMSV and BioNano Solve, OMIndel uniquely called 62 (13%) deletions and 87 (12%) insertions (Venn diagram in Fig. 6a and b), in which 37 (60%) deletions and 77 (89%) insertions are also called by either parents (NA12891 and NA12892) or overlap with orthogonal sequencing-based calls. We construct the orthogonal sequence-based calls such that it is a deduplicated union set of indels from Delly [32], PacBio calls generated in [30], 10X calls (ftp://ftp-trace.ncbi.nlm.nih.gov/giab/ftp/data/NA12878) and hybrid methods including HySA [10], svclassify [29] and metaSV [26]. We found here that only 25 (5.2%) deletions and 10 (1.4%) insertions can be validated neither by the parents' calls nor by orthogonal sequence-based calls. The estimated precision of OMIndel is therefore 94.8% for deletion and 98.6% for insertion, compared with 93.4% and 99.7% for OMSV and 93.0% and 95.4% for BioNano Solve. It was observed that among those numbers, only OMSV's insertion detection has around 1% advantage of precision over OMIndel. However, it is at the cost of missing 127 (22.8%) validated insertions shared by both OMIndel and BioNanoSolve. In addition, Fig. 6e and f show that OMIndel can potentially complement or even outperform OMSV and BioNano Solve in terms of detecting novel calls validated by sequence-based method (13 for deletion and 26 for insertion, compared with 13 and 11 for OMSV, and 53 and 62 for BioNano Solve).

We further looked into OMIndel unique and validated calls (named set A), and compared with the number of supporting reads between those are shared (set B). We found set A has a much smaller number of variant supporting read number (Table 2) than that of set B. Along with the recalls from simulation, OMIndel has been proven to be advantageous in calling indels with weak signals.

The CEU trio data provided us the opportunity to evaluate our genotype accuracy compared with the other two methods. Table 3 listed the accuracy of proband's genotype given parents' calls and corresponding genotypes according to the mendelian inheritance rule. Except on deletion, when OMIndel's genotype accuracy ties with that of BioNano Solve, OMIndel outperforms the two methods on both insertion and deletion. Particularly, we observed that OMSV's genotype accuracy is pretty low, as compared with its high precision in making a call. This particularly demonstrates OMSV's disadvantages in extracting supporting variant and reference reads corresponding to a variant call when mis-alignments are involved.

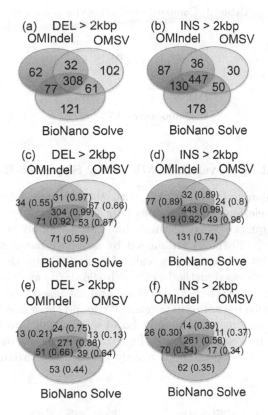

**Fig. 6.** Venn diagram comparing the indels of OMIndel, OMSV and BioNano Solve. (a) and (b) show the numbers of deletion and insertion in the Venn diagram. (c) and (d) show the number of calls that can be validated by the parents' OM calls for deletion and insertion, respectively. (e) and (f) show the number of calls that can be validated by the orthogonal sequenced-based method for deletion and insertion, respectively. The corresponding percentages over the total call are in parenthesis.

**Table 2.** Comparison of mean variant supporting read number between unique validated calls and shared calls. In the parenthesis are the total call number within the category.

|     | Mean variant read # (Total calls) | |
| --- | --- | --- |
|     | OMIndel unique | OMIndel shared |
| DEL | 16.48 (37) | 26.72 (417) |
| INS | 18.79 (77) | 28.73 (613) |

**Table 3.** Comparison of genotype accuracy.

|  | GT accuracy | |
|---|---|---|
|  | DEL | INS |
| OMIndel | 0.75 | 0.83 |
| OMSV | 0.63 | 0.73 |
| BioNano solve | 0.75 | 0.77 |

### 3.3   Investigating Novel Calls Missed by Sequence-Based Methods

We further compared OMIndel calls with the sequence-based calls described above for both deletions and insertions (Fig. 7). We investigated the novel calls missed by all sequence-based methods but can be found in parents' calls. Of the 479 deletions, 120 (25%) are missed by sequence-based methods, of which 86 (72%) are validated by parents' calls. Of the 700 insertions, 329 (47%) are missed by sequence-based methods, of which 296 (90%) are validated by parents' calls. We randomly selected 20 deletions and 20 insertions that are novel to sequence-based calls but called in parents. We found these novel indels are missed by sequencing-based methods mainly because they fall into repetitive regions (Fig. 8) or they are complex events (Fig. 9). In all, these events are missed by Illumina or PacBio based methods mainly because the variant signals are very weak.

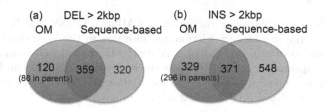

**Fig. 7.** Venn diagram comparing sequence-based indels and OMIndel calls on OM. The numbers in parenthesis are those that are also called by parents.

## 4   Discussion

With the long reads provided by OM data, it is of most interest to know what OM can deliver as compared with other technologies. While this study is focused on answering this question, we found the existing tools on OM are limited in handling the indels with weak signals. Our proposed OM-based targeted assembly approach, OMIndel, falls in the same paradigm of that in TIGRA [5] and HySA [10].

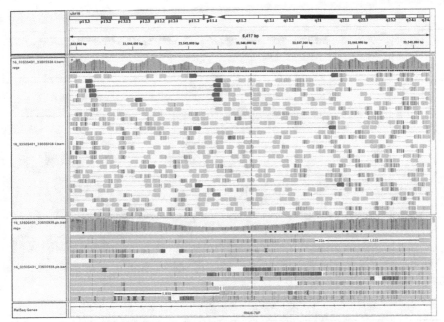

(a) A deletion on 16:33505401-33605938 in GRCh37

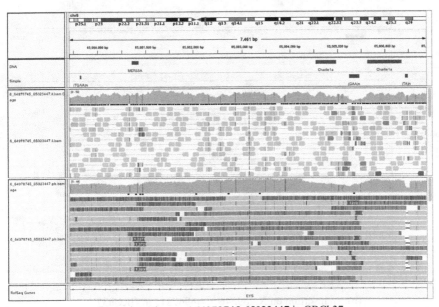

(b) An insertion on 6:64978745-65023447 in GRCh37.

**Fig. 8.** IGV of complex indels that are also called by parents but missed by sequence-based methods. In all IGVs, the upper and bottom panels show the alignment of Illumina and PacBio reads, respectively.

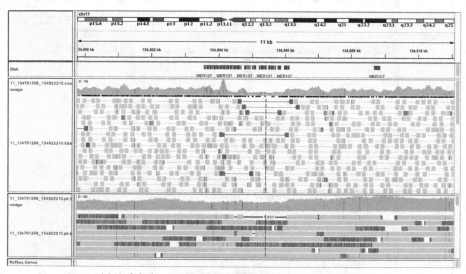

(a) A deletion on 11:134791298-134822210 in GRCh37

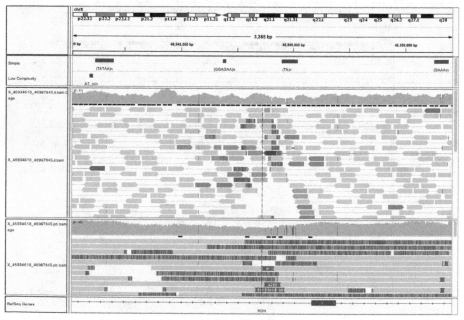

(b) An insertion on X:46934610-46967845 in GRCh37.

**Fig. 9.** IGV of indels overlapping with repetitive regions. These indels are also called by parents but missed by sequence-based methods.

In simulations, we found that BioNano Solve is only advantageous over OMIndel on high VAFs and insertions when measuring recall. This is consistent with our prior assumption that a *de novo* assembly-based approach is better at detecting large insertions than alignment-based or local assembly-based approach when there are enough reads representing the variant. The recall corresponding to different insertion sizes is yet to be summarized for alignment-based or assembly-based approaches.

We summarized some major categories where OM has its unique advantage of detecting indels over sequencing-based technologies. However, further investigation is needed so that such list of categories can be comprehensive as a reference for sequencing. We also observed that there are quite a few sequence-based calls (47.1% for deletions and 59.6% for insertions) missed by our OM-based method. Further investigation will facilitate exploring why OM missed the indels called by sequence-based methods. This may further help to increase the recall of OM-based algorithms. It is also valuable to identify OM's limitations, i.e., which indels are beyond OM's detection ability. This could be done by looking into the calls unique to sequence-based methods.

We acknowledge that the error profiles unique to OM data have not been fully investigated in this study. As discussed by [22], missing and additional labels are indicative of small indels, whereas sizing differences are indicative of large indels. In this study, we focused only on large indels (indels > 2000 bp) by modeling sizing difference distribution in our genotyping algorithm. Detecting smaller indels requires further investigation where the missing and additional label errors need to be modeled.

Finally, this paper's scope is limited only to indel detection. OM's long reads have advantages in detecting inversions, translocations, or complex events such as chromothripsis and chromoplexy. With all of these explored, a comprehensive characterization of the human genome could potentially be achieved.

## 5    Conclusions

We proposed OMIndel, a method that utilizes OM reads to detect large indels. It differs from the previous two methods in that it is alignment-based but follows a local assembly-like fashion, so that it can simultaneously detect indels with weak signal as well as maintain a low FDR. We applied OMIndel to both simulated and real data, and found that it is advantageous over the other two OM-based methods, OMSV and BioNano Solve, by detecting those indels that have weak signals while maintaining a higher or comparable precision. We also manually inspected the indels unique to OM but missed by sequence-based methods. We found that they fall into either a category of complex events or at repetitive regions. OMIndel is freely downloadable online, and we expect that with the increasing availability of samples having OM data and the decreasing cost of OM technology, this tool can be widely used for SV detection.

**Acknowledgements.** We thank Drs. Ken Chen at MD Anderson Cancer Center and Feng Yue at Pennsylvania State University for discussions and data sharing. We would like to thank the Human Genome Structural Variation Consortium. Special thanks to Dr. Charles Lee, Dr. Evan Eichler, Dr. Jan Korbel and Dr. Mark Chaisson for their comments on OM analysis. We used the Extreme Science and Engineering Discovery Environment (XSEDE), which is supported by National Science Foundation grant number ACI-1548562.

# References

1. Alkan, C., Coe, B.P., Eichler, E.E.: Genome structural variation discovery and genotyping (2011)
2. Baca, S.C., et al.: Punctuated evolution of prostate cancer genomes. Cell **153**(3), 666–677 (2013)
3. Cao, H., et al.: Rapid detection of structural variation in a human genome using nanochannel-based genome mapping technology. GigaScience **3**(1), 34 (2014)
4. Chaisson, M.J.P., et al.: Resolving the complexity of the human genome using single-molecule sequencing. Nature **517**(7536), 608–611 (2015)
5. Chen, K., et al.: TIGRA: a targeted iterative graph routing assembler for breakpoint assembly. Genome Res. **24**(2), 310–317 (2014)
6. Chen, K., et al.: BreakDancer: an algorithm for high-resolution mapping of genomic structural variation. Nat. Methods **6**(9), 677–681 (2009)
7. Chin, C.-S., et al.: Nonhybrid, finished microbial genome assemblies from long-read SMRT sequencing data. Nat. Methods **10**(6), 563–569 (2013)
8. Chong, Z., et al.: NovoBreak: local assembly for breakpoint detection in cancer genomes. Nat. Methods **14**(1), 65–67 (2016)
9. Durbin, R., Eddy, S., Krogh, A., Mitchison, G.: Biological sequence analysis. J. Electrochem. Soc. **129**, 2865 (2002)
10. Fan, X., Chaisson, M., Nakhleh, L., Chen, K.: HySA: a hybrid structural variant assembly approach using next-generation and single-molecule sequencing technologies. Genome Res. **27**(5), 793–800 (2017)
11. Feuk, L., Carson, A.R., Scherer, S.W.: Structural variation in the human genome. Nat. Rev. Genet. **7**(2), 85–97 (2006)
12. Hastie, A.R., et al.: Rapid automated large structural variation detection in a diploid genome by nanochannel based next-generation mapping. bioRxiv **NA**, 1–14 (2017)
13. Hormozdiari, F., et al.: Alu repeat discovery and characterization within human genomes. Genome Res. **21**(6), 840–849 (2011)
14. Howe, K., Wood, J.M.D.: Using optical mapping data for the improvement of vertebrate genome assemblies (2015)
15. Jaratlerdsiri, W., et al.: Next generation mapping reveals novel large genomic rearrangements in prostate cancer. Oncotarget **8**(14), 23588–23602 (2017)
16. Korbel, J.O., et al.: PEMer: a computational framework with simulation-based error modelsfor inferring genomic structural variants from massive paired-end sequencingdata. Genome Biol. **10**(2), R23 (2009)
17. Lam, E.T., et al.: Genome mapping on nanochannel arrays for structural variation analysis and sequence assembly. Nat. Biotechnol. **30**(8), 771–776 (2012)
18. Layer, R.M., Chiang, C., Quinlan, A.R., Hall, I.M.: LUMPY: a probabilistic framework for structural variant discovery. Genome Biol. **15**(6), R84 (2014)
19. Leung, A.K.-Y., Jin, N., Yip, K.Y., Chan, T.-F.: OMTools: a software package for visualizing and processing opticalmapping data. Bioinformatics **33**(18), 2933–2935 (2017)

20. Leung, A.K.-Y., et al.: OMBlast: alignment tool for optical mapping using a seed-and-extend approach. Bioinformatics **33**(3), 311–319 (2016)
21. Levy-Sakin, M., Ebenstein, Y.: Beyond sequencing: optical mapping of DNA in the age of nanotechnology and nanoscopy. Curr. Opin. Biotechnol. **24**(4), 690–698 (2013)
22. Li, L., et al.: OMSV enables accurate and comprehensive identification of large structural variations from nanochannel-based single-molecule optical maps. Genome Biol. **18**(1), 230 (2017)
23. Lupski, J.R.: Genomic rearrangements and sporadic disease. Nat. Genet. **39**(7S), S43–S46 (2007)
24. Mak, A.C.Y., et al.: Genome-wide structural variation detection by genome mapping onnanochannel arrays. Genetics **202**(1), 351–362 (2016)
25. Mardis, E.R., Wilson, R.K.: Cancer genome sequencing: a review. Hum. Mol. Genet. **18**(R2), R163–R168 (2009)
26. Mohiyuddin, M., et al.: An accurate and integrative structural-variant caller fornext generation sequencing. Bioinformatics **31**(16), 2741–2744 (2015)
27. Mostovoy, Y., et al.: A hybrid approach for de novo human genome sequence assembly and phasing. Nat. Methods **13**(7), 587–590 (2016)
28. Muggli, M.D., Puglisi, S.J., Ronen, R., Boucher, C.: Misassembly detection using paired-end sequence reads and optical mapping data. Bioinformatics **31**(12), i80–i88 (2015)
29. Parikh, H., et al.: svclassify: a method to establish benchmark structural variant calls. BMC Genom. **17**(1), 64 (2016)
30. Pendleton, M., et al.: Assembly and diploid architecture of an individual human genome viasingle-molecule technologies. Nat. Methods **12**(8), 780–786 (2015)
31. Quinlan, A.R., Hall, I.M.: Characterizing complex structural variation in germline and somatic genomes. Trends Genet. **28**(1), 43–53 (2012)
32. Rausch, T., Zichner, T., Schlattl, A., Stütz, A.M., Benes, V., Korbel, J.O.: DELLY: structural variant discovery by integrated paired-end and split-read analysis. Bioinform. Oxf. Engl. **28**(18), i333–i339 (2012)
33. Samad, A., Huff, E.F., Cai, W., Schwartz, D.C.: Optical mapping: a novel, single-molecule approach to genomic analysis. Genome Res. **5**(1), 1–4 (1995)
34. Scott, E.C., Gardner, E.J., Masood, A., Chuang, N.T., Vertino, P.M., Devine, S.E.: A hot L1 retrotransposon evades somatic repression and initiates human colorectal cancer. Genome Res. **26**(6), 745–755 (2016)
35. Sedgewick, R., Wayne, K.: Algorithms. Addison-Wesley Professional, Boston (2011)
36. Seo, J.S., et al.: De novo assembly and phasing of a Korean human genome. Nature **538**(7624), 243–247 (2016)
37. Sharp, A.J., Cheng, Z., Eichler, E.E.: Structural variation of the human genome. Annu. Rev. Genomics Hum. Genet. **7**, 407–442 (2006)
38. Shelton, J.M., et al.: Tools and pipelines for BioNano data: molecule assembly pipeline and FASTA super scaffolding tool. BMC Genomics **16**(1), 34 (2015)
39. Stephens, P.J., et al.: Massive genomic rearrangement acquired in a single catastrophic event during cancer development. Cell **144**(1), 27–40 (2011)
40. Sudmant, P.H., et al.: An integrated map of structural variation in 2,504 human genomes. Nature **526**(7571), 75–81 (2015)
41. Wang, J., et al.: CREST maps somatic structural variation in cancer genomes withbase-pair resolution. Nat. Methods **8**(8), 652–654 (2011)

# Applied Comparative Genomics

# mClass: Cancer Type Classification with Somatic Point Mutation Data

Md Abid Hasan$^{(\boxtimes)}$ and Stefano Lonardi$^{(\boxtimes)}$

University of California Riverside, 900 University Ave, Riverside, CA 92507, USA
mhasa006@ucr.edu, stelo@cs.ucr.edu

**Abstract.** Cancer is a complex disease associated with abnormal DNA mutations. Not all tumors are cancerous and not all cancers are the same. Correct cancer type diagnosis can indicate the most effective drug therapy and increase survival rate. At the molecular level, it has been shown that cancer type classification can be carried out from the analysis of somatic point mutation. However, the high dimensionality and sparsity of genomic mutation data, coupled with its small sample size has been a hindrance in accurate classification of cancer. We address these problems by introducing a novel classification method called mClass that accounts for the sparsity of the data. mClass is a feature selection method that ranks genes based on their similarity across samples and employs their normalized mutual information to determine the set of genes that provide optimal classification accuracy. Experimental results on TCGA datasets show that mClass significantly improves testing accuracy compared to DeepGene, which is the state-of-the-art in cancer-type classification based on somatic mutation data. In addition, when compared with other cancer gene prediction tools, the set of genes selected by mClass contains the highest number of genes in top 100 genes listed in the Cancer Gene Census. mClass is available at https://github.com/mdahasan/mClass.

**Keywords:** Cancer classification · Somatic point mutation
Genetic variation

## 1 Introduction

Cancer is a complex disease that results from an accumulation of DNA mutations and epigenetic modifications in somatic cells. Remarkable scientific progress has shed light on almost every biological aspect of this disease. Despite this progress, cancer is still one of the most challenging disease of our time with an increasing numbers of new cases and resulting in 14.6% of all human death each year [1]. Not all tumors are cancerous and not all cancers are the same. There is no single test that can diagnose cancer type with perfect accuracy. The diagnosis process requires careful examination and extensive testing to determine whether a person has cancer and which type. Traditional cancer diagnosis method involves lab tests, genetic tests, tumor biopsies, etc. The effective differentiation of cancers

© Springer Nature Switzerland AG 2018
M. Blanchette and A. Ouangraoua (Eds.): RECOMB-CG 2018, LNBI 11183, pp. 131–145, 2018.
https://doi.org/10.1007/978-3-030-00834-5_7

with similar histopathological appearance can indicate the most effective drug treatment and increase survival rates (see, e.g., [2, 6, 8, 9]).

Technological advancements in sequencing technologies has resulted in a dramatic increase in the quantity and quality of sequencing data related to cancer, now available in databases such as The Cancer Genome Atlas [4] and the International Cancer Genome Consortium [3]. These vast repositories provide genomic data from thousands of patients across different cancer subtypes [5]. The abundance of this data has enabled researchers to devise new statistical approaches for the accurate identification of cancer types and subtypes. Cancer classification methods use gene expression data and/or somatic point mutation such as copy number variation, translocations and small insertions and deletions. Several methods have been proposed to accurately predict cancer types and subtypes (see, e.g., [2, 11–13]). The classification of cancer based on the somatic point mutation data can be challenging because of the high dimensionality and sparsity of the data. In cancer patients only a few genes are mutated with high frequency, while most of the genes have a low rate of mutation [10].

The literature on cancer classification methods is extensive. For instance, in [7] the authors proposed a pan-cancer classification method based on gene expression data. They used over nine thousand samples for 31 cancer types to train a method in which a genetic algorithm carries out the gene selection and a nearest neighbor method is used as a classifier.

The authors of [23] proposed to find discriminatory gene sets by measuring the relevance of individual genes using mean and standard deviation of each sample to the class centroid. In [24] the authors introduced new scoring functions to design a stable gene selection method. Their method scores genes based on the assumption that discriminatory genes have different mean values across different classes, small intra-class variation and relatively large inter-class variation.

The authors of [14] combined the clustering gene selection with statistical tests such as T-test and F-test and the gene selection method proposed in [23] to deal the high dimensionality in gene expression data. Genes are assigned to clusters if they are close to the centroids after applying $k$-means clustering.

In [2], the authors proposed a deep neural network for the classification of multiple cancer types from somatic point mutation data, called DeepGene. To the best of our knowledge, DeepGene is the state-of-the-art for multiple cancer classifications using somatic point mutation data. DeepGene clusters genes based on mutation occurrence and uses a sparse representation to index non-zero elements. The data is then fed into a fully connected deep neural network that learns specific cancer types.

In this paper, we address the shortcomings of existing methods dealing with the sparsity and high-dimensionality of somatic point mutation data by proposing an efficient feature selection method based on information theory. A logistic regression model demonstrates the effectiveness of our approach for cancer type classification. Although in a medical setting the task of predicting cancer type from somatic point mutation data might not be practical, here we investigate the fundamental question on whether somatic point mutation data has sufficient discriminative power to allow for cancer type classification.

## 2   Methods

Given $m$ individuals affected by cancer, the input to our feature selection method is composed of the class labels, i.e., the cancer type for the $m$ individuals, and the mutation frequency of all genes for the $m$ individuals. Selected features are then fed into a classifier as described below.

Let $n$ be the number of human genes for which somatic point mutation data is available. Let $C \in \{1 \ldots l\}^m$ be the vector containing the class labels where $l$ is the number of cancer types, and let $G \in \{0 \ldots k\}^{m \times n}, k \in \mathbb{N}$ be the matrix representing the number of mutations observed in each gene (i.e., $G(i,j) = k$ if gene $i$ has $k$ mutations in sample $j$).

The significance of a gene being involved in a particular type of cancer depend on its mutation frequency. Genes with higher mutations are expected to be more relevant for the causation of cancer [16]. In our method, we disregard genes that contain less than $t\%$ mutations across all samples. This filtering step removes non-significant genes from further consideration thus reduce the adverse impact of the data sparsity. Our feature selection model has two steps. First, we cluster genes based on their pairwise similarity. Then, we rank genes using a normalized mutual information criterion [15].

### 2.1   Gene Clustering

Grouping similar genes into clusters allows our method to identify and eliminate redundant genes within a cluster without compromising the efficiency of the feature selection. The reduction of data also reduces the complexity of downstream steps. Since $G$ is a sparse matrix, we use the cosine similarity because of its good mathematical properties on sparse vectors. Given two $n$-dimensional vectors $X$ and $Y$ the cosine similarity is defined as

$$s(X,Y) = \frac{\sum_{i=1}^{n} X_i Y_i}{\sqrt{\sum_{i=1}^{n} X_i^2} \sqrt{\sum_{i=1}^{n} Y_i^2}}$$

where $X_i$ and $Y_i$ are the $i$-th components of vector $X$ and $Y$. Gene $p$ is assigned to the cluster of gene $q$ if the cosine similarity between row vectors $G[:,p]$ and $G[:,q]$ is higher than a predefined threshold $e$. According to this procedure, it is possible that the same gene could end up in multiple clusters. To select unique genes out of these clusters, we rank the genes based on mutation count and mutual information with the class label within the cluster as described next.

### 2.2   Normalized Mutual Information

Our gene selection method relies on an information theoretic measure that evaluates the predictive ability of each gene. Let $X$ be a discrete random variable where each event $x \in X$ occurs with probability $p(x)$. The *entropy* $H(X)$ of variable $X$ is the sum of the information content of each discrete event weighted by the individual event probability, that is $H(X) = -\sum_{x \in X} p(x) \log_2 p(x)$.

Given two discrete random variables $X$ and $Y$ with joint probability $p(x, y)$ and marginal probabilities $p(x)$ and $p(y)$, the *conditional entropy* of variable $Y$ conditioned on variable $X$ is defined as $H(Y|X) = \sum_{x \in X, y \in Y} p(x, y) \log_2(p(x)/p(x, y))$. Similarly, $H(X|Y) = \sum_{x \in X, y \in Y} p(x, y) \log_2(p(y)/p(x, y))$. We have that $H(Y|X) = H(Y)$ iff $X$ and $Y$ are independent random variables. The *mutual information* $I(X, Y)$ is the gain of information about random variable $X$ due to additional information from random variable $Y$, that is

$$I(X, Y) = H(X) - H(X|Y) = \sum_{x \in X} \sum_{y \in Y} p(x, y) \log_2 \left( \frac{p(x, y)}{p(x)p(y)} \right)$$

Given a set $F$ of features (the set of genes in $G$ in this case) and class variables $C$, the feature selection based on mutual information finds a subset $S \subset F$ such that the mutual information $I(C, S)$ is maximized. In order to achieve that goal we use the Normalized Mutual Information based Feature Selection (NMIFS) technique. NMIFS is a heuristic algorithm that selects one feature at a time. NMIFS differs from other mutual information based feature selection technique such as MIFS [17], MIFS-U [18] and mRMR [19] in that it does not depend on the parameter used to control the redundancy penalization. Also NMIFS does not assume that the random variables have uniform probability distribution.

Given features $f_i \in F - S$ and $f_s \in S$ we express the mutual information as

$$I(f_i, f_s) = H(f_i) - H(f_i|f_s) = H(f_s) - H(f_s|f_i) \tag{1}$$

where $H(f_i)$ and $H(f_s)$ are the entropies and $H(f_i|f_s)$ and $H(f_s|f_i)$ are conditional entropies.

The mutual information $I(f_i, f_s)$ is non-negative, and attains its maximum at $\min\{H(f_i), H(f_s)\}$. We can define the normalized mutual information between $f_i$ and $f_s$ as

$$\text{norm}I(f_i, f_s) = \frac{I(f_i, f_s)}{\min\{H(f_i), H(f_s)\}} \tag{2}$$

The *average normalized mutual information* is a measure of redundancy between $f_i$ and $f_s \in S$ for $s = 1, \ldots, |S|$ and it defined as

$$\frac{1}{|S|} \sum_{f_s \in S} \text{norm}I(f_i, f_s)$$

where $|S|$ is the cardinality of subset $S$. Our gene selection criterion selects a gene $f_i \in F - S$ that maximizes

$$J(C, f_i) = I(C, f_i) - \frac{1}{|S|} \sum_{f_s \in S} \text{norm}I(f_i, f_s) \tag{3}$$

where $I(C, f_i)$ is the mutual information between feature $f_i$ and class variable $C$.

## 2.3  Feature Selection

A sketch of mClass' algorithm is shown as Algorithm 1. The algorithm first determines the number of mutations of each gene from the input matrix $G$. Then it computes the cosine similarity between all pairs of genes that have a mutation percentage across all sample of at least $t\%$. Genes are assigned to the same cluster when their similarity exceeds threshold $e$. The process assigns each gene to one or more clusters. The top $v$ genes from each clusters are selected into a representative list $R'$.

Next, mClass collects the unique set of genes $U$ from the representative set $R'$. It then calculates the mutual information between all features/genes $f_i \in U$ and the class variable $C$. To calculate Eqs. (2) and (3) mClass discretizes the gene mutation values into $d$ equal-width bins. The gene $\hat{f_i}$ which has the maximum mutual information with the class variable $C$ is selected as the first feature in $S$ ($S$ is the final set of ranked genes). That gene is then removed from $U$. For all the other genes in $U$ mClass first calculates the normalized mutual information between all pair of genes in $U$ and $S$ using Eq. (2). A gene $f_i \in U$ is selected when it maximizes Eq. (3). The gene is then added to $S$ and removed from $U$. This process is repeated until all genes are given a rank in the ordered set $S$. Instead of deciding on a predefined number of features *a priori* to be used in the classifier, we select a variable number of genes in $S$ based on their ability to classify the data.

## 2.4  Cancer Type Classifier

As said, we employ a logistic regression (LG) multi-class classifier for a given number of genes in the ranked set $S$. The linear model describes the probabilities describing the possible outcome of a single trial using logistic function. Here we use a One-vs-Rest (OvR) for the multi-class classification implementation with $L_2$ regularization. For the binary case, the $L_2$-regularized logistic regression optimizes the following cost function

$$\text{minimize}_w \sum_{x,y} \log(1 + \exp(-w^T x.y)) + \lambda w^T w) \tag{4}$$

The objective is to find the feature weights $(w)$ that minimizes the cost function in Eq. (4). Here $x$ is the feature vector (genes) and $y$ is the class label. The hyper-parameter $\lambda$ used to control the strength of regularization was left as the default value (as defined by `scikit-learn`). As said, the classifier is fed the genes in $S$ incrementally. To determine the final set of features we select genes based on their ability to accurately classify the dataset. The model decomposes the optimization problem in Eq. (4) in a OvR fashion so that the binary classifier can be trained on all classes.

# 3  Experimental Results

In this section, we describe the experimental setup, i.e., datasets and the parameters used in the feature selection and classification, as well as other implemen-

**Data:** Gene mutation data $G \in \{0, k\}^{m \times n}$, similarity measure threshold $e$,
mutation count threshold $t$, discretization value $d$, $v$, class variable $C$
**Result:** Ordered set of genes $S$
set $R \leftarrow \emptyset$;
**for** *each gene* $f_i \in G$ **do**
 **if** *number of mutation of* $f_i > t$ **then**
  |  $R \leftarrow R \cup \{f_i\}$;
 **end**
**end**
set $CL \leftarrow \emptyset$;
**for** *each gene* $f_i \in R$ **do**
 create a new cluster in $CL$ for $f_i$;
 **for** *each gene* $f_j \in R$, $j \neq i$ **do**
  **if** *cosine similarity* $s(f_i, f_j) > e$ **then**
   |  assign $f_i$ and $f_j$ to same cluster in $CL$
  **end**
 **end**
**end**
set $R' \leftarrow \emptyset$;
**for** *each cluster* $cl \in CL$ **do**
 |  set $R' \leftarrow R' \cup \{$top $v$ genes in $cl\}$
**end**
collect unique genes $U \leftarrow set(R')$;
discretize gene mutation values in $d$ equal-width bins;
select the first feature $\hat{f}_i = \text{argmax}_{f_i \in U}\{I(C; f_i)\}$ ;
set $U \leftarrow U - \{\hat{f}_i\}$;
set $S \leftarrow \{\hat{f}_i\}$;
**for** *each gene* $f_i$ *in* $U$ **do**
 calculate $I(f_i; f_s)$ for all pairs $(f_i, f_s)$ with $f_i \in U$ and $f_s \in S$;
 select feature $f_i \in U$ that maximizes $J$ in Equation (3);
 set $U \leftarrow U - \{f_i\}$;
 set $S \leftarrow S \cup \{f_i\}$;
**end**
**return** ordered set $S$;

**Algorithm 1.** mClass feature selection algorithm

tation details. Data preprocessing, feature selection and classification evaluation steps were implemented in Python. All tested classifiers are available from the Python package `scikit-learn`.

## 3.1   Datasets

We used two cancer datasets to test mClass. The first dataset is a twelve-type cancer dataset from The Cancer Genome Atlas (TCGA) [4]. The dataset was assembled by selecting the genes across all samples for all cancer types that contain mutations. Table 1 shows the basic statistics of each cancer type. Observe that the number of samples and the number of mutations varies significantly

**Table 1.** Sample and mutation statistics for the twelve-type cancer dataset

| Cancer type | Number of samples | Number of mutations |
|-------------|-------------------|---------------------|
| ACC | 90 | 18,272 |
| BLCA | 130 | 37,948 |
| BRCA | 982 | 83,360 |
| CESC | 194 | 45,293 |
| HNSC | 279 | 49,264 |
| KIRP | 161 | 13,640 |
| LGG | 286 | 9,228 |
| LUAD | 230 | 68,270 |
| PAAD | 150 | 30,123 |
| PRAD | 332 | 11,802 |
| STAD | 289 | 130,050 |
| UCS | 57 | 10,129 |
| Total | 3,180 | 507,379 |

across cancer types. After removing samples that have less than five mutations across all genes, the dataset contained 3,151 samples and 23,236 genes. The second dataset from TCGA contains four cancer types, namely COAD, SKCM, LAML and KIRC. It contains 1,043 samples with a total of 363,285 mutations across 25,286 genes. Details about this dataset and the corresponding experimental results are discussed in Sect. 3.5.

## 3.2 Parameters

mClass' feature selection uses four parameters: the similarity measure threshold $e$ for the clustering step, the minimum mutation count threshold $t$ to eliminate non-informative genes, the number $v$ of top genes selected from each cluster and the number of bins $d$ used for discretizing gene mutation values (see Algorithm 1).

In our experiments, parameter $t$ was set to 1 which has the effect of disregarding genes with less that 1% mutation across the samples. As said, the pairwise gene similarity is calculated using the cosine similarity measure and genes are assigned into same cluster if the similarity between them is greater than the similarity threshold $e$. The algorithm then selects the top $v$% genes from each cluster for gene ranking step. The values for $e$, $t$, $v$ and $d$ were selected experimentally based on ability of the method to accurately classify the datasets using the selected number of features. For instance, Table 2 shows the classification accuracy of mClass+LG (mClass's feature selection followed by logistic regression) on the twelve-type cancer dataset, for various choices of $e$. Based on this analysis, we selected $e = 0.55$. Similarly, we tested the values of $v$ in the range 5%–25%, and we obtained the highest classification accuracy with $v = 10$%.

**Table 2.** Classification accuracy of mClass+LG as a function of similarity threshold $e$ on the twelve-type cancer dataset

| Similarity threshold (e) | Classification Accuracy |
|---|---|
| 0.50 | 0.708 |
| 0.55 | **0.718** |
| 0.60 | 0.715 |
| 0.65 | 0.715 |
| 0.70 | 0.715 |
| 0.75 | 0.715 |

**Table 3.** Ten-fold cross validation accuracy for mClass+LG and DeepGene (three configurations) on the twelve-type cancer dataset

| Method | Cross-validation Accuracy |
|---|---|
| DeepGene (CGF + ISR) | 0.655 |
| DeepGene (CGF) | 0.638 |
| DeepGene (ISR) | 0.649 |
| mClass+LG | **0.675** |

A similar experimental analysis (not shown) indicated that $d = 5$ was the optimal choice for these datasets. Incidentally, the same value of $d$ was used in [22].

## 3.3   Evaluation Metrics and Comparison with DeepGene

We have used the evaluation metrics introduced in [2] to compare the results. All evaluation experiments were performed by randomly selecting 90% of the input data as training data and 10% of the input as testing data. We compared the ten-fold cross validation accuracy of mClass+LG (mClass's feature selection followed by logistic regression) and testing accuracy against state-of-the-art DeepGene [2].

As said, mClass selects the optimal number of features in a forward selection fashion. We compared mClass' cross-validation results with DeepGene, which employs a convolutional neural network (CNN) as the classifier. The performance of DeepGene was calculated in three different configuration: clustered gene filter and indexed sparsity reduction, only cluster gene filter and only indexed sparsity reduction.

The ten-fold cross-validation results between mClass and three configuration of DeepGene on the twelve-type cancer dataset is shown in Table 3. Observe that the classification accuracy of mClass outperformed all three configurations of DeepGene proposed in [2]. The classification accuracy of mClass is more than 3% higher than the best configuration of DeepGene.

We also compared the testing accuracy of mClass with (i) the best configuration of DeepGene and (ii) LG on the full dataset (i.e., no feature selection).

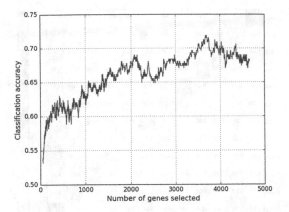

**Fig. 1.** Classification accuracies as a function of the number of feature (genes) selected

**Table 4.** Testing accuracies of mClass+LG, DeepGene and LG on full dataset (twelve-type cancer dataset)

| Method | Classification accuracy |
| --- | --- |
| Full dataset (no feature selection) | 0.677 |
| DeepGene (CGF+ISR) | 0.655 |
| mClass+LG | **0.718** |

The logistic regression classifier in mClass uses balanced weights to counter the imbalance in the number of samples in the dataset. Using the forward feature selection technique described in Algorithm 1, the testing accuracy of the classifier was measured by adding ranked gene one at a time. Figure 1 shows the progression of forward feature selection. mClass obtains the best testing accuracy $(TP + TN)/(TP + TN + FP + FN)$ of 0.718 using a collection of top 3,676 genes which is 9.6% higher than the accuracy obtained by the best configuration of DeepGene with an average precision $TP/(TP + FP)$ of 0.74, recall $TP+(TP+FN)$ of 0.718 and F-Score $(2 \times precision \times recall)/(precision + recall)$ of 0.711 as shown in Table 5. Figure 2 illustrates the confusion matrix for the twelve-type cancer dataset. Observe that with mClass + LG, false positives rate is highest for BRCA while BLCA has the highest rate of false negatives. Table 4 summarizes the testing accuracy of these three methods.

## 3.4 Testing Other Classifiers

As said, mClass+LG uses a logistic regression as the classifier for the cancer classification datasets. We have tested the classification accuracies of other classifiers following mClass' feature selection. We employed Support Vector Machine (SVM) both with the linear and RBF kernel, $k$-nearest neighbor (KNN), Naive Bayes and Random Forest. All the classifiers were available from the Python package `scikit-learn`.

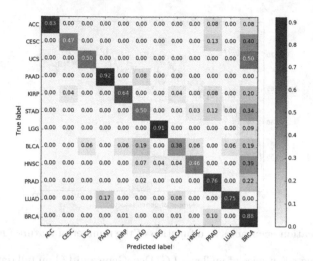

**Fig. 2.** Normalized confusion matrix for the twelve-type cancer dataset

**Table 5.** Classification results on twelve-type cancer dataset

| Cancer type | Precision | Recall | F-Score | Support |
|---|---|---|---|---|
| ACC | 1.00 | 0.83 | 0.91 | 12 |
| CESC | 0.88 | 0.47 | 0.61 | 15 |
| UCS | 0.33 | 0.50 | 0.40 | 2 |
| PAAD | 0.80 | 0.92 | 0.86 | 13 |
| KIRP | 0.89 | 0.64 | 0.74 | 25 |
| STAD | 0.70 | 0.50 | 0.58 | 32 |
| LGG | 0.95 | 0.91 | 0.93 | 23 |
| BLCA | 0.67 | 0.38 | 0.48 | 16 |
| HNSC | 0.81 | 0.46 | 0.59 | 28 |
| PRAD | 0.68 | 0.78 | 0.73 | 50 |
| LUAD | 0.90 | 0.75 | 0.82 | 12 |
| BRCA | 0.62 | 0.88 | 0.71 | 88 |
| Average/Total | 0.74 | 0.72 | 0.71 | 316 |

To classify the data using SVM with the RBF kernel, we optimized the parameter $C$ and $\gamma$ using 10-fold cross validation (keeping other parameters to default). The highest accuracy was obtained with $C = 2e^2$ and $\gamma = 2e^{-5}$. We have used the same parameter $C$ for the linear kernel version of the SVM. The classification with KNN employed Euclidean distance and Pearson correlation coefficient. The 10-fold cross validation showed an optimal accuracy of 0.316 for Euclidean distance using a threshold of 3 and an accuracy of 0.436 with the Pearson correlation coefficient using a neighborhood size of 4. The ensemble

Random Forest classifier's employed a maximum of 1,000 trees in the forest. We set the minimum number of samples required to split an internal node to 9. All other parameters were set to default.

The performance of the various classifier is shown in Fig. 3. The experimental results show a significant advantage of LG over all other classifiers. mClass+LG achieves (i) a 9.6% testing classification improvement over the best configuration of DeepGene (ii) a 24.6% improvement over the linear kernel SVM, (iii) a 29.6% improvement over the RBF kernel SVM, (iv) a 106.9% improvement over KNN with Euclidean distance, (v) a 64.6% improvement over the KNN with Pearson correlation coefficient, (vi) a 83.6% improvement over Naive Bayes and (vii) a 30.3% improvement over Random Forest.

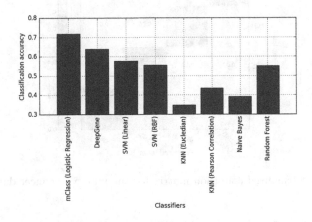

**Fig. 3.** Classification accuracy of mClass+LG, DeepGene and other classifiers applied to the features selected by mClass

## 3.5   Experimental Results on the Four-Type Dataset

As mentioned above, we used a second dataset consisting four type of cancers, namely COAD, SKCM, LAML and KIRC. After removing genes with less than 1% mutations across all samples, the dimension of the dataset was reduced to $1043 \times 25286$. The dataset contains 154 samples for COAD, 345 samples for SKCM, 158 samples for LAML and 386 samples for KIRC. Total number of mutations in this dataset is 363,285. We used the same parameter values for $e$, $t$, $v$ and $d$ as in the previous experiment. The 10-fold cross-validation peaked with an accuracy score of 89.5% with 1,132 genes. For testing accuracy, the dataset was divided into training and testing dataset of size 698 (67%) and 345 (33%), respectively. Using 1,132 features, mClass+LG achieves an accuracy of 87.5% on this dataset. Table 6 shows the average precision and f1-score for each class in this dataset. Figure 4 shows the normalized confusion matrix for our classifier. We could not compare the performance of mClass with DeepGene on

**Table 6.** Testing accuracies on the four-type cancer dataset using mClass

| Cancer Type | Precision | Recall | F-score | Support |
|---|---|---|---|---|
| COAD | 0.93 | 0.75 | 0.83 | 51 |
| SKCM | 0.97 | 0.88 | 0.92 | 101 |
| LAML | 0.65 | 0.98 | 0.79 | 56 |
| KIRC | 0.94 | 0.88 | 0.91 | 137 |
| Avg/Total | 0.90 | 0.87 | 0.88 | 345 |

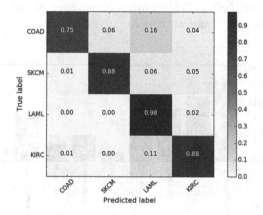

**Fig. 4.** Normalized confusion matrix for the four-type cancer dataset

this second dataset because, according to the authors, the data pre-processing code necessary to feed the training model for DNN is not available anymore.

### 3.6  Comparisons of Predicted Genes

We compared the genes selected by mClass+LG using the 12-types dataset with genes from Cancer Gene Census (CGC). At the time of writing the CGC database contains 719 genes. About 90% of these genes contain somatic mutations, 20% contain germline mutation and 10% contain both types of mutations. We compared mClass' selected genes against the selection carried out by Mutsig 2.0, Mutsig CV [20], MutationAccessor [21] and Muffin [16]. These latter methods predicts cancer genes by analyzing cancer somatic mutation data from 18 types of cancer. We examined the top 100, 500 and 1000 genes produced by these methods, and counted how many of these genes were annotated in the CGC database.

Figure 5 shows these counts for mClass, Mutsig 2.0, Mutsig CV, MutationAccessor and Muffin. Observe that for the top 100 genes, mClass identifies about 50% more CGC genes than MutSig 2.0, MutSig CV and MutationAccessor. mClass identifies more CGC genes than Mutsig 2.0, Mutsig CV and MutationAccessor for the 500 and 1000 case. However, mClass falls short by 18% and 14%

than Muffin in identifying CGC genes in top 500 and top 1000 genes. Although the purpose of mClass was not identifying driver genes, it is remarkable that the top ranked genes selected by mClass contains a large proportion of cancer driver genes.

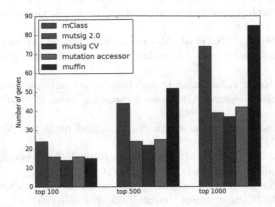

**Fig. 5.** Number of CGC genes produced by mClass, Mutsig 2.0, Mutsig CV, MutationAccessor and Muffin in their top 100, 500 and 1000 selection

## 4 Conclusions

In this paper we proposed a gene selection method based on clustering and normalized mutual information to rank genes for multiple cancer classification using somatic point mutation data. A logistic regression classifier in an one-vs-rest configuration is applied for multiple cancer classification using the selected genes. Experimental results on two TCGA datasets shows significant improvements in classification accuracy. We also showed that our feature selection method ranked genes that match CGC-annotated genes. Moreover, the model can be extended by including other genomic data that could further improve the overall classification performance. For instance, one could use mutation signature associated with specific cancer types to improve the overall accuracy.

**Funding.** This work was supported in part by the US National Science Foundation [IOS-1543963, IIS-1526742].

## References

1. Stewart, B., Wild, C.P.: World Cancer Report 2014. World Health Organization (2015)
2. Yuchen, Y., Yi, S., et al.: DeepGene: an advanced cancer type classifier based on deep learning and somatic point mutations. BMC Bioinf. **17**(Suppl 17), 243–303 (2016)

3. Zhang, J., Baran, J., et al.: International cancer genome consortium data portal - a one-stop shop for cancer genomics data. Database: J. Biol. Databases Curation **2011**, bar026 (2015). https://doi.org/10.1093/database/bar026

4. Tomczak, K., Czerwinska, P., Wiznerowicz, M.: The cancer genome atlas (TCGA): an immeasurable source of knowledge. Contemp. Oncol. **19**(1A), A68–A77 (2011)

5. Amar, D., Izraeli, S., et al.: Utilizing somatic mutation data from numerous studies for cancer research: proof of concept and applications. Oncogene **36**, 3375–3383 (2017)

6. Golub, T.R., Slonim, D.K., et al.: Molecular classification of cancer: class discovery and class prediction by gene expression monitoring. Science **286**(5439), 531–537 (1999)

7. Yuanyuan, L., Kang, K., et al.: A comprehensive genomic pan-cancer classification using The Cancer Genome Atlas gene expression data. BMC Genomics **18**, 508 (2017)

8. Long, D.L.: Tumor heterogeneity and personalized medicine. New Engl. J. Med. **366**(10), 956–957 (2012)

9. Gudeman, J., Jozwiakowski, M., Chollet, J., Randell, M.: Potential risks of pharmacy compounding. Drugs R D **13**(1), 1–8 (2013)

10. Michael, S.L., Stojanov, P., Craig, H.M.: Discovery and saturation analysis of cancer genes across 21 tumor types. Nature **505**, 495–501 (2014)

11. Browne, R.P., McNicholas, P.D., Sparling, M.D.: Model-based learning using a mixture of mixtures of gaussian and uniform distributions. IEEE Trans. Pattern Anal. Mach. Intell. **34**(4), 814–817 (2012)

12. Chicco, D., Sadowski, P., Baldi, P.: Deep autoencoder neural networks for gene ontology annotation prediction. In: Proceedings of ACM Conference in Bioinformatics and Computational Biology, 2014, pp. 533–540. Newport Beach: Health Informatics (2014)

13. Chow, C.K., Zhu, H., Lacy, J., et al.: A cooperative feature gene extraction algorithm that combines classification and clustering. In: IEEE International Conference on Bioinformatics and Biomedicine Workshop (BIBMW), pp. 197–202 (2009)

14. Cai, Z., Xu, L., Shi, Y., et al.: Using gene clustering to identify discriminatory genes with higher classification accuracy. In: IEEE Symposium on Bioinformatics and BioEngineering (BIBE), pp. 235–242. Arlington (2006)

15. Pablo, A.E., Michel, T., Claudio, A.P., Jacek, M.Z.: Normalized mutual information feature selection. IEEE Trans. Neural Netw. **20**(2), 189–201 (2009)

16. Cho, A., Shim, J.E., Kim, E., et al.: MUFFIN: cancer gene discovery via network analysis of somatic mutation data. Genome Biol. **17**, 129 (2016)

17. Battiti, R.: Using mutual information for selection features in supervised neural network. IEEE Trans. Neural Netw. **5**(4), 537–550 (1994)

18. Kwak, N., Choi, C.H.: Input feature selection for classification problems. IEEE Trans. Neural Netw. **13**(1), 143–159 (2002)

19. Peng, H., Long, F., Ding, C.: Feature selection based on mutual information: criteria of max-dependency, max-relevance and min-redundancy. IEEE Trans. Pattern Anal. Mach. Intell. **27**(8), 1226–1238 (2005)

20. Lawrence, M.S., Stojanov, P., Polak, P., et al.: Mutational heterogeneity in cancer and the search for new cancer-associated genes. Nature **499**(7454), 214–218 (2013)

21. Reva, B., Antipin, Y., Sander, C.: Predicting the functional impact of protein mutations: application to cancer genomics. Nucleic Acids Res. **39**(17), e118 (2011)

22. Carter, H.: Cancer-specific high-throughput annotation of somatic mutations: computational prediction of driver missense mutations. Cancer Res. **69**(176), 6660–6667 (2009)

23. Ji-Hoon, C., Dongkwon, L., Jin Hyun, P., In-Beum, L.: New gene selection method for classification of cancer subtypes considering within-class variation. FEBS Lett. **551**, 3–7 (2003)
24. Kun, Y., Zhipeng, C., Jianzhong, L., Gouhui, L.: A stable gene selection in microarray data analysis. BMC Bioinf. **7**, 228 (2006)

# Speciation and Rate Variation in a Birth-and-Death Account of WGD and Fractionation; the Case of Solanaceae

Yue Zhang, Chunfang Zheng, and David Sankoff[(✉)] [ID]

University of Ottawa, Ottawa, Canada
sankoff@uottawa.ca

**Abstract.** We derive the mixture of distributions of sequence similarity for duplicate gene pairs generated by repeated episodes of whole genome doubling. This involves integrating sequence divergence and gene pair loss through fractionation, using a birth-and-death process and a mutational model. We account not only for the timing of these events in terms of local modes, but also the amplitude and variance of the component distributions. This model is then extended to orthologous gene pairs, applied to the evolution of the Solanaceae, focusing on the genomes of economically important crops. We assess how consistent or variable fractionation is from species to species and over time.

**Keywords:** Birth-and death process · Whole genome doubling Fractionation · Solanaceae

## 1 Introduction

A major source of data for the study of genomic evolution is the distribution of some measure of the similarity or difference either between pairs of paralogous genes, generated by a series of whole genome doubling, tripling, etc. (all subsumed under the acronym WGD), or between pairs of orthologous genes, generated by speciation. These distributions have been routinely analyzed in comparative genomics by finding peaks or local modes, in order to estimate each of the WGD or speciation times. We have previously shown how to model the random processes of paralogous gene pair divergence, by mutation, and by gene pair loss through *fractionation*–duplicate gene deletion, in terms of a birth-and-death process integrated with a mutational model. This accounts not only for the timing of peaks, but also their amplitude and how spread out or concentrated they are [1–3]. In this paper, our goal is to extend this model to the study of orthologous gene pairs, so that we can apply it to the evolution of the Solanaceae, focusing on the genomic comparisons among tomato, potato, eggplant, pepper, tobacco and petunia genomes. We aim to systematically elucidate the process of fractionation, using this family as an example, to assess how consistent or variable it is from species to species and how it changes over time.

© Springer Nature Switzerland AG 2018
M. Blanchette and A. Ouangraoua (Eds.): RECOMB-CG 2018, LNBI 11183, pp. 146–160, 2018.
https://doi.org/10.1007/978-3-030-00834-5_8

We first present the details of our discrete-time birth-and-death model for generating populations of paralogs in Sect. 2.2, as summarized from [2], as well as expected counts of present-day paralogous pairs with most recent common ancestors at each ancestral time in Sect. 2.3. These results are then reduced to simpler expressions (no summations, no factorials) for several important cases in Sect. 2.4. In Sect. 2.5, we extend our model to introduce speciation and explore a simplification that allows us to derive the expected number of orthologous pairs with most recent common ancestors at each ancestral time.

In order to account for genomic data, we can observe all the paralogous pairs, as well as the orthologous pairs if two species are involved, but we cannot directly observe at which WGD or speciation time each pair originated. Here is where the mutational model plays a role. The set of pairs generated by a single WGD or speciation event displays a distribution of similarities, whose mean is directly related to the time of that event and whose variance reflects the degree of randomness of the process of similarity decay, as discussed in Sect. 3. The similarities of all the pairs originating from all the events thus constitute a mixture of distributions. The means of the component distributions can be identified as local modes in the distribution of gene pair similarities, as discussed in Sect. 3.1. Maximum likelihood methods can then fill out the remaining information about the variances of each component distribution and their proportions in the mixture.

In Sect. 4.1, we apply our model and methodology to six genomes from the Solanaceae ("nightshade") family of flowering plants using the grapevine genome as an outgroup. We compare all the genomes to each other (21 comparisons) and five of the six to themselves, using the SYNMAP tool on the COGE platform [4,5] to obtain the distribution of paralogous and orthologous gene pair similarities, resulting from WGD and speciation event. The goal is to estimate rates of fractionation, based on the information previously derived about the component distributions. Results from the 27 distributions are then compared for consistency and for variation between genomes.

## 2   Methods

### 2.1   The Birth-and-Death Model

We treat the generation and loss of paralogous genes due to WGD as a discrete time birth-and-death process. At the $i$-th step, each gene in the population is independently replaced by a "litter" of size $r_i \geq 2$, of whom a number $j \geq 1$ survive, with probability $u_j^{(i)}$. The *ploidy* of the event at time is $r_i$, and $r_i - j \geq 1$ is the effect of *fractionation* on the litter. The final population size is observed at the $n$-th step. In practice, what is observed is not the number of genes, but the number of pairs of genes, as calculated in Sect. 2.3, together with a statistic, for each pair, which helps determine at which $t_i$ the two separate lineages, represented by the pair, originated. This statistic, and how we use it will be discussed in Sect. 3. Before that, however, we will discuss how our model may be extended to study orthologous gene pairs.

## 2.2  The Evolution of Population Size

Denote by $m_1, \ldots, m_n$ the total number of individuals (genes) in the population at times $t_1 < \cdots < t_n$. At each time $t_i$, $i = 1, \ldots n-1$, each of the population of $m_i$ genes is replaced by $r_i \geq 2$ progeny. Independently for each gene's progeny, any or all of the $r_i$ survive until time $t_{i+1}$, but at least one does ("no lineage extinction"). We denote by $u_j^{(i)}$ the probability that $j$ of the $r_i$ progeny survive from time $t_i$ to time $t_{i+1}$. This model of fractionation may be termed "sibling rivalry"; there is no constraint on the survival of "cousins". Motivations for the "no lineage extinction" and "sibling rivalry" assumptions are given in [2,3].

Let $a_1^{(i)}, \ldots, a_{r_i}^{(i)}$ be the number of genes at time $t_i$, of which $1, \ldots, r_i$, respectively, survive until $t_{i+1}$, so that

$$m_i = \sum_{j=1}^{r_i} a_j^{(i)}, \qquad m_{i+1} = \sum_{j=1}^{r_i} j a_j^{(i)}. \tag{1}$$

The probability distribution of the evolutionary histories represented by $\mathbf{r} = \{r_i\}_{i=1\ldots n-1}$ and the variable $\mathbf{a} = \{a_j^{(i)}\}_{j=1\ldots r_i}^{i=1\ldots n-1}$ is

$$P(\mathbf{r}; \mathbf{a}) = \prod_{i=1}^{n-1} \left[ \binom{m_i}{a_1^{(i)}, \ldots, a_{r_i}^{(i)}} \prod_{j=1}^{r_i} (u_j^{(i)})^{a_j^{(i)}} \right], \tag{2}$$

as can be proved by induction on $i$. The expected number of genes at time $t_n$ is

$$\mathbf{E}(m_n) = \sum_{\mathbf{a}} P(\mathbf{r}; \mathbf{a}) m_n. \tag{3}$$

Similarly, for the events starting at time $t_j$ with $m_j$ genes, up to $t_k$, we write

$$P^{(j,k)}(\mathbf{r}; \mathbf{a}) = \prod_{i=j}^{k-1} \left[ \binom{m_i}{a_1^{(i)}, \ldots, a_{r_i}^{(i)}} \prod_{h=1}^{r_i} (u_h^{(i)})^{a_h^{(i)}} \right]$$

$$\mathbf{E}^{(j,k)}(m_k) = \sum_{\mathbf{a}} P^{(j,k)}(\mathbf{r}; \mathbf{a}) m_k. \tag{4}$$

## 2.3  Paralogous Gene Pairs

Having described the origin and survival of individual genes, we now summarize the analysis in [2] of the *pairs* of genes observed at time $t_n$ whose most recent common ancestor was replaced by $r_i$ progeny at some time $t_i$. For each of the $a_j^{(i)}$ genes with $j \geq 2$ surviving copies, there are $\binom{j}{2}$ surviving pairs of genes at time $t_{i+1}$. The total number of pairs created at time $t_i$ and surviving to time $t_{i+1}$ is thus

$$d^{(i,i+1)} = \sum_{j=2}^{r_i} \binom{j}{2} a_j^{(i)}. \tag{5}$$

These are called the $t_i$-pairs at time $t_{i+1}$. The expected number of such pairs is

$$\mathbf{E}(d^{(i,i+1)}) = \sum_{a} P^{(1,i+1)}(\mathbf{r};a) \sum_{j=2}^{r_i} \binom{j}{2} a_j^{(i)}. \tag{6}$$

At time $t_j$, for $i + 1 \leq j \leq n$, any two descendants of the two genes making up a $t_i$-pair *with no more recent common ancestor* is also called a $t_i$-pair (at time $t_j$).

For a given $t_i$-pair $g'$ and $g''$ at time $t_{i+1}$, where $i < n - 1$, the expected number of pairs of descendants $d^{(i,n)}$ having no more recent common ancestor is

$$\mathbf{E}(d^{(i,n)}) = \mathbf{E}(d^{(i,i+1)}) \big(\mathbf{E}^{(i+1,n)}(m_n)\big)^2 \tag{7}$$

where $m_{i+1} = 1$ in both factors representing the descendants of a $t_i$-pair. This follows from the independence among the fractionation process between times $t_i$ and $t_{i+1}$ and both processes starting with $g'$ and $g''$.

Of the $m_n$ genes in Eq. (3), the expected number of unpaired genes is

$$\mathbf{E}(m^*) = m_1 \prod_{i=1}^{n-1} u_1^{(i)}. \tag{8}$$

## 2.4   Reductions to Simple Form

Though Eq. (7) would seem to entail an increasing complexity of formulae as $n$ increases, in many important cases this reduces to simple expressions.

**Successive Doublings (Tetraploidizations).** For example if all $r_i = 2$ for $1 \leq i \leq n - 1$, we have by induction that Eq. (7) reduces to

**Theorem 1.**

$$E(t_1) = u_2^{(1)} \Pi_{j=2}^{n-1}(1 + u_2^{(j)})^2$$
$$E(t_i) = \Pi_{j=1}^{i-1}(1 + u_2^{(j)}) u_2^{(i)} \Pi_{j=i+1}^{n-1}(1 + u_2^{(j)})^2 \tag{9}$$
$$E(t_{n-1}) = u_2^{(n-1)} \Pi_{j=1}^{n-2}(1 + u_2^{(j)}),$$

**Corollary 1.** *If all the* $u_2^{(j)} = u$, *then for* $1 \leq i \leq n - 1$,

$$E(t_i) = u(1 + u)^{2n-i-1}. \tag{10}$$

**Successive Triplings (Hexaploidizations).** In the case all $r_i = 3$ for $1 \leq i \leq n - 1$,

$$E(t_1) = (3u_3^{(1)} + u_2^{(1)}) \Pi_{j=2}^{n-1}(1 + 2u_3^{(j)} + u_2^{(j)})^2$$
$$E(t_i) = \Pi_{j=1}^{i-1}(1 + 2u_3^{(j)} + u_2^{(j)})(3u_3^{(i)} + u_2^{(i)}) \Pi_{j=i+1}^{n-1}(1 + 2u_3^{(j)} + u_2^{(j)})^2 \tag{11}$$
$$E(t_{n-1}) = (3u_3^{(n-1)} + u_2^{(n-1)}) \Pi_{j=1}^{n-2}(1 + 2u_3^{(j)} + u_2^{(j)}).$$

**General $r$.** For $r \geq 2$ the same at all $t_i$, and $u_j^{(i)} = u_j$ for $j = 1, \ldots, r$ and $i = 1, \ldots n - 1$, there will be $K \geq 0$ and $K' \geq 0$, depending on the distribution of $u_j$, such that

$$E(t_i) = K'K^{2n-i-1}. \tag{12}$$

## 2.5   Introducing Speciation into the Model

When two populations of a species evolve into two daughter species, we may assume that they initially have the same gene complement, and share identical paralog trees. Instead of observing the state of the paralog tree at time $t_n$, however, we observe a set of orthologous gene pairs at time $t_{n+1}$. Obviously, if such a tree has $m_n$ genes at time $t_n$, this will create $m_n$ different $t_n$ orthologous pairs at time $t_{n+1}$, the time of observation, putting aside consideration of fractionation between $t_n$ and $t_{n+1}$ for the moment.

Under this assumption, we can also calculate the number of $t_i$ orthologous pairs, for $i = 1, \ldots, n-1$. Any $t_i$ paralogous pair creates two $t_i$ orthologous pairs, namely the first gene in the paralogous pair in one species together with the second gene in the other species, and vice versa. For any $i < n$, the number of orthologous $t_i$ pairs is twice the number of paralogous $t_i$ pairs. If, however, we allow fractionation to continue beyond the speciation event, the modeling problem becomes more complicated. We can extend the birth-and-death process, treating speciation as another WGD event, though the counting of orthologs is necessarily different than the counting of $t_i$ paralogs as illustrated in Fig. 1.

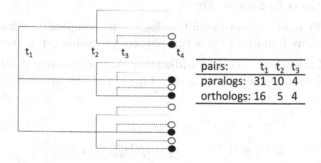

**Fig. 1.** A gene tree produced by two triplications at times $t_1$ and $t_2$, followed by a speciation at time $t_3$, showing the number of paralogous and orthologous $t_1, t_2$ and $t_3$ pairs. (The time of origin of a pair is that of its most recent common ancestor.)

For this sequence of events, the same logic behind Eqs. (9-12) allows us to write

$$E(t_1) = 0.5(3u_3^{(1)} + u_2^{(1)})(1 + 2u_3^{(2)} + u_2^{(2)})^2(1 + u_2^{(3)})^2$$
$$E(t_2) = 0.5(1 + 2u_3^{(1)} + u_2^{(1)})(3u_3^{(2)} + u_2^{(2)})(1 + u_2^{(3)})^2$$
$$E(t_3) = u_2^{(3)}(1 + 2u_3^{(1)} + u_2^{(1)})(1 + 2u_3^{(2)} + u_2^{(2)}), \tag{13}$$

and there is a lengthy associated expression for the expected number of unpaired genes. This approach is more general than simply counting two pairs of orthologs for every pair of paralogs required by the no fractionation assumption, since $u_2^{(3)}$ can be less than 1. However, even this is not really satisfactory, since it may incur a lineage extinction in one of the two genomes created at time $t_n$. The "correct" way of proceeding would be to allow the sibling rivalry fractionation regime operative between $t_{n-1}$ and $t_n$ to continue independently in each of the two genomes until the time of observation $t_{n+1}$. Implementing this is the subject of ongoing research.

# 3    The Distribution of Similarities

The goal of this work is to understand fractionation, so that if at the time of observation we could count the $t_i$ pairs for $i \geq 1$, we could use Equations like (9–13) as a basis for making inferences about the $u_j^{(i)}$. But although we can observe all the paralogous pairs, as well as the orthologous pairs if two species are involved, we cannot *directly* observe at which WGD or speciation time each pair originated. Instead, what we observe at time $t_n$ (or $t_{n+1}$ in the case of orthology) is a measure $p$ of similarity (e.g., the proportion of identical nucleotides in the aligned coding sequences) between each pair of genes in the population. Because of how sequence similarity decays by random substitutions of nucleotides, we can expect an approximately exponential decline in $p$.

Thus if the distribution of gene pair similarities clusters around values $p_1 < p_2 < \cdots < p_{n-1}$, we can infer that these correspond to WGD events at some time $t_1 < t_2 < \cdots < t_{n-1}$. And assuming a large sample of gene pairs, each of these clusters can be modeled by a normal distribution. The distribution of gene pairs is thus a mixture of $n - 1$ normals.

Previous work assumed that the variance of the similarity of a gene pair was proportional to $p(1-p)$, but this did not provide a very good fit in practice. In the present paper, we do not assume any such relationship. Indeed, our strategy will be to identify the $t_i$ by a combination of techniques described in Sect. 3.1, and fix these in a standard maximum likelihood estimate of the variance and amplitude of each component of the mixture. This enables us to calculate the proportion of the sample in each component. We use these proportions, or frequencies derived by multiplying by the sample size, as the numbers of $t_i$ pairs, from which we can estimate the survival proportions using Eqs. (9–13).

## 3.1    The Mode as an Estimator of $t_i$

There are well-established methods for decomposing a mixture of normals (or other predetermined distributions) into their component distributions [6]. Experience shows, however, that these methods, despite their built-in validation criteria, are not robust against non-normality, especially with genomic data, and tend to deliver spurious extra components, and components located in unlikely places. We will nevertheless make use of these methods, but in a way constrained to give appropriate results.

We will compare several genomes to each other. Our strategy is first to locate the $t_i$ in each comparison by picking out local modes in the distribution of similarities, guided by the knowledge that some of these $t_i$ are shared among several genome comparisons, since they reflect the same events. Then for each comparison, some of these estimates are refined by maximum likelihood methods, which also produce the amplitude and variance of the component. From these we can directly partition all the gene pairs into sets of $t_1$ pairs, $t_2$ pairs, etc. Finally, the numbers of genes in these sets can be used to produce estimates of the $u_j^{(i)}$.

Why use the mode? Because of overlapping tails, reminiscent of the mixing of generations, i.e., the decay of synchrony, in initially synchronized population, studied in the antediluvian literature [7], the means of the component distributions cannot be estimated by averaging, but can be identified as local modes in the overall distribution of gene pair similarities.

Estimating the local modes of an underlying distribution by using the modes of the sample involves a trade-off between precision and a proliferation of misleading modes. With gene pair similarities grouped into large bins, or averaged among moving windows of large size, the empirical distribution will be relatively smooth, and bonafide modes will be easily noticed. But a large bin size only indicates that the mode is somewhere in a large interval. With small bin sizes, or sliding window sizes, the position of the nodes are more precisely determined, but more subject to a proliferation of spurious nodes due to statistical fluctuation. Again, we control this problem by considering several related comparisons at a time.

## 4    Results

### 4.1    The Evolution of the Family Solanaceae

The Solanaceae is a family of plants in the asterid order Solanales. This family is distinguished biologically by its early whole genome tripling, as indicated in Fig. 2, and scientifically by the fact that many of its species boast sequenced genomes, namely all the economically most important ones (cf [8]).

### 4.2    The Genomes

We use the SYNMAP software on COGE, and thus have direct access to most of the data, in an appropriate format, among those available on the COGE platform. Those genome data gathered elsewhere (cited below) were uploaded to a temporary private account on COGE for purposes of the present research.

The tomato (*Solanum lycopersicum*) genome sequence and annotation [9] are considered the gold standard among the asterid genome projects. Although there is a recent update to version 3, we used the more familiar (from previous work) version 2.40.

The potato (*Solanum tuberosum*) genome [10] is also a high quality sequence has now been fully assembled into pseudomolecules (version 4.03).

The pepper genome (*Capiscum annuum* version 1.55) [11] is drawn from a genus closely related to *Solanum*.

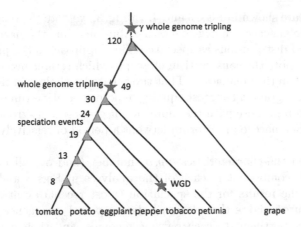

**Fig. 2.** Phylogenetic relationships among the Solanaceae, showing WGD and speciation events. Numbers indicate millions of years from the event to the present, drawn from Fig. 3 in [13], except for the interpolated age of eggplant speciation.

The tobacco (*Nicotiana benthamiana*) genome was sequenced some years ago [12], but its sequence and annotation have been updated and made available for comparative purposes, together with the petunia (*Petunia hybrida* genome [13], both via SGN–the Sol Genomics Network https://solgenomics.net. Among the Solanaceae genomes studied here, only tobacco has undergone a WGD since the original Solanaceae tripling.

A draft version of the eggplant genome (*Solanum melongena*) has also been available for some time [14], and this is what we use here, although a new version is available for browsing via SGN, with restrictions against comparative use awaiting the writing up and publication of the project.

As an outgroup, we use the grapevine (*Vitis vinifera*) genome [15], one of the first flowering plant genomes to be sequenced (in 2007), and one that has proven to be extraordinarily conservative, both with respect to mutational rate and to rearrangement of chromosomal structure. Indeed, the structure of the 19 grape chromosomes resembles in large measure that of the 21 chromosomes of the ancestor of the core eudicots, resulting from a tripling of a seven-chromosome precursor [16]. This is known as the "γ" tripling. Over half of the known flowering plants, including the Solanaceae, belong to this group.

### 4.3   The Comparisons

We applied SynMap to all pairs of the seven genomes and also compared each genome with itself (with the exception of eggplant, because of technical difficulties). We used the default parameters, which are fairly strict in ensuring that all pairs were part of a syntenic block, and thus created at the same time. This excluded duplicate gene pairs that may have been created individually, at some time other than during a WGD event.

The results are shown in Figs. 3 and 4. In Fig. 3, we note the relative stability of the $\gamma$ and Solanaceae tripling-based distributions, but the narrowing of the speciation-based distributions as speciation time approaches the present.

In Fig. 4, we note the conservatism of grape, which retains higher similarities for $\gamma$ paralogs than the Solanaceae. That the $\gamma$-based orthologs in the Solanaceae comparisons with grape all suggest equally remote speciation times, rather than manifesting a compromise with the more recent grape-versus-grape values indicates that the Solanaceae ancestor underwent a period of relatively rapid evolution.

We compiled the characteristics - $p, \sigma$, number (and overall proportion) of pairs - for each component in each of the analyses in Figs. 3 and 4. Of those in Fig. 3, only the results for the speciation (most recent) event are displayed in Table 1. Figure 5 shows the relation between $p$ and divergence time for the speciation event pertinent to each pair of genomes, and their common earlier WGD.

On the left of Fig. 5, the cluster of points around 120 My represents the gene pairs generated by the $\gamma$ tripling event pre-dating all core eudicots, too remote in time to be distinguished from the speciation of the ancestor of grape and the ancestor of the Solanaceae. Points near the centre represent the Solanaceae tripling. Scattered points at more recent times indicate the speciation events among the six Solanaceae species.

The trend line in the figure is $p = 1.2e^{-0.09t}$, which fits well, although the coefficient of the exponential is greater than expected (i.e., 1.0). The right of Fig. 5 suggests that the standard deviation of the component normals are linearly related to their modes (and hence their means). The speciation data for modal values unequivocally support the phylogeny in Fig. 2, e.g., as calculated by neighbour joining (not shown).

## 4.4    Fractionation Rates

We calculated maximum likelihood estimates for $u_2^{(1)}, u_2^{(2)}$ and $u_2^{(3)}$, based on component proportions like those in the bottom section of Table 1. Because there are only two independent proportions per comparison, pertaining to $t_1, t_2$ and $t_3$, and an estimate of the number of unpaired genes (predicted by the model in Eq. (8)), we could not also infer the $u_3^{(i)}$, and simply assumed $u_3^{(1)} = (u_2^{(1)})^2$ and $u_3^{(2)} = (u_2^{(2)})^2$, on the premise that the small probability of two additional progeny surviving (beyond the one essential to avoid extinction) would be approximately the product of their individual probabilities. These event-specific and species-specific survival parameters $u_j^{(i)}$ on the left of Table 2 are directly estimable from the distribution statistics, and reveal much about the difference between the event and the species pairs, but our ultimate interest is in fractionation rates, which we denote $\rho$, and their consistency or variability. In general,

$$u(t) = e^{-\rho t}$$
$$\rho = \frac{-\ln u(t)}{t}. \tag{14}$$

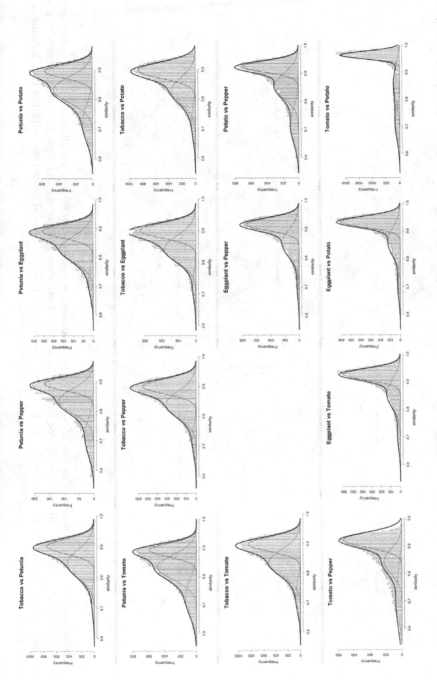

**Fig. 3.** Distribution of ortholog similarities in comparisons among six Solanaceae genomes, with normal distributions fitted to similarities generated by each WGD and speciation event.

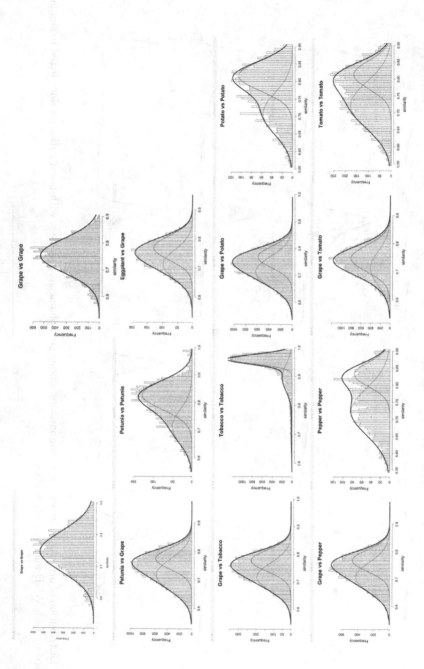

**Fig. 4.** Distribution of paralog similarities in five* Solanaceae genomes and in grape, with normal distributions fitted to similarities generated by each WGD. This is compared to ortholog similarities in each Solanaceae genome versus grape. Two grape panels represent two slightly different fits to the data. Note the Y-axis in the tobacco self-comparison is out of proportion with the rest, because of its recent WGD. (*We were unable to run SYNMAP for eggplant self-comparison.)

**Table 1.** Characteristics inferred for speciation event distributions.

| modal similarity p | Grape | Tomato | Potato | Eggplant | Pepper | Tobacco | Petunia |
|---|---|---|---|---|---|---|---|
| Grape | | 0.769 | 0.770 | 0.772 | 0.771 | 0.779 | 0.777 |
| Tomato | 0.769 | | 0.966 | 0.930 | 0.925 | 0.901 | 0.895 |
| Potato | 0.770 | 0.966 | | 0.936 | 0.927 | 0.900 | 0.898 |
| Eggplant | 0.772 | 0.930 | 0.936 | | 0.920 | 0.904 | 0.894 |
| Pepper | 0.771 | 0.925 | 0.927 | 0.920 | | 0.902 | 0.896 |
| Tobacco | 0.779 | 0.901 | 0.900 | 0.904 | 0.902 | | 0.904 |
| Petunia | 0.777 | 0.895 | 0.898 | 0.894 | 0.896 | 0.904 | |

| component σ | Grape | Tomato | Potato | Eggplant | Pepper | Tobacco | Petunia |
|---|---|---|---|---|---|---|---|
| Grape | | 0.039 | 0.040 | 0.041 | 0.038 | 0.039 | 0.037 |
| Tomato | 0.039 | | 0.014 | 0.021 | 0.027 | 0.024 | 0.023 |
| Potato | 0.040 | 0.014 | | 0.020 | 0.020 | 0.027 | 0.023 |
| Eggplant | 0.041 | 0.021 | 0.020 | | 0.021 | 0.023 | 0.024 |
| Pepper | 0.038 | 0.027 | 0.020 | 0.021 | | 0.024 | 0.022 |
| Tobacco | 0.039 | 0.024 | 0.027 | 0.023 | 0.024 | | 0.024 |
| Petunia | 0.037 | 0.023 | 0.023 | 0.024 | 0.022 | 0.024 | |

| component proportion | Grape | Tomato | Potato | Eggplant | Pepper | Tobacco | Petunia |
|---|---|---|---|---|---|---|---|
| Grape | | 0.45 | 0.40 | 0.51 | 0.44 | 0.47 | 0.40 |
| Tomato | 0.45 | | 0.57 | 0.59 | 0.53 | 0.44 | 0.31 |
| Potato | 0.40 | 0.57 | | 0.61 | 0.41 | 0.54 | 0.33 |
| Eggplant | 0.51 | 0.59 | 0.61 | | 0.55 | 0.48 | 0.42 |
| Pepper | 0.44 | 0.53 | 0.41 | 0.55 | | 0.46 | 0.36 |
| Tobacco | 0.47 | 0.44 | 0.54 | 0.48 | 0.46 | | 0.45 |
| Petunia | 0.40 | 0.31 | 0.33 | 0.42 | 0.36 | 0.45 | |

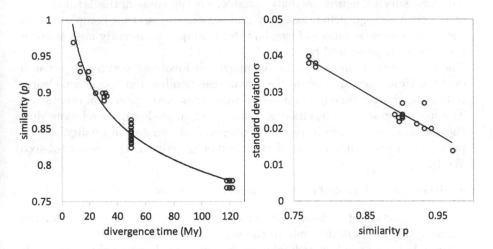

**Fig. 5.** Left: Similarity of orthologs as a function of speciation time. Divergence times taken from Fig. 3 in [13]. Right: Relation of standard deviation to component mean.

When we apply this rule to the survival rates in the table, using the time intervals derived from [13], we derive the fractionation rates on the right of the table. From the sections of Table 2 on survival we observe:

- the 15 estimates of survival between $\gamma$ and the Solanaceae tripling are systematically much lower than the survival between the latter tripling and speciation, and after speciation.

**Table 2.** Estimates of survival (left) and of fractionation rates (right).

γ survival x 100

|          | Tomato | Potato | Eggplant | Pepper | Tobacco | Petunia |
|----------|--------|--------|----------|--------|---------|---------|
| Tomato   |        | 5.12   | 1.64     | 8.34   | 4.05    | 7.89    |
| Potato   | 5.12   |        | 1.42     | 7.13   | 2.68    | 6.58    |
| Eggplant | 1.64   | 1.42   |          | 0.98   | 1.05    | 1.93    |
| Pepper   | 8.34   | 7.13   | 0.98     |        | 1.95    | 3.23    |
| Tobacco  | 4.05   | 2.68   | 1.05     | 1.95   |         | 1.87    |
| Petunia  | 7.89   | 6.58   | 1.93     | 3.23   | 1.87    |         |

γ fractionation rate

|          | Tomato | Potato | Eggplant | Pepper | Tobacco | Petunia |
|----------|--------|--------|----------|--------|---------|---------|
| Tomato   |        | 0.042  | 0.058    | 0.035  | 0.045   | 0.036   |
| Potato   | 0.042  |        | 0.060    | 0.037  | 0.051   | 0.038   |
| Eggplant | 0.058  | 0.060  |          | 0.065  | 0.064   | 0.056   |
| Pepper   | 0.035  | 0.037  | 0.065    |        | 0.055   | 0.048   |
| Tobacco  | 0.045  | 0.051  | 0.064    | 0.055  |         | 0.056   |
| Petunia  | 0.036  | 0.038  | 0.056    | 0.048  | 0.056   |         |

Solanaceae survival x100

|          | Tomato | Potato | Eggplant | Pepper | Tobacco | Petunia |
|----------|--------|--------|----------|--------|---------|---------|
| Tomato   |        | 15.6   | 7.22     | 12.32  | 26.79   | 22.92   |
| Potato   | 15.6   |        | 6.43     | 16.6   | 13.59   | 20.97   |
| Eggplant | 7.22   | 6.43   |          | 6.79   | 8.92    | 11.78   |
| Pepper   | 12.32  | 16.6   | 6.79     |        | 13.15   | 8.65    |
| Tobacco  | 26.79  | 13.59  | 8.92     | 13.15  |         | 16.83   |
| Petunia  | 22.92  | 20.97  | 11.78    | 8.65   | 16.83   |         |

Solanaceae rate

|          | Tomato | Potato | Eggplant | Pepper | Tobacco | Petunia |
|----------|--------|--------|----------|--------|---------|---------|
| Tomato   |        | 0.045  | 0.073    | 0.070  | 0.053   | 0.078   |
| Potato   | 0.045  |        | 0.076    | 0.060  | 0.080   | 0.082   |
| Eggplant | 0.073  | 0.076  |          | 0.090  | 0.097   | 0.113   |
| Pepper   | 0.070  | 0.060  | 0.090    |        | 0.081   | 0.129   |
| Tobacco  | 0.053  | 0.080  | 0.097    | 0.081  |         | 0.094   |
| Petunia  | 0.078  | 0.082  | 0.113    | 0.129  | 0.094   |         |

Speciation survival x100

|          | Tomato | Potato | Eggplant | Pepper | Tobacco | Petunia |
|----------|--------|--------|----------|--------|---------|---------|
| Tomato   |        | 27.52  | 8.53     | 20.66  | 23.84   | 11.68   |
| Potato   | 27.52  |        | 7.95     | 13.56  | 15.55   | 10.61   |
| Eggplant | 8.53   | 7.95   |          | 5.97   | 5.80    | 6.53    |
| Pepper   | 20.66  | 13.56  | 5.97     |        | 9.31    | 4.09    |
| Tobacco  | 23.84  | 15.55  | 5.80     | 9.31   |         | 12.02   |
| Petunia  | 11.68  | 10.61  | 6.53     | 4.09   | 12.02   |         |

Speciation rate

|          | Tomato | Potato | Eggplant | Pepper | Tobacco | Petunia |
|----------|--------|--------|----------|--------|---------|---------|
| Tomato   |        | 0.161  | 0.189    | 0.083  | 0.060   | 0.072   |
| Potato   | 0.161  |        | 0.195    | 0.105  | 0.078   | 0.075   |
| Eggplant | 0.189  | 0.195  |          | 0.148  | 0.119   | 0.091   |
| Pepper   | 0.083  | 0.105  | 0.148    |        | 0.099   | 0.107   |
| Tobacco  | 0.060  | 0.078  | 0.119    | 0.099  |         | 0.071   |
| Petunia  | 0.072  | 0.075  | 0.091    | 0.107  | 0.071   |         |

- The early survival figures are quite variable; a major cause of this is the quality of the genome sequencing, assembly and annotation, so that comparisons of the draft genome sequence of eggplant, for example, apparently miss many of the gene pairs generated by $\gamma$.
- The high rates of survival in the comparisons involving petunia or tobacco over the time interval between the Solanaceae tripling and speciation clearly reflect the shorter time interval before their respective speciation events.
- The speciation survival results reflect, as expected, phylogenetic relationships, though imperfectly, due in part to sequence and annotation quality, and in part due to the amplification of the number of pairs in the recent tobacco WGD.

From the sections of Table 2 on fractionation rates we observe:

- A large reduction of variability (compared to survival) in the results for the inter-tripling interval, due only to the logarithmic transform.
- A large, but not complete, reduction in the difference between the two periods of fractionation, due to the normalization by the time span. This is compatible with the idea that fractionation rates may be universally constrained to a relatively narrow range of values.
- The high rates of post-speciation ortholog loss within *Solanum*, and the relatively low rates for the comparisons involving petunia or tobacco, suggest that the process initially proceeds more quickly than fractionation, or levels off after a certain point, or both.

# 5   Conclusions

We modeled the process of fractionation to account for the distribution of similarities between paralog or ortholog gene pairs after a number of whole genome doublings, triplings, etc., each followed by a period of duplicate gene loss. The model is a discrete-time birth-and-death process, with synchronous birth across the population and non-independent death probabilities constrained by two biologically-motivated conditions: no lineage extinction and (metaphorical) sibling rivalry, i.e, independence of "cousin" death.

The observations of gene pair similarities consist of a mixture of normals, each component generated by one event, with the event time estimated by the sequence divergence from the event to the present. Despite the overlapping distributions, we can estimate the mean (*via* a local mode), standard deviation and proportion of the sample.

We then use these parameters to estimate survival probabilities for gene pairs from one event to the next, according to the birth-and-death model. From the survival data we can then estimate fractionation rates, the number of gene pairs lost per unit time.

We apply our ideas to six genomes from the family Solanaceae and outlier grape. The SynMap program on the CoGe platform produces the distribution of similarities of syntenically validated paralogs and orthologs to feed into our analysis. The 21 pairwise genome comparisons produce a highly consistent picture of the creation and loss of duplicate gene pairs. The survival probabilities and fractionation rates are eminently interpretable in terms of phylogenetic considerations.

Based on our methods and results, we can accurately characterize fractionation rates, something first attempted some years ago [17]. Indeed, we are now in a position to question to what extent fractionation embodies clocklike behaviour.

**Acknowledgements.** Research supported in part by grants from the Natural Sciences and Engineering Research Council of Canada. DS holds the Canada Research Chair in Mathematical Genomics.

# References

1. Zhang, Y., Zheng, C., Sankoff, D.: Evolutionary model for the statistical divergence of paralogous and orthologous gene pairs generated by whole genome duplication and speciation. IEEE/ACM Trans. Comput. Biol. Bioinf. 1545–5963 (2017)
2. Zhang, Y., Zheng, C., Sankoff, D.: Pinning down ploidy in paleopolyploid plants. BMC Genomics **19**(Suppl 5), 287 (2018)
3. Sankoff, D., Zheng, C., Zhang, Y., Meidanis, J., Lyons, E., Tang, H.: Models for similarity distributions of syntenic homologs and applications to phylogenomics. IEEE/ACM Trans. Comput. Biol. Bioinf. (2018). https://doi.org/10.1109/TCBB.2018.2849377
4. Lyons, E., Freeling, M.: How to usefully compare homologous plant genes and chromosomes as DNA sequences. Plant J. **53**, 661–673 (2008)

5. Lyons, E., Pedersen, B., Kane, J., Freeling, M.: The value of non-model genomes and an example using SynMap within CoGe to dissect the hexaploidy that predates rosids. Trop. Plant Biol. **1**, 181–190 (2008)
6. McLachlan, G.J., Peel, D., Basford, K.E., Adams, P.: The EMMIX software for the fitting of mixtures of normal and t-components. J. Stat. Softw. **4**, 1–14 (1999)
7. Sankoff, D.: Duration of detectible synchrony in a binary branching process. Biometrika **58**, 77–81 (1971)
8. Vallée, G.C., Santos Muńoz, D., Sankoff, D.: Economic importance, taxonomic representation and scientific priority as drivers of genome sequencing projects. BMC Genomics **17**, (Suppl 10), 782 (2016)
9. The Tomato Genome Consortium: The tomato genome sequence provides insights into fleshy fruit evolution. Nature **485**, 635–641 (2012)
10. The Potato Genome Sequencing Consortium: Genome sequence and analysis of the tuber crop potato. Nature **475**, 189–195 (2011). DNA Research **21**, 649–C660 (2014)
11. Kim, S., Park, M., Yeom, S.I., Kim, Y.M., Lee, J.M.: Genome sequence of the hot pepper provides insights into the evolution of pungency in *Capsicum* species. Nature Genet. **46**, 270–278 (2014)
12. Bombarely, A., Rosli, H.G., Vrebalov, J., Moffett, P., Mueller, L.A., Martin, G.B.: A draft genome sequence of *Nicotiana benthamiana* to enhance molecular plant-microbe biology research. Mol. Plant-Microbe Interact. **25**, 1523–1530 (2012)
13. Bombarely, A., Moser, M., Amrad, A., Bao, M., et al.: Insight into the evolution of the Solanaceae from the parental genomes of *Petunia hybrida*. Nature Plants 2 (16074) (2016)
14. Hirakawa, H., Shirasawa, K., Miyatake, K., Nunome, T., Negoro, S., et al.: Draft genome sequence of eggplant (*Solanum melongena* L.): the representative *Solanum* species indigenous to the Old World. DNA Research **21**, 649–60 (2014)
15. Jaillon, O., Aury, J.M., Noel, B., Policriti, A., Clepet, C., et al.: The grapevine genome sequence suggests ancestral hexaploidization in major angiosperm phyla. Nature **449**, 463–467 (2007)
16. Zheng, C., Chen, E., Albert, V.A., Lyons, E., Sankoff, D.: Ancient eudicot hexaploidy meets ancestral eurosid gene order. BMC Genomics **14**, S7:S3 (2013)
17. Sankoff, D., Zheng, C., Zhu, Q.: The collapse of gene complement following whole genome duplication. BMC Genomics **11**, 313 (2010)

# Reconciliation and Coalescence

# Detecting Introgression in Anopheles Mosquito Genomes Using a Reconciliation-Based Approach

Cedric Chauve[1]([✉]), Jingxue Feng[2], and Liangliang Wang[2]

[1] Department of Mathematics, Simon Fraser University,
8888 University Drive, Burnaby, BC, Canada
`cedric.chauve@sfu.ca`
[2] Department of Statistics and Actuarial Sciences, Simon Fraser University,
8888 University Drive, Burnaby, BC, Canada

**Abstract.** Introgression is an important evolutionary mechanism in insects and animals evolution. Current methods for detecting introgression rely on the analysis of phylogenetic incongruence, using either statistical tests based on expected phylogenetic patterns in small phylogenies or probabilistic modeling in a phylogenetic network context. Introgression leaves a phylogenetic signal similar to horizontal gene transfer, and it has been suggested that its detection can also be approached through the gene tree/species tree reconciliation framework, which accounts jointly for other evolutionary mechanisms such as gene duplication and gene loss. However so far the use of a reconciliation-based approach to detect introgression has not been investigated in large datasets. In this work, we apply this principle to a large dataset of *Anopheles* mosquito genomes. Our reconciliation-based approach recovers the extensive introgression that occurs in the gambiae complex, although with some variations compared to previous reports. Our analysis also suggests a possible ancient introgression event involving the ancestor of *An. christyi*.

## 1 Introduction

Introgression is the transfer of genetic material between sympatric species, a donor and a receptor species, through hybridization between individuals of both species. It is an important evolutionary mechanism, that plays a key role in the evolution of eukaryotic genomes [15], especially toward the adaptation to a changing environment, a phenomenon known as adaptive introgression (reviewed in [16]). Among recent examples, the evolution of a group of African *Anopheles* mosquitoes, known as the *gambiae complex*, is of interest. This species complex includes most African vectors for the disease malaria, although not all species of the complex are malaria vectors. In 2015, Fontaine *et al.* demonstrated that there is extensive introgression within the gambiae complex, with possible implications related to the rapid acquisition of enhanced vectorial capacities [9]. The extent of introgression within the gambia complex was later confirmed by another

© Springer Nature Switzerland AG 2018
M. Blanchette and A. Ouangraoua (Eds.): RECOMB-CG 2018, LNBI 11183, pp. 163–178, 2018.
https://doi.org/10.1007/978-3-030-00834-5_9

work [29], using a different methodology, although the suggested introgression events were not in full agreement with Fontaine *et al.* The present paper follows this line of work, aiming at detecting traces of introgression within a larger group of *Anopheles* mosquito genomes, covering African and Asian mosquitoes.

There exist several methods that have been designed specifically for detecting footprints of introgression from genomic data, that can be classified into two main groups: methods based on summary statistics and methods based on evolutionary models. Among the first group, specific methods target the detection of introgression between two closely related sister lineages, relying on population genomics data for detecting haplotype blocks at a genetic distance lower than the expected distance if no introgression was involved; we refer to [20] for a recent discussion on these methods. When four species are considered, the most common summary statistic method is the *D statistics* [7], also called the *ABBA BABA statistics*. This method records, over several loci, the frequency of evolutionary trees that are incongruent with a given species phylogeny, and tests if the imbalance between the observed incongruent topologies is significant against a null hypothesis assuming that phylogenetic incongruence is solely due to Incomplete Lineage Sorting (ILS). There exist related methods that consider other invariants [3] or extend it to handle more than four taxa [8,18], although at a significant computational cost. A common feature of these methods is that they aim at disentangling two evolutionary mechanisms that result in discordant gene trees compared to a given species tree: ILS and introgression. Another line of work is based on modeling introgression, that results from hybridization events, using phylogenetic networks, with evolutionary models that account for both ILS and hybridization. This model-based approach has been implemented in combinatorial [11,31] and probabilistic frameworks [14,28,30,33]. We refer the reader to [6] for a recent perspective on model-based approaches. These methods are highly parameterized, and generally their computational complexity grows exponentially with the number of reticulate edges considered in the phylogenetic network, and they have mostly been used with data sets of relatively moderate size so far, although recent pseudo-likelihood methods have shown promising improvements in computation time [21,32].

An important drawback of the methods outlined above is that they rely on the analysis of orthologous loci, thus disregarding gene duplication and gene loss. While this can be a reasonable approach for small data sets, it does exclude many gene families for larger data sets. Moreover, as observed in [17], introgression through hybridization leaves a phylogenetic signal similar to Horizontal Gene Transfer (HGT), although both are very different from a biological point of view. HGT is an evolutionary mechanism that is well handled by several efficient gene tree/species tree reconciliation algorithms [12,23–25] that scale well to large data sets. This suggests that the framework of reconciling gene trees with a known species tree could be used for detecting introgression without the need to filter out paralogous genes.

In the present work, we explore this idea, and apply a reconciliation-based method to detect signals of introgression in a large data set of 14 *Anopheles* genomes covering both African and Asian mosquitoes and including the gambiae complex. We use a combination of published methods to sample reconciled gene trees in an evolutionary model accounting for gene duplication, gene loss and HGT using almost the full complement of genes in our data set. In order to disentangle ILS and introgression, we rely on the hypothesis that introgression acts on larger genome segments, as discussed in [22], and we develop a statistical test to detect genome segments with significantly more genes whose evolution shows a signal of HGT than expected if such genes were located at random along chromosomes. Our approach recovers a strong signal for several introgression events within the gambiae complex, confirming the extensive level of introgression within this group of species, although with some differences related to specific introgressed segments. We also find support for a potential ancient introgression event involving the *An. christyi* lineage and the most common ancestor of the clade of Asian *Anopheles* mosquitoes.

## 2   Data and Methods

### 2.1   Data

Our starting data are the full genome sequences of 14 *Anopheles* species:

- the gambiae complex composed of *An. gambia* (AGAMB), *An. coluzzi-* (ACOLU), *An. arabiensis* (AARAB), *An. quadriannulatus* (AQUAD), *An. melas* (AMELA), *An. merus* (AMERU);
- two outgroups to the gambiae complex, *An. christyi* (ACHRI, an African mosquito) and *An. epiroticus* (AEPIR, an Asian mosquito);
- a clade of Asian mosquitoes, *An. stephensi India* (ASTEI), *An. stephensi sensu stricto* (ASTES), *An. maculatus* (AMACU), *An. culicifacies* (ACULI), *An. minimus* (AMINI), also including the African mosquito *An. funestus* (AFUNE) related to Asian vectors [10]; from now we call this group the *Asian clade*.

The species tree relating these species is given in Fig. 1; it is the so-called X-phylogeny used in [1]. In our experiments, we consider this tree as undated, i.e. with no given branch length. The branching pattern within the gambiae complex, a highly debated question, follows [9,26].

The 14 genomes contain a single fully assembled genome, *An. gambia*, while some others are assembled at the contig level; we refer the reader to [1] for a precise discussion on the assembly of these genomes. The considered genomes contain from 10,000 to above 14,000 genes, that have been clustered prior to our study into homologous gene families using the OrthoDB algorithm [27] and represent an improvement of the set of gene families used in [1]. Figure 2 below illustrates the distribution of the number of genes per genome and the sizes of the gene families.

An important observation is the large number of very small gene families, likely due to errors in assembling genes or in clustering genes into homologous

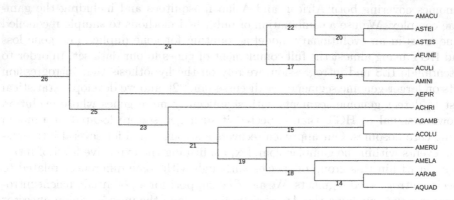

**Fig. 1.** Species tree of the 14 considered *Anopheles* species. Numbers on the internal branches identify ancestral species.

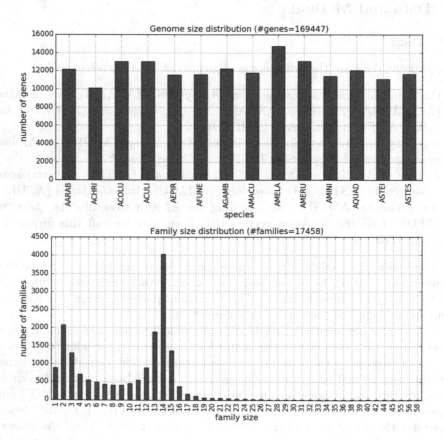

**Fig. 2.** (Top) Distribution of the number of genes per species. (Bottom) Distribution of the size (number of genes) of gene families.

families, an expected issue with large-scale multi-species genomic data sets. For each family, a multiple sequences alignment (MSA) of the coding sequences of the genes belonging to the family was obtained using the method described in [1].

## 2.2   Methods

Our analysis of this data set contains three main steps. In a first step, we sample reconciled gene trees for each homologous family, in a model including gene duplication, gene loss and HGT. Then we evaluate the consistency of these inferred HGTs to verify that they do not contain a high level of noise. Last we rely on robust HGTs to detect potential introgression events and we apply a statistical test of the co-clustering of the involved genes to detect genome segments potentially involved in these introgression events.

*Reconciled Genes Trees.* For each gene family, we ran MrBayes, a software package for Bayesian phylogenetics inference [19], using the family MSA as input and computing two independent MCMC (Markov-Chain Monte-Carlo) chains per family. MrBayes was run using the General Time Reversible model of sequence evolution with a proportion of invariable sites and a $\Gamma$-shaped distribution of rates across sites. The MCMC ran for $10,000,000$ generations and both chains started from different random trees. Since the average standard deviation of split frequencies (ASDSF) approaching 0 indicates convergence [13,19], we used 0.01 as the threshold of this statistic to determine if the MCMC chain has converged or not. The ASDSF was examined every $5,000$ iterations, and, after convergence, tree samples were saved every 500 MCMC iterations, leading to a maximum of $20,000$ sampled gene trees. Families for which at least one chain generated less than $5,000$ samples were discarded from further analysis. We refer to these sampled trees as the *MrBayes trees*, with two sets of MrBayes trees being generated for each gene family.

Next, for each selected family, the MrBayes trees were provided as input to ALE [24], a method for the exploration of the space of reconciled gene trees, accounting for gene duplication, gene loss and HGT[1]. Reconciled gene trees are gene trees augmented with a mapping of each internal node to a species of the species tree (extant or ancestral) and an annotation of the node as either a speciation or a duplication or a HGT; in the latter case the receptor species of the HGT is also indicated. Given a set of MrBayes trees, ALE extracts the clades observed in these trees and their frequencies, and explores the space of reconciled gene trees that can be assembled from these clades (a process called *gene tree amalgamation*) while maximizing the likelihood of observing the reconciled gene tree. The result is a maximum likelihood amalgamated reconciled gene tree; moreover, when used with its Bayesian MCMC mode, ALE determines the rates of gene duplication, loss and HGT and can sample reconciled gene trees.

---

[1] As mentioned previously, introgression and HGT are different evolutionary mechanisms; however, for expository reasons, we refer to the transfer of genetic material between two *Anopheles* species as an HGT.

For each homologous gene family, ALE was run independently on the two sets of MrBayes trees resulting of the two MrBayes MCMC chains, using its Bayesian MCMC mode. For each run, an amalgamated reconciled gene tree was computed, together with a sample of $1,000$ reconciled gene trees, sampled every 100 iterations of the MCMC chain. We call these two sets of sampled reconciled gene trees the *ALE trees*. Gene families for which the two amalgamated reconciled gene trees were not identical were excluded from further analysis. For a given family, the frequency of an HGT, defined by a donor species $d$ and a receptor species $r$ and denoted by the ordered pair $(d, r)$, is obtained by averaging, over the two independent runs of ALE, the frequency of observing this HGT in the ALE trees; note that $d$ and $r$ can both be either an extant or an ancestral species. The final output of this step is a list of quadruples (donor $d$, receptor $r$, family $g$, frequency $f$): each such quadruple records that, for the given family $g$, an HGT from species $d$ to species $r$ was observed in the sampled reconciled gene trees with frequency $f$. In the rest of this work, we analyze the observed HGT to detect traces of introgression.

*Consistency of HGTs.* It is well known that the accurate detection of HGT is challenging, especially when using an undated species tree. It is then important to evaluate the noise due to likely erroneous HGTs. To do so, we rely on the recent method MaxTiC [5]. MaxTiC aims at ranking the internal nodes of a species tree provided with weighted ranking constraints derived from a set of HGTs, in order to maximize the total weight of the satisfied constraints. In our case, constraints are obtained from HGTs as described in [5]: a given HGT from a donor species $d$ to a receptor species $r$ defines a ranking constrain that the ancestor $a$ of $d$ should be older than $r$. Note that, by definition, an HGT whose donor is an extant species does not create a constraint that can conflict with a ranking of the internal nodes; as a consequence, we excluded such constraints from the input of MaxTiC. The weight of a ranking constraint is the sum of the frequencies of the HGTs defining it that are observed across all selected gene families. We applied MaxTiC with inputs composed of ranking constraints derived from several sets of HGTs, obtained by filtering out inferred HGTs whose frequency is below a threshold $t$, ranging from 0.20 to 0.95 by steps of 0.05.

The result of MaxTiC, for a given value of the threshold $t$, is composed of two sets of ranking constraints, the constraints consistent with the computed ranking of the internal nodes of the species tree, and the constraints in conflict with this ranking. We define the *consistency ratio* as the ratio between the weight of the consistent constraints divided by the weight of all considered constraints at frequency threshold $t$. Intuitively, a high consistency ratio points at a low proportion of erroneous HGTs.

*Detecting Potentially Introgressed Segments.* Gene duplication and HGT are two mechanisms that can cause incongruence between a gene tree and a species tree, that are accounted for in ALE. However, ILS is a third common cause of phylogenetic incongruence, that is not considered in the ALE model. A crucial question toward detecting introgression is to distinguish inferred HGTs likely due to introgression to HGTs that could be due to ILS. To do so we rely on the hypothesis that, unlike ILS, introgression is more likely to impact blocks of contiguous genes [22]. Based on this hypothesis, for a given pair of species $(d, r)$, we aim at detecting genome regions where the concentration of genes belonging to families whose evolutionary history as given by sampled reconciled trees involves HGT from $d$ to $r$ is significantly higher compared to a null hypothesis that such families are scattered randomly along chromosomes. As *An. gambia* is the only fully assembled genome in our data set, we perform all tests using *An. gambia* chromosomes; we discuss the impact of this approximation in the Discussion section.

We designed our analysis as follows. Consider an *An. gambiae* genome segment (called *window* from now) containing $n$ genes. Let $p$ denote the probability of observing, within a given window, a gene from a family whose evolution involves a $(d, r)$ HGT, and $p_0$ be the average of the $(d, r)$ HGT frequencies for all the genes on the whole genome, where HGTs are the ones inferred from the ALE results. A statistical hypothesis testing is conducted to test the null hypothesis, $p = p_0$, versus the alternative hypothesis, $p > p_0$. Note that prior to this test, tandem arrays, i.e. segments of consecutive genes from the same family, were reduced to a single gene. Let $X_i$ be the number of observed HGTs from $d$ to $r$ in the $s$ ALE sampled reconciled gene trees for the $i$-th gene in the window, where $i = 1, \ldots, n$. We assume the distribution of $X_i$ to be Binomial$(s, p)$. An unbiased estimator for $p$ is $\hat{p} = \frac{\sum_{i=1}^{n} X_i}{sn}$. Under the assumption that $X_i$'s are independent for simplicity, we have $var(\hat{p}) = \frac{p(1-p)}{sn}$. Consequently, the test statistics $Z = \frac{\hat{p} - p_0}{\sqrt{p_0(1-p_0)/(sn)}}$ is approximately distributed as a standard Normal distribution. Let $z$ be the observed value of $Z$ given the ALE results. The $p$-value of the hypothesis testing can then be obtained by computing $P(Z \geq z)$.

For the multiple tests over all the windows on each chromosome, we used the Benjamini-Yekutieli (BY) [2] method to control the False Discovery Rate (FDR) in a multiple testing setting with dependencies between the tests, which is the case in our experiment as adjacent windows are not independent.

The result of this analysis is a list of windows for which we detected a significantly higher density of genes supporting an HGT from $d$ to $r$, under an FDR of 1%; we selected a window size of $n = 20$ genes, although results were similar with $n = 10$ or $n = 30$.

## 3   Results

*Reconciled Gene Trees.* After running MrBayes and ALE, and filtering out gene families for which both MrBayes chains did not generate at least 5,000 sampled gene trees and gene families for which the two amalgamated gene trees generated by ALE were not identical, there are 11,589 gene families containing a total of 137,180 genes left, with each species "losing" roughly 2,000 genes. Figure 3 illustrates the impact of this filtering on homologous families sizes.

Comparing with Fig. 2, we observe a significant decrease in the number of gene families with 12 or more genes, indicating that many of these families do

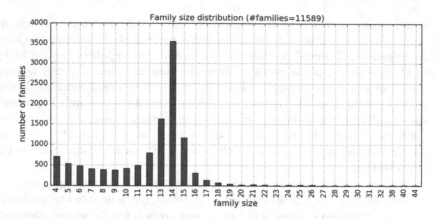

**Fig. 3.** Distribution of the number of genes per homologous family after filtering out families for consistency of the two runs of the MrBayes+ALE pipeline.

not generate consistent amalgamated gene trees under our relatively stringent filtering criteria. We can also observe that the majority of the gene families passing our filtering step are not composed of one-to-one orthologous genes, motivating an approach based on an evolutionary model accounting for gene duplication and gene loss.

*Horizontal Gene Transfers.* Next we consider the inferred HGTs that can suggest potential introgression events. After filtering out, for each gene family, all HGT that do not appear in at least 20% of both sets of ALE trees, the total number of conserved HGTs is 16, 210, leading to an average number of inferred HGT per gene family slightly above 1. Figure 4 shows that low-frequency HGTs dominate the landscape, although there are 4, 771 (resp. 1, 778) HGTs observed with frequency at least 50% (resp. 80%).

Next, the MaxTiC results suggest that the inferred HGTs do not show an apparent high level of noise, measured in terms of conflicting HGTs. The consistency ratio increases steadily from 0.908 at $t = 0.2$ to 0.938 at $t = 0.5$ and 0.973 at $t = 0.8$, indicating a low level of conflict among HGTs with frequency at least 0.2. The most interesting finding is that, at threshold $t = 0.7$, only two constraints with a significant weight are discarded, constraints $(18, 15)$ and $(14, 15)$ – where $(x, y)$ means that node $x$ should be ranked before node $y$ – while the reversed constraints $(15, 18)$ and $(15, 14)$ are among the conserved constraints, although with a weight an order of magnitude higher. The constraints $(15, 14)$ and $(14, 15)$ originate respectively from HGTs from *An. arabiensis* and *An. quadriannulatus* to ancestral species 15 and from *An. gambia* and *An. coluzzi* to ancestral species 14. This observed time inconsistent HGTs between these two groups of species is discussed in Sect. 4.

*Potential Introgression Events.* In order to classify inferred HGTs as potential introgression events from a donor species $d$ to a receptor species $r$, we used

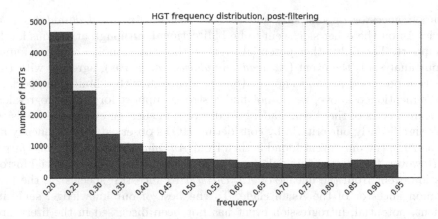

**Fig. 4.** Distribution of the frequency of observed HGTs appearing with frequency at least 20% in ALE trees.

the following stringent criteria: the HGT must be observed in at least 50 gene families, at frequency at least 50%, with an accumulated frequency over all such gene families at least 50. These criteria are based on the results of the MaxTiC analysis. Figure 5 shows the potential introgression events detected using these criteria.

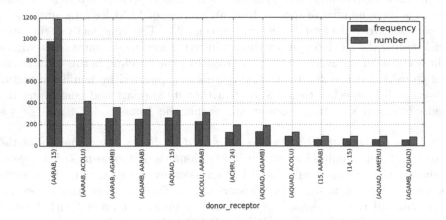

**Fig. 5.** Potential introgression events based on sets of at least 50 inferred HGTs of frequency 0.5 or above and accumulated frequency at least 50.

As expected, most potential introgression events are recent and concern the gambiae complex, in agreement with the extensive amount of introgression seen in this group [9]; in particular, we retrieve the major introgression from *An. arabiensis* to the common ancestor of *An. gambiae* and *An. coluzzi* (ancestral species 15) that was found in [9,29]. We can also observe that almost all potential introgression between the two groups of *An. gambia, An. coluzzi* and their common ancestor on one side and *An. arabiensis*, *An. quadriannulatus* and species 14 on the other side seem to be bidirectional, although at various levels of support. The only other potential event within the gambiae complex found by our analysis is the event (*A. quadriannulatus*,*An. merus*), agreeing with the direction proposed in [29] as opposed to [9], although with a limited support.

As mentioned above, we do not find a strong support for any introgression between species of the Asian clade. In order to find such an event, one needs to relax significantly our criteria by considering HGTs observed with frequency as low as 20%; the only event found is then from *An. maculatus* to *An. culicifaces*.

However, the most striking observation is the hypothesis of a potential introgression event from the lineage of *An. christyi* to ancestral species 24, the last common ancestor of the Asian clade. To the best of our knowledge, such an ancient, potential, introgression event has not been discussed in the literature so far. This potential introgression is supported by 195 HGTs with an average frequency of 0.65, comparable to likely introgression events, such as the one from *An. quadriannulatus* to *An. gambia* (193 HGT, average frequency 0.70), discussed in [29].

In order to assess further the level of support for these various potential introgression events, we considered, for each such event, the taxon coverage of the gene families whose evolution involves an HGT supporting the event. The rationale is that for HGTs supported by gene families with low taxon coverage, the identification of the donor and receptor species could lack precision. Overall, we find that all potential introgression events are supported by gene families covering a large number of species, from an average of 12.51 for (*An. arabiensis*, *An. gambiae*) to 13.89 for (*An. christyi*, species 24). The same analysis repeated after lowering the HGT frequency threshold to 0.2 lead to similar results, with a slight decrease of the average taxon coverage by gene families; in this context the event from *An. maculatus* to *An. culicifaces* is supported by families covering on average 6.12 species, thus lowering further its support and confirming the absence of signal for introgression events within the Asian clade.

*Spatial Distribution of Gene Families Involved in HGTs.* Our analysis of the clustering of gene families involved in HGTs along the chromosomes of *An. gambia* showed that for all the potential introgression events shown on Fig. 5, some genome regions contain significant clusters of gene families supporting the event. We provide all the corresponding chromoplots images at https://github.com/cchauve/Anopheles_introgression_RECOMBCG_2018 and discuss below some interesting observations.

First, considering the three events with *An. arabiensis* as donor species and the chromosome arm 2L, one of the 4 autosomal arms of all *Anopheles* species, the pattern of potentially introgressed genes is very different, as shown in Fig. 6. It is interesting to observe that there seems to be close to no introgression signal toward *An. gambia* on this arm, while the introgression to *An. coluzzi* is centered around the region of the so-called 2La polymorphic inversion. A similar pattern showing very specific regions of the X chromosome being introgressed can be observed, illustrated on Fig. 7.

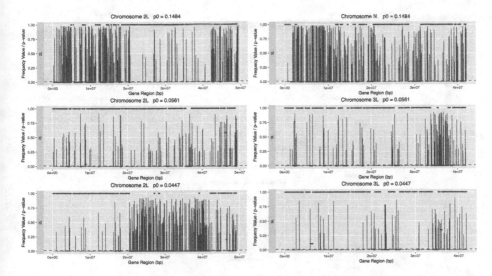

**Fig. 6.** Chromoplots for chromosome arms 2L and 3L for introgression from *An. arabiensis* to species 15 (Top), *An. gambia* (Middle) and *An. coluzzi* (Bottom). Blue vertical bars indicate genes with their HGT frequency, the red dotted line the FDR of 1% and green dots the BY corrected p-value. (Color figure online)

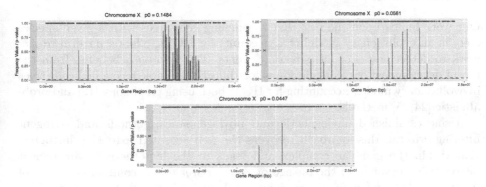

**Fig. 7.** Similar to Fig. 6 for chromosome X

Within the gambiae complex, we retrieve patterns observed in other works. We see for example that the introgression from *An. quadriannulatus* to *An. gambia* involves mostly the 2La inversion again, as was discussed in [29]. We can also see that the signal for an introgression event from *An. quadriannulatus* to *An. merus* involves limited regions of chromosomal arms 3R and 3L.

Last, looking at the chromoplots obtained from the HGTs observed between the lineage of *An. christyi* and species 24 (Fig. 8), we can see a level of support similar to the potential events located within the gambiae complex, although with a much stronger signal for introgression located on the X chromosome.

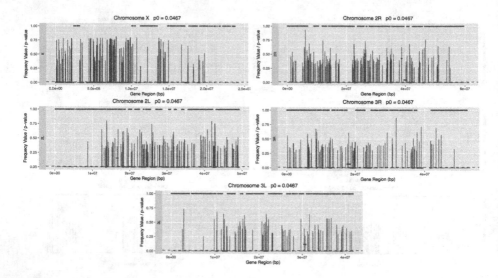

**Fig. 8.** Chromoplots for the potential introgression event from *An. christyi* to species 24.

## 4   Discussion

Our work builds upon the fact that reconciling gene trees with a given species tree, in an evolutionary model that accounts for HGTs, offers a natural framework to detect traces of introgression events. This approach benefits from the recent development of reconciliation algorithms that sample reconciled gene trees in evolutionary models accounting for HGT, both using parsimony [12] and probabilistic [24,25] methods.

Using established phylogenetic and phylogenomics methods and stringent filtering criteria, this approach recovers the well accepted extensive introgression within the gambiae complex and leads to the hypothesis of an ancient introgression event from the lineage of *An. christyi* to the common ancestor of Asian *Anopheles* mosquitoes, involving predominantly the X chromosome. This hypothesis is interesting as the phylogenetic placement of *An. christyi* was questioned in [3] – that uses the same species tree as in the present work – where it was suggested that the significant level of phylogenetic incongruence is indeed due to biological causes (ILS or hybridization); nevertheless, the branching pattern of the gambiae complex, the Asian mosquitoes, *An. christyi* and *An. epiroticus*, deserves further investigation related to ILS and introgression.

Compared to methods based on summary statistics of gene trees, such as the D-statistics, the approach we suggest has several advantages, that are well illustrated by our work. First, the reconciliation framework can handle gene families with gene duplication and gene loss events. Therefore, we can use almost the full complement of genes in larger data sets, unlike methods based on the analysis of one-to-one orthologous loci. Moreover, for all considered gene families, evolutionary trees are computed using a more comprehensive evolutionary model that actually accounts for hybridization events; this contrasts with summary statistics methods that rely on the analysis of gene or loci trees computed without accounting for such evolutionary events. Finally, sampling reconciled gene trees is important toward providing a more nuanced view of evolutionary processes at play within a group of species; the impact of filtering out HGTs sampled with low frequency in our work illustrates this important feature of our approach. Finally, summary statistics methods are limited to data sets with few species, that are in general assumed to be closely related, unlike our approach, a feature which is a crucial toward raising the hypothesis of an ancient introgression along the lineage of *An. christyi*.

Despite the promising results we obtain on a well studied data set, the idea of using a reconciliation-based approach for detecting footprints of introgression requires to be evaluated very carefully. Indeed, our work can certainly not be considered sufficient to claim that HGTs inferred from reconciliations can capture accurately introgression events. Such an evaluation would require to assess its accuracy on simulated datasets, especially toward measuring the impact of using homologous gene families instead of orthologous gene families. The impact of errors in gene families, gene trees and the considered species tree, among other factors, should also be assessed in these simulations. It would also allow to evaluate different reconciliation algorithms, including recently developed algorithms that account for ILS [4,23]. Regarding these two algorithms, it would be interesting to see if they could be extended to sample reconciled gene trees; also, both consider a parsimony framework and the ability of ALE to sample reconciled gene trees in a probabilistic framework was key in our decision to use it for our study.

Regarding the question of the species tree, it is very natural to argue that, given the level of introgression observed in the gambiae complex for example, an approach based on a starting species tree is questionable and phylogenetic networks could provide a better principled framework. However, current phylogenetic networks methods do not scale well to the number of species we consider in this work and the number of potential introgression events. Moreover, to maintain a reasonable computational complexity, they often require either prior potential reticulate edges or an upper bound on the number of reticulations to be given. It would be interesting to test more efficient pseudo-likelihood methods [21,32] and methods jointly computing a species networks and gene trees [28,33]. We nevertheless believe that an interesting feature of our approach is the ability to propose introgression events, that could be tested in a network framework. Last, the consistency analysis using MaxTiC also suggests that current models of phylogenetic networks might need further developments to account for extensive bidirectional and repeated hybridizations events that take place within a short amount of time, as is the case in the gambiae complex. For example, considering the significant level of introgression observed between the clades of *An. arabiensis* and *An. quadriannulatus* of *An. gambia* and *An. coluzzi* suggests that the speciations leading to these four species could have taken place over an extended period of time – which conflicts with the MaxTiC principle of ranking speciation events – during which extensive bidirectional introgression occurred.

The main issue with our approach concerns disentangling introgression from ILS. We followed an indirect approach based on synteny. This approach is further weakened by the fact that we use only the chromosomes of *An. gambia* for all potential introgression events, as it is the only fully assembled genome. This is especially questionable for the potential introgression involving *An. christyi*; but in this precise case, within the clade of Asian mosquitoes the best available assembled genome is *An. minimus*, which is fragmented in around one hundred scaffolds [1], which likely reduces its effectiveness as a support for a synteny analysis. Our synteny-based approach would then naturally benefit from better assembled genomes, including ancestral genomes [1].

Along the lines of making the most out of the reconciliation framework, the ability to model missing species, either because they are extinct or unsampled (called *ghost species*), is an intriguing avenue; the hypothesis that an ancient introgression event along the lineage of *An. christyi* would involve a ghost species is reasonable we believe. Two reconciliation methods, ALE and ecceTERA [12] can handle ghost species, although they require a dated species tree; this is another interesting future avenue to explore.

To conclude, we believe that our work demonstrates that a reconciliation-based approach to study introgression in larger data sets is worth exploring and several interesting methodological questions require further work such as integrating better ILS, networks and unresolved species phylogenies. From an applied point of view, the hypothesis of an introgression event between *An. christyi* and the Asian mosquitoes clade is an interesting case to study further.

**Acknowledgments.** CC is supported by Natural Science and Engineering Research Council of Canada (NSERC) Discovery Grant RGPIN-2017-03986. Most computations were done on the Cedar system of ComputeCanada through a resource allocation to CC. We thank Luay Nakhleh for useful feedback on an early draft of this work.

# References

1. Anselmetti, Y., Duchemin, W., Tannier, E.: Phylogenetic signal from rearrangements in 18 anopheles species by joint scaffolding extant and ancestral genomes. BMC Genomics **19**(2), 96 (2018)
2. Benjamini, Y., Yekutieli, D.: The control of the false discovery rate in multiple testing under dependency. Ann. Stat. **29**(4), 1165–1188 (2001)
3. Blischak, P.D., Chifman, J., Wolfe, A.D., Kubatko, L.S.: HyDe: a python package for genome-scale hybridization detection. Syst. Biol. (2018). https://doi.org/10.1093/sysbio/syy023
4. Chan, Y.-B., Ranwez, V., Scornavacca, C.: Inferring incomplete lineage sorting, duplications, transfers and losses with reconciliations. J. Theor. Biol. **432**, 1–13 (2017)
5. Chauve, C., Rafiey, A., Davin, A.A., Scornavacca, C., et al.: Maxtic: fast ranking of a phylogenetic tree by maximum time consistency with lateral gene transfers (2017). biorxiv: https://doi.org/10.1101/127548. Reviewed https://doi.org/10.24072/pci.evolbiol.100037

6. Degnan, J.H.: Modeling hybridization under the network multispecies coalescent. Syst. Biol. (2018). Advance access. https://doi.org/10.1093/sysbio/syy040

7. Durand, E.Y., Patterson, N., Reich, D., Slatkin, M.: Testing for ancient admixture between closely related populations. Mol. Biol. Evol. **28**(8), 2239–2252 (2011)

8. Elworth, R., Allen, C., Benedict, T., Dulworth, P., Nakhleh, L.: DGEN: a test statistic for detection of general introgression scenarios (2018). https://doi.org/10.1101/348649

9. Fontaine, M.C., Pease, J.B., Steele, A.: Extensive introgression in a malaria vector species complex revealed by phylogenomics. Science **347**(6217), 1258524 (2015)

10. Garros, C., Koekemoer, L., Coetzee, M., Coosemans, M., Manguin, S.: A single multiplex assay to identify major malaria vectors within the African Anopheles funestus and the Oriental An. minimus groups. Am. J. Trop. Med. Hyg. **70**, 583–590 (2004)

11. Holland, B.R., Benthin, S., Lockhart, P.J., Moulton, V., Huber, K.T.: Using super-networks to distinguish hybridization from lineage-sorting. BMC Evol. Biol. **8**(1), 202 (2008)

12. Jacox, E., Chauve, C., Szöllősi, G.J., Ponty, Y., Scornavacca, C.: ecceTERA: comprehensive gene tree-species tree reconciliation using parsimony. Bioinformatics **32**(13), 2056–2058 (2016)

13. Lakner, C., Van Der Mark, P., Huelsenbeck, J.P., Larget, B., Ronquist, F.: Efficiency of Markov chain Monte Carlo tree proposals in bayesian phylogenetics. Syst. Biol. **57**(1), 86–103 (2008)

14. Liu, K.J., Dai, J., Truong, K., Song, Y., Kohn, M.H., Nakhleh, L.: An HMM-based comparative genomic framework for detecting introgression in eukaryotes. PLOS Comput. Biol. **10**(6), 1–13 (2014)

15. Mallet, J., Besansky, N., Hahn, M.W.: How reticulated are species? BioEssays **38**(2), 140–149 (2015)

16. Martin, S.H., Jiggins, C.D.: Interpreting the genomic landscape of introgression. Curr. Opin. Genet. Dev. **47**, 69–74 (2017)

17. Nakhleh, L.: Computational approaches to species phylogeny inference and gene tree reconciliation. Trends Ecol. Evol. **28**(12), 719–728 (2013)

18. Pease, J.B., Hahn, M.W.: Detection and polarization of introgression in a five-taxon phylogeny. Syst. Biol. **64**(4), 651–662 (2015)

19. Ronquist, F., Teslenko, M., van der Mark, P., Ayres, D.L., Darling, A.: Mrbayes 3.2: efficient bayesian phylogenetic inference and model choice across a large model space. Syst. Biol. **61**(3), 539–542 (2012)

20. Rosenzweig, B.K., Pease, J.B., Besansky, N.J., Hahn, M.W.: Powerful methods for detecting introgressed regions from population genomic data. Mol. Ecol. **25**(11), 2387–2397 (2016)

21. Solìs-Lemus, C., Ané, C.: Inferring phylogenetic networks with maximum pseudo-likelihood under incomplete lineage sorting. PLOS Genet. **12**(3), 1–21 (2016)

22. Sousa, F., Bertrand, Y.J.K., Doyle, J.J.: Using genomic location and coalescent simulation to investigate gene tree discordance in Medicago l. Syst. Biol. **66**(6), 934–949 (2017)

23. Stolzer, M., Lai, H., Xu, M., Sathaye, D., Vernot, B., Durand, D.: Inferring duplications, losses, transfers and incomplete lineage sorting with nonbinary species trees. Bioinformatics **28**(18), i409–i415 (2012)

24. Szöllősi, G.J., Rosikiewicz, W., Boussau, B., Tannier, E., Daubin, V.: Efficient exploration of the space of reconciled gene trees. Syst. Biol. **62**(6), 901–912 (2013)

25. Szöllosi, G.J., Davín, A.A., Tannier, E., Daubin, V., Boussau, B.: Genome-scale phylogenetic analysis finds extensive gene transfer among fungi. Philos. Trans. Royal Soc. B Biol. Sci. **370**(1678), 20140335 (2015)

26. Wang, Y., Zhou, X., Yang, D., Rokas, A.: A genome-scale investigation of incongruence in culicidae mosquitoes. Genome Biol. Evol. **7**(12), 3463–3471 (2015)

27. Waterhouse, R.M., Tegenfeldt, F., Li, J.: OrthoDB: a hierarchical catalog of animal, fungal and bacterial orthologs. Nucleic Acids Res. **41**(D1), D358–D365 (2012)

28. Wen, D., Nakhleh, L.: Coestimating reticulate phylogenies and gene trees from multilocus sequence data. Syst. Biol. **67**(3), 439–457 (2018)

29. Wen, D., Yu, Y., Hahn, M.W., Nakhleh, L.: Reticulate evolutionary history and extensive introgression in mosquito species revealed by phylogenetic network analysis. Mol. Ecol. **25**(11), 2361–2372 (2016)

30. Wen, D., Yu, Y., Zhu, J., Nakhleh, L.: Inferring phylogenetic networks using PhyloNet. Syst. Biol. **67**(4), 35–40 (2018)

31. Yu, Y., Barnett, R.M., Nakhleh, L.: Parsimonious inference of hybridization in the presence of incomplete lineage sorting. Syst. Biol. **62**(5), 738–751 (2013)

32. Yu, Y., Nakhleh, L.: A maximum pseudo-likelihood approach for phylogenetic networks. BMC Genomics **16**(10), S10 (2015)

33. Zhang, C., Ogilvie, H.A., Drummond, A.J., Stadler, T.: Bayesian inference of species networks from multilocus sequence data. Mol. Biol. Evol. **35**(2), 504–517 (2018)

# Reconstructing the History of Syntenies Through Super-Reconciliation

Mattéo Delabre[1], Nadia El-Mabrouk[1(✉)], Katharina T. Huber[2],
Manuel Lafond[3], Vincent Moulton[2], Emmanuel Noutahi[1],
and Miguel Sautie Castellanos[1]

[1] Département d'informatique (DIRO), Université de Montréal,
Montréal, Québec, Canada
`mabrouk@iro.umontreal.ca`
[2] School of Computing Sciences, University of East Anglia, Norwich, UK
[3] Department of Computer Science, Université de Sherbrooke, Sherbrooke, Canada

**Abstract.** Classical gene and species tree reconciliation, used to infer
the history of gene gain and loss explaining the evolution of gene families,
assumes an independent evolution for each family. While this assumption is reasonable for genes that are far apart in the genome, it is
clearly not suited for genes grouped in syntenic blocks, which are more
plausibly the result of a concerted evolution. Here, we introduce the
*Super-Reconciliation* model, that extends the traditional Duplication-
Loss model to the reconciliation of a set of trees, accounting for segmental duplications and losses. From a complexity point of view, we show
that the associated decision problem is NP-hard. We then give an exact
exponential-time algorithm for this problem, assess its time efficiency on
simulated datasets, and give a proof of concept on the opioid receptor
genes.

**Keywords:** Gene tree · Reconciliation · Duplication · Loss · Synteny

## 1 Introduction

Gene gain and loss is known as a major force driving evolution. Assuming the
gene and species trees are known and correspond to the true evolution, incongruence between the two trees can be explained by gene gain and loss events,
and "reconciling" the two trees allows recovering these events.

Tree reconciliation can be performed through different biological models of
evolution, the most common being the Duplication-Loss (DL) [15,32,33] or
Duplication-Loss and Transfer [5,10,30] models. While most reconciliation methods are based on the parsimony principle of minimizing the number or cost of

**Electronic supplementary material** The online version of this chapter (https://
doi.org/10.1007/978-3-030-00834-5_10) contains supplementary material, which is
available to authorized users.

M. Blanchette and A. Ouangraoua (Eds.): RECOMB-CG 2018, LNBI 11183, pp. 179–195, 2018.
https://doi.org/10.1007/978-3-030-00834-5_10

operations, probabilistic models seeking for a reconciliation with maximum likelihood or maximum posterior probability have also been developed [4,26,29].

Regardless of the model, current algorithms for reconciliation take each gene family individually, assuming an independent evolution through single duplications and losses. Although this hypothesis holds for genes that are far apart in the genome, it is clearly too restrictive for those organized in syntenic blocks or paralogons, i.e. sets of homologous chromosomal regions, among one or many genomes, sharing the same genes (e.g. neuropeptide Y-family receptors [21], the Homeobox gene clusters [1,13,14], the FGFR fibroblast growth factor receptors [3,16] or the genes of the opioid system [11,27,28]). These genes are more plausibly the result of an evolution from a common ancestral region, rather than from a set of independent gene duplications that would have converged to the same organization in different genomic regions.

The purpose of this paper is to generalize the DL reconciliation model from a unique gene tree to a set of gene trees, accounting for segmental duplications and losses. As far as we know, this problem has never been considered before. The closest algorithms are DeCo [6] and DeCoStar [31] which, given a set of gene families, a set of adjacencies between genes, a set of gene trees and a species tree, compute an adjacency forest reflecting the evolution of each adjacency. However, adjacencies are taken independently, and only single duplications and losses are considered. A correction strategy that adjusts the computation of the evolutionary cost to favour co-evolution events, hence grouping seemingly individual events into single segmental ones was latter proposed in [12]. Another related problem asks for the reconciliation of a set of gene trees leading to a minimum number of duplication episodes, referring to possible whole genome duplication events, defined as sets of single duplications mapped to the same node in the species tree [9,23]. However the considered model does not account for gene orders and duplications involving a set of neighboring genes.

Here, we consider the *Super-Reconciliation* problem in which, given a set of gene families, a set of syntenies, a gene tree for each gene family and a species tree, we seek an evolutionary history of the set of syntenies that is in agreement with the individual gene trees whilst minimizing the number of segmental duplications and losses. Our proposed model is a direct generalization of the reconciliation of a single gene tree. As such, it ignores tandem duplications, rearrangements and assumes that the input set of gene trees is consistent.

After defining the new Super-Reconciliation model in the next section, we begin, in Sect. 3, by characterizing the conditions under which a Super-Reconciliation exists for a set of syntenies and a set of gene trees. We prove, in Sect. 4, that the associated decision problem is NP-hard and gave a dynamic programming algorithm in Sect. 5. An application on simulated datasets and a proof of concept on the genes of the opioid system are then presented in Sect. 6. We conclude with a discussion on future work in Sect. 7.

# 2    Trees, Reconciliation and Problem Statement

A *string* or a *sequence* is an ordered set of characters. Given a string $X = x_1 \cdots x_n$, a *substring* of $X$ is a consecutive set of characters from $X$ in the same order as in $X$, and a *subsequence* is a set of characters of $X$ in the same order, but not necessarily consecutive in $X$ ($X$ is a substring and a subsequence of $X$).

All trees are considered rooted. Given a tree $T$, we denote by $r(T)$ its root, by $V(T)$ its set of nodes and by $\mathcal{L}(T) \subset V(T)$ its leafset. We say that $T$ *is a tree for* $L = \mathcal{L}(T)$. A node $v$ is an *ancestor* of $v'$ if $v$ is on the path from $r(T)$ to $v'$; $v$ is the *father* of $v'$ if it directly precedes $v'$ on this path. In this latter case, $v'$ is called the *child* of $v$. We denote by $E(T)$ the set of edges of $T$, where an edge is represented by its two terminal nodes $(v, v')$, with $v$ being the father of $v'$. Two nodes $v$ and $v'$ are *separated* in $T$ iff neither one is an ancestor of the other. A node is said to be *unary* if it has a single child and *binary* if it has two children. Given a node $v$ of $T$, the subtree of $T$ rooted at $v$ is denoted $T[v]$.

A *binary tree* is a tree with all internal (i.e. non-leaf) nodes being binary. If internal nodes have one or two children, then the tree is said *partially binary*.

*Creating a unary root* consists in creating a new node $v$, a new edge $(v, r(T))$ and assigning $v$ as the new root of $T$. *Grafting* a leaf $w$ consists of subdividing an edge $(v, v')$ of $T$, thereby creating a new node $v''$ between $v$ and $v'$, then adding a leaf $w$ with parent $v''$. If $W$ is a rooted tree, *grafting $W$ to $T$* corresponds to grafting a leaf $w$, then replacing $w$ by the root of $W$.

The *lowest common ancestor* (LCA) in $T$ of a subset $L'$ of $\mathcal{L}(T)$, denoted $lca_T(L')$, is the ancestor common to all nodes in $L'$ that is the most distant from the root. The restriction $T|_{L'}$ of $T$ to $L'$ is the tree with leafset $L'$ obtained from the subtree of $T$ rooted at $lca_T(L')$ by removing all leaves that are not in $L'$ and all unary nodes. Let $T'$ be a tree such that $\mathcal{L}(T') = L' \subseteq \mathcal{L}(T)$. We say that $T$ *displays* $T'$ iff $T|_{L'}$ is label-isomorphic to $T'$ (i.e., isomorphic with preservation of leaf labels). We also say that $T$ is an *extension* of $T'$.

*Species, Gene and Synteny Trees:* (See Fig. 1) The *species tree* $S$ for a set $\Sigma$ of species represents an ordered set of speciation events that have led to $\Sigma$.

A *gene family* is a set $\Gamma$ of genes where each gene $g$ belongs to a given species $s(g)$ of $\Sigma$. If $\Gamma' \subseteq \Gamma$ is a subset of genes, we denote $s(\Gamma') = \{s(g) : g \in \Gamma'\}$.

A *synteny* is an ordered sequence of genes. We consider that genes of a synteny all belong to different gene families (tandem duplications are ignored). More precisely, let $\mathcal{F} = \{\Gamma_1, \Gamma_2, ..., \Gamma_t\}$ be a set of gene families, and $\lambda_{\mathcal{F}} = \{(g, \Gamma) : g \in \Gamma \wedge \Gamma \in \mathcal{F}\}$ be a function. We say that an ordered sequence of genes $X = g_1 g_2 ... g_k$ is a synteny on $\mathcal{F}$ iff $\lambda_{\mathcal{F}}$ is well-defined for all genes of $X$, $\lambda_{\mathcal{F}}$ is injective, and all genes in $X$ belong to the same species. If $X$ is a synteny, then $s(X)$ simply denotes the genome containing $X$.

A *synteny family* is a set $\mathcal{X}$ of syntenies. We say that a set $\mathcal{F}$ of gene families are *organized into a set $\mathcal{X}$ of syntenies* iff there is a bijection between the genes of $\mathcal{F}$ and the genes in $\mathcal{X}$ (each gene of $\mathcal{F}$ belongs to exactly one synteny of $\mathcal{X}$).

A tree $T$ is a *gene tree* for a gene family $\Gamma$ (respec. a *synteny tree* for a synteny family $\mathcal{X}$) if its leafset is in bijection with $\Gamma$ (respec. $\mathcal{X}$).

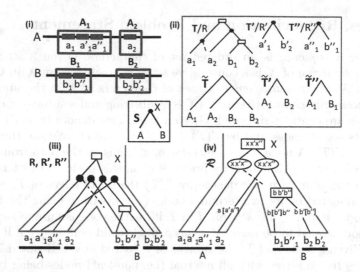

**Fig. 1.** (i) Two genomes $A$ and $B$; three gene families (red, green and blue) grouped into two syntenies $A_1, A_2$ in $A$ and two syntenies $B_1, B_2$ in $B$. (ii) Ignoring node labels and dotted lines, $T$, $T'$ and $T''$ are the corresponding gene trees and $\tilde{T}$, $\tilde{T}'$ and $\tilde{T}''$ are the corresponding synteny trees. The reconciled gene trees $R$, $R'$ and $R''$ are the same trees but including node labels and dotted lines. Nodes identified by circles are speciations, those represented by rectangles are duplications, and dotted lines represent lost branches. (iii) The reconciled trees embedded in the species tree $S$. (iv) A Super-Reconciliation $\mathcal{R}$, representing a more realistic evolutionary history from a common ancestral synteny. Each ancestral node is identified by the synteny, the event and the segment of the synteny affected by the event. Square nodes represent $Dup$ events, round nodes $Spe$ events, brackets $pLoss$ events and dotted lines $fLoss$ (see text). (Color figure online)

Given a gene tree $T$, the *corresponding synteny tree* is the tree $\tilde{T}$ obtained from $T$ by replacing each leaf of $T$ by the synteny containing the considered genes.

Given a tree $T$ (either gene tree or synteny tree), we extend the mapping $s$ to internal nodes $v$ of $T$ by defining $s(v) = lca_S(\{s(l) : l \in \mathcal{L}(T[v])\})$.

An evolutionary history is represented by a *labeled* tree, where the label of a node is its corresponding event. In the case of gene families, an event is entirely determined by its type, either a duplication, a speciation or a loss. The labels of a gene tree are obtained through reconciliation, as described below.

## 2.1   Reconciliation

**Definition 1 (Reconciled gene tree).** *Let $T$ be a binary gene tree and $S$ be a binary species tree. A DL reconciliation (or simply reconciliation) $R(T, S)$ of $T$ with $S$ is a labeled extension of $T$ obtained by grafting new leaves satisfying: for each internal node $v$ of $R(T, S)$ with two children $v_l$ and $v_r$, either $s(v_l) =$*

$s(v_r) = s(v)$, *or* $s(v_l)$ *and* $s(v_r)$ *are the two children of* $s(v)$. *The node* $v$ *is a duplication in* $s(v)$ *in the former case and a speciation in the latter case. A grafted leaf on a newly created node* $v$ *corresponds to a loss in* $s(v)$. *All other leaves are labeled by the default event "extant".*

*The cost of a reconciliation* $R(T, S)$ *is the number of induced duplications and losses.*

Given a gene tree $T$ and a species tree $S$, a *minimum reconciliation*, i.e. a reconciliation of minimum cost, is obtained from the LCA-mapping which consists in setting $s(v) = lca_S(s(\mathcal{L}(T[v])))$ for each $v \in V(T)$, and labeling each internal node $v$ of $T$ as a speciation if and only if $s(v_l)$ and $s(v_r)$ are separated in $S$, and as a duplication otherwise. Observe that in any case, if $s(v_l)$ and $s(v_r)$ are not separated, then it is impossible for $v$ to be a speciation. We denote by *LCA-reconciliation* the reconciliation labeled by means of the LCA-mapping.

Before extending the reconciliation concept to a set of gene trees, we need to specify an evolutionary model for syntenies. In this paper, syntenies are considered to have evolved from a single ancestral synteny through speciation (defined as for single genes), segmental duplication and segmental loss, where:

- a speciation $Spe(X, [1, l])$ acting on a synteny $X = g_1 \cdots g_l$ belonging to a genome $s(X)$ has the effect of reproducing $X$ in the two genomes $s_l$ and $s_r$ children of $s(X)$ in $S$.
- a (segmental) duplication $Dup(X, [i, j])$ acting on a synteny $X$ belonging to a genome $s(X)$ is an operation that copies a substring $g_i \cdots g_j$ of size $j - i + 1$ of $X = g_1 g_2 \cdots g_i \cdots g_j \cdots g_l$ somewhere else into the genome $s(X)$, creating a new *copied synteny* $X' = g'_i \cdots g'_j$ where each $g'_k$, for $i \le k \le j$ belongs to the same gene family as $g_k$;
- a (segmental) loss $Loss(X, [i, j])$ acting on a synteny $X = g_1 \cdots g_i \cdots g_j \cdots g_l$ is an operation that removes a substring $g_i \cdots g_j$ of size $j - i + 1$ of $X$, leading to the *truncated synteny* $X' = g_1 \cdots g_{i-1} g_{j+1} \cdots g_l$. A loss is called *full* if $X'$ is the empty string (i.e. all genes of $X$ are removed) and *partial* otherwise. We may denote full loss events as $fLoss$ and partial loss events as $pLoss$.

An evolutionary history of a set of syntenies can thus be represented as a partially binary tree where leaves correspond to extant syntenies and lost syntenies (resulting from full losses), and each internal node $v$ corresponds to an event $\mathcal{E}(X, [i, j])$ with $\mathcal{E} \in \{Spe, Dup, pLoss\}$ (and leaves correspond to either extant genes or $fLoss$ events). Thus, in contrast to a single gene family, a tree representing the evolution of a set of syntenies is not only labeled by the type of event corresponding to each internal node, but also by the segment of the synteny affected by the event (see the bottom-right tree in Fig. 1). If $\mathcal{E}$ is:

1. $Spe$, then $v$ is a binary node with two children corresponding to syntenies $Y$ and $Z$ such that $X = Y = Z$ and $s(Y)$ and $s(Z)$ being the two children of $s(X)$ in $S$.
2. $Dup$, then $v$ is a binary node with two children corresponding to syntenies $X$ and $X' = X[i, j]$, where $s(X) = s(X')$.

3. *pLoss*, then $v$ is a unary node with a child corresponding to the truncated synteny $X' = X[1, i - 1]X[j + 1, l]$, and $s(X) = s(X')$.

The topology of a tree representing the evolution of a set of syntenies differs from that of a single gene family since the former may contain unary nodes, resulting from partial losses, while the latter only contains binary nodes.

Our goal is to infer an evolutionary history of a set of syntenies which is a reconciliation of a set of individual gene trees, formally defined below.

**Definition 2 (Super-Reconciliation).** *Let $\mathcal{G} = \{T_1, T_2, \cdots, T_n\}$ be a set of binary gene trees for the gene families $\mathcal{F} = \{\Gamma_1, \Gamma_2, \cdots, \Gamma_t\}$ organized into a set $\mathcal{X}$ of syntenies belonging to a set $\Sigma$ of taxa, and let $S$ be a binary species tree for $\Sigma$. For each $i$, $1 \leq i \leq n$, let $\tilde{T}_i$ be the synteny tree corresponding to $T_i$.*

*A Super-Reconciliation $R(\mathcal{G}, S)$ of $\mathcal{G}$ with $S$ is a labeled synteny tree which is an extension of the trees $\tilde{T}_i$, for $1 \leq i \leq n$, representing a valid history for $\mathcal{X}$.*

*The cost of a Super-Reconciliation $R(\mathcal{G}, S)$ is the number of induced Dup, fLoss and pLoss events.*

For example, the cost of the Super-Reconciliation in Fig. 1 is 6. Notice that, although this cost is higher than that obtained by considering each gene family independently (cost of 3), the induced history is much more realistic as it is unlikely that independent gene duplications would have led to the same gene organization in different genomic regions.

We are now ready to state the optimization problem considered in this paper.

SUPER-RECONCILIATION problem:
**Input:** A set $\Sigma$ of species and a species tree $S$ for $\Sigma$; a set of gene families $\mathcal{F} = \{\Gamma_1, \Gamma_2, \cdots, \Gamma_t\}$ organized into a set of syntenies $\mathcal{X}$; a set of gene trees $\mathcal{G} = \{T_1, T_2 \cdots, T_t\}$ one for each family of $\mathcal{F}$.
**Output:** A Super-Reconciliation $R(\mathcal{G}, S)$ of minimum cost.

## 3    Existence Conditions

As a synteny is represented by a gene order and can only be modified through losses (duplications create new syntenies but do not modify existing syntenies), an evolutionary history does not always exist for a set of syntenies $\mathcal{X}$, regardless of the trees linking them. If this holds, the syntenies are said to be *order consistent*. Due to space constraints, we leave the details on order consistency constraints in the supplementary material.

In addition, in contrast to the reconciliation of a single gene tree which always exists, this is not the case for a Super-Reconciliation as different gene trees may exhibit inconsistent speciation histories for the same syntenies. A set of trees on subsets of $\mathcal{X}$ is said *consistent* iff, for any triplet $Trp = \{X_1, X_2, X_3\}$ of disjoint elements of $\mathcal{X}$, all trees containing $Trp$ as a sub-leafset exhibit the same topology for $Trp$.

**Lemma 1 (Tree consistency condition).** *Let $\mathcal{G} = \{T_1, T_2, \cdots, T_t\}$ be a set of gene trees for a set of gene families organized into a set $\mathcal{X}$ of syntenies, and let $S$ be the species tree. If a Super-Reconciliation $R(\mathcal{G}, S)$ exists, then the set of corresponding synteny trees $\{\tilde{T}_1, \tilde{T}_2, \cdots \tilde{T}_t\}$ is consistent.*

*Proof.* By definition, a Super-Reconciliation $R(\mathcal{G}, S)$ displays $\tilde{T}_i$, for all $1 \leq i \leq t$, as $R(\mathcal{G}, S)$ is an extension of each tree. Thus, for any triplet $Trp = \{X_1, X_2, X_3\}$ of $\mathcal{X}$, if $\tilde{T}_i$ and $\tilde{T}_j$ contain the triplet $Trp$ as a sub-leafset, then $R(\mathcal{G}, S)$ displays both $\tilde{T}_i|_{Trp}$ and $\tilde{T}_j|_{Trp}$. In other words, $\tilde{T}_i|_{Trp}$ and $\tilde{T}_j|_{Trp}$ are label-isomorphic. □

The consistency problem of rooted trees has been widely studied. The BUILD algorithm [2] can be used to test, in polynomial-time, whether a collection of rooted trees is consistent, and if so, construct a compatible, not necessarily fully resolved, supertree, i.e. a tree displaying them all. This algorithm has been generalized to output all compatible minimally resolved supertrees [8,22,25], which may be exponential in the number of genes.

The following theorem makes the link between a supertree and a reconciliation.

**Theorem 1.** *Let $\mathcal{G} = \{T_1, T_2 \cdots, T_t\}$ be a set of trees for a set of families organized in an order consistent set of syntenies $\mathcal{X}$, and $S$ be the species tree. Let $\tilde{\mathcal{G}} = \{\tilde{T}_1, \tilde{T}_2 \cdots, \tilde{T}_t\}$ be the set of synteny trees corresponding to those in $\mathcal{G}$. If $\tilde{\mathcal{G}}$ is a consistent set of trees then:*

1. *A Super-Reconciliation $R(\mathcal{G}, S)$ is an extension of a supertree for $\tilde{\mathcal{G}}$;*
2. *Any supertree is the "backbone" of a Super-Reconciliation. Namely, for any supertree $\tilde{T}$ for $\tilde{\mathcal{G}}$, there is a Super-Reconciliation $R(\mathcal{G}, S)$ which is an extension of $\tilde{T}$.*

The first statement of Theorem 1 follows from Lemma 1. As for the second statement, we will prove it implicitly in Sect. 5 by providing an algorithm that yields a minimum cost reconciliation on any supertree.

Following Theorem 1, the problem reduces to finding a supertree for the set of synteny trees minimizing the number of segmental duplications and losses. A natural algorithm for the SUPER-RECONCILIATION problem follows:

1. Explore the space of all order consistent ancestral syntenies $A$ for $\mathcal{X}$;
2. Explore the space of all supertrees $\tilde{T}$ for $\tilde{\mathcal{G}}$;
3. Find a Super-Reconciliation of minimum cost which is an extension of $\tilde{T}$ with $A$ as an ancestral synteny;
4. Select the Super-Reconciliations leading to the minimum cost.

Step 1 is discussed in Supplementary material and Step 2 has been discussed in this section. Before developing an algorithm for Step 3, which is the purpose of Sect. 5, we begin by analyzing the theoretical complexity of the SUPER-RECONCILIATION problem.

# 4   Complexity of the Super-Reconciliation Problem

We have recently considered the problem of finding a supertree of a set of gene trees minimizing the classical single gene duplication and single gene duplication and loss distances. The problem has been shown NP-hard for the duplication distance, and exponential-time algorithms have been developed for both distances. For segmental duplications only, the hardness of SUPER-RECONCILIATION is almost immediate from the results of [20]. For both duplications and losses, the problem remains NP-hard, although the proof is far more technical. Here we give the simpler proof of hardness for minimizing duplications only, and refer the reader to the Supplementary material for the NP-hardness proof for minimizing segmental duplications *and* losses.

**Theorem 2.** *The* SUPER-RECONCILIATION *problem is NP-hard for the duplication cost. Furthermore, the minimum number of duplications is hard to approximate within a factor $n^{1-\epsilon}$ for any $0 < \epsilon < 1$, where n is the number of syntenies in the input.*

*Proof.* The hardness follows from that of the MINDUP-SUPERTREE problem, defined as follows. Given a species tree $S$ and a set of gene trees $T_1, \ldots, T_k$, possibly with overlapping leafsets, MINDUP-SUPERTREE asks for a supertree $T$ that displays $T_1, \ldots, T_k$ such that the LCA-reconciliation of $T$ and $S$ yields a minimum number $d$ of duplications. It was shown in [20] that it is NP-hard to approximate $d$ within a factor $n^{1-\epsilon}$ for any $0 < \epsilon < 1$, where here $n$ is the number of genes in $\Gamma = \bigcup_{i=1}^{k} \mathcal{L}(T_i)$.

To reduce MINDUP-SUPERTREE to the SUPER-RECONCILIATION problem, it essentially suffices to exchange the roles of genes and syntenies. More precisely, given an instance of MINDUP-SUPERTREE consisting of a species tree $S$ and gene trees $T_1, \ldots, T_k$, we compute an instance of SUPER-RECONCILIATION as follows. The species tree is the same as $S$, and for each gene $g \in \Gamma$, we have a synteny $X_g$ with $s(X_g) = s(g)$. Moreover for each gene tree $T_i$, we create an identical gene tree $T_i'$, but in which each gene $g \in \mathcal{L}(T_i)$ is replaced by a unique gene $g_{T_i}$ that belongs to synteny $X_g$ (and hence $s(g) = s(g_{T_i}) = s(X_g)$). Thus the synteny tree $\tilde{T}_i$ for $T_i'$ is obtained by replacing each leaf $g$ of $T_i$ by $X_g$. In particular, there are $n$ syntenies. The order of the genes on the syntenies is arbitrary (since we are not counting segmental losses).

It only remains to show the correspondence between the solutions for the two problem instances. Suppose that the MINDUP-SUPERTREE instance admits a supertree $T$ with $d$ duplications when reconciled. Let $\tilde{T}$ be the synteny tree obtained from $T$ by replacing each gene $g \in \mathcal{L}(T)$ by $X_g$. Because $s(g) = s(X_g)$, both $T$ and $\tilde{T}$ have the same duplications under the LCA reconciliation, which is $d$. Conversely, if our SUPER-RECONCILIATION instance admits a synteny tree $\tilde{T}$ with $d$ duplications, replacing each leaf $X_g$ by $g$ yields a supertree for the MINDUP-SUPERTREE instance with $d$ duplications. Because the value of the solutions are preserved and $n = |\Gamma|$ corresponds to the number of syntenies, this reduction is approximation-preserving and the hardness result follows.    □

We state our second hardness result formally here.

**Theorem 3.** *The* SUPER-RECONCILIATION *problem is NP-hard for the Dup, fLoss and pLoss cost.*

## 5   A Super-Reconciliation for a Supertree

In this section, we are given a set $\mathcal{G} = \{T_1, T_2, \cdots, T_t\}$ of consistent gene trees for a set of families $\mathcal{F} = \{\Gamma_1, \Gamma_2, \cdots, \Gamma_t\}$ organized in an order consistent set of syntenies $\mathcal{X}$, and a species tree $S$ for the set $\Sigma$ of taxa containing the genes. In addition, we are given a supertree $\tilde{T}$ for the synteny trees $\tilde{\mathcal{G}} = \{\tilde{T}_1, \tilde{T}_2, \cdots, \tilde{T}_t\}$ corresponding to those in $\mathcal{G}$, and an order consistent ancestral synteny $A$ for $\mathcal{X}$.

Given a Super-Reconciliation $R(\mathcal{G}, S)$ ($R$ for short), because $R$ is obtained from $\tilde{T}$ by grafting leaves, each node of $\tilde{T}$ is present in $R$. Hence we say that $v \in V(\tilde{T})$ has a *corresponding node* $v'$ in $R$. More precisely, if $l \in \mathcal{L}(\tilde{T})$, then $l \in \mathcal{L}(R)$ also and the correspondence is immediate. If $v$ is an internal node of $V(\tilde{T})$, the node $v'$ of $R$ corresponding to $v$ is $lca_R(\{l : l \in \mathcal{L}(\tilde{T}[v])\})$. We show that, as in the traditional reconciliation setting, the nodes of $R$ that are also in $\tilde{T}$ should be mapped to the lowest species possible. To simplify the argument, we will call an internal node a full loss if it is the parent of a $fLoss$ event.

**Lemma 2.** *Let $R(\mathcal{G}, S)$ be a Super-Reconciliation of minimum cost which is an extension of $\tilde{T}$. Let $v \in V(\tilde{T})$ and let $v'$ be the node corresponding to $v$ in $R(\mathcal{G}, S)$. Then $s(v') = lca_S(s(\mathcal{L}(\tilde{T}[v])))$.*

*Proof.* First, observe that the statement is clearly true for the leaves. Assume that the statement is false. Now, let $v$ be a node of $\tilde{T}$ such that its corresponding node $v'$ does not satisfy the statement - moreover, choose $v$ to be a minimal node with this property (meaning that for the children $v_l$ and $v_r$ of $v$, the corresponding nodes $v_l'$ and $v_r'$ in $R(\mathcal{G}, S)$ satisfy $s(v_l') = lca_S(s(\mathcal{L}(\tilde{T}[v_l])))$ and $s(v_r') = lca_S(s(\mathcal{L}(\tilde{T}[v_r])))$). Note that $v$ must exist, since the statement is true for the leaves.

Now, we may assume that $s(v') \neq lca_S(s(v_l'), s(v_r'))$, as otherwise $v'$ satisfies the lemma. Thus in $S$, there are at least $k$ edges on the path from $s(v')$ to $lca_S(s(v_l'), s(v_r'))$, where here $k > 0$. It is not hard to verify that in this case, $v'$ must be a duplication node, according to the definition of a reconciliation. This implies that there are at least $k$ full losses on the path from $v'$ to $v_l'$ and at least $k$ full losses on the path from $v'$ to $v_r'$. Consider the Super-Reconciliation $R'$ that is identical to $R(\mathcal{G}, S)$, with the exception that $s(v') = lca_S(s(v_l'), s(v_r'))$. Then the $2k$ losses on the paths between $v'$ and $v_l'$ and between $v'$ and $v_r'$ are not needed anymore, although if $v'$ is not the root, $k$ losses become necessary on the path between $v'$ and $w'$, where $w'$ is the node corresponding to the parent $w$ of $v$ in $\tilde{T}$. Remapping $v'$ cannot increase the number of duplications, and so we have saved $k$ losses.

It remains to argue that the number of partial losses remains the same. But this is easy to see. We keep the same synteny assignment at nodes $v'$, $v_l'$ and $v_r'$

(and $w'$ if $v'$ is not the root) as in $R(\mathcal{G}, S)$. If $v'$ was a segmental duplication in $R(\mathcal{G}, S)$, we set $v'$ to be a segmental duplication in $R'$ as well. The number of partial losses on the paths between $v'$ and $v'_l, v'_r$ (and $w'$) therefore remains the same as in $R(\mathcal{G}, S)$.    □

We now show that speciation and duplication nodes are easy to identify. Essentially, we may set the events of internal nodes as in the classical LCA-mapping reconciliation. In what follows, assume that $\tilde{T}$ is reconciled under the LCA-mapping, and put $s(v) = lca_S(\mathcal{L}(s(\tilde{T}[v])))$ for every $v \in V(\tilde{T})$.

**Lemma 3.** *Let $R(\mathcal{G}, S)$ be a Super-Reconciliation of minimum cost which is an extension of $\tilde{T}$. Let $v \in V(\tilde{T})$ be an internal node of $\tilde{T}$ and let $v'$ be its corresponding node in $R(\mathcal{G}, S)$. Moreover let $v_l$ and $v_r$ be the children of $v$. If $s(v_l)$ and $s(v_r)$ are separated in $S$, then $v'$ is a speciation, and otherwise $v'$ is a duplication.*

*Proof.* Let $v'_l$ and $v'_r$ be the nodes corresponding to $v_l$ and $v_r$, respectively, in $R(\mathcal{G}, S)$. First, if $s(v_l)$ and $s(v_r)$ are not separated, then by Lemma 2, $s(v'_l)$ and $s(v'_r)$ are not separated, hence it is not possible for $v'$ to be a speciation. Therefore $v'$ must be a duplication.

Suppose instead that $s(v_l)$ and $s(v_r)$ are separated in $S$, but that $v'$ is labeled by a duplication event $Dup(X, [i, j])$, where $X$ is the synteny assigned at $v'$. On the path from $v'$ to $v'_l$, there may be some $pLoss$ events and some nodes that were grafted owing to full losses. We may assume that all full loss events, if any, have occurred before the $pLoss$ events on this path (i.e., nodes grafted from full losses are closer to $v'$). This is without loss of generality, as this does not change the resulting synteny in $v'_l$. We shall make the same assumption with the path from $v'$ to $v'_r$. Now, by Lemma 2, $s(v') = lca_S(s(v_l), s(v_r))$. Because $v'$ is a duplication, the two children $w_l, w_r$ of $v'$ in $R(\mathcal{G}, S)$ must satisfy $s(w'_l) = s(w'_r) = s(v')$. Since $s(v'_l) \neq s(v') \neq s(v'_r)$, we have that $\{w_l, w_r\} \cap \{v'_l, v'_r\} = \emptyset$, and therefore $w_l$ and $w_r$ were grafted on $\tilde{T}$ due to full losses. If we label $v'$ as a speciation $Spe(X, [1, |X|])$, these two full losses are not needed anymore, and by doing so we have one duplication less and two full losses less. Let $Y_l$ and $Y_r$ be the two syntenies that were assigned at $w_l$ and $w_r$ in $R(\mathcal{G}, S)$, respectively. Then $Y_l = X$ and $Y_r = X[i, j]$ or vice-versa (assume the former, without loss of generality). Suppose that $w_r$ was an ancestor of $v'_r$ in $R(\mathcal{G}, S)$, again without loss of generality. The substring $X[i, j]$ can be obtained from $X$ by adding at most two partial losses on the path from $v'$ to $v'_r$. The rest of the reconciliation can remain the same. To sum up, we have removed one duplication and two full losses, and inserted at most two partial losses to reproduce the effect of the segmental duplication. This contradicts that $R(\mathcal{G}, S)$ is a reconciliation of minimum cost.    □

From Lemma 3, it follows that we know the event-type (Dup or Spe) of each internal node of the supertree $\tilde{T}$. It then remains to extend the tree with losses and infer the actual event at each node (i.e., the corresponding synteny and segment being duplicated or lost). It is easy to see that losses and segments affected by the events are fully determined by gene orders assigned to internal nodes. Therefore, the problem reduces to the classical "small phylogeny problem" generally defined as follows: Given an alphabet $\Sigma$ (nucleotides or amino-acids or genes), a distance on the set of words of $\Sigma$ (edit distance for gene sequences or rearrangement distances for gene orders) and a tree $T$ with leaves being words on $\Sigma$ (extant gene sequences or gene orders), find the labeling of ancestral nodes (ancestral sequences or orders) minimizing the total cost of the tree. This cost is the sum of costs of each branch, which is the distance between the two words connected by the branch.

Here, we are given a synteny tree $\tilde{T}$ for a set $\mathcal{X}$ of syntenies on a set of gene families $\mathcal{F}$, and an ancestral synteny $A$ which is an order of $\mathcal{F}$. We want to find a *synteny assignment* attributing a partial order on $\mathcal{F}$ to each node of $V(\tilde{T})$. We assume that the root $r$ of $\tilde{T}$ is assigned the synteny $A$. It follows from the considered evolutionary model that, for two nodes $u$ and $v$ of $\tilde{T}$ with $u$ being an ancestor of $v$, the synteny $X_v$ assigned to $v$ should be a subsequence of the string $X_u$ assigned to $u$. A synteny assignment verifying this condition is called a *valid synteny assignment* for $\tilde{T}$.

For $v \in V(\tilde{T})$, define $d(v, X)$ as the minimum number of segmental duplications and losses induced by a synteny assignment on $\tilde{T}[v]$ with $X$ being the assignment at $v$. The problem SMALL-PHYLOGENY FOR SYNTENIES is to find an optimal assignment, i.e. an assignment leading to $d(\tilde{T}) = \min_X d(r(\tilde{T}), X)$ for $X$ belonging to the set of syntenies that are order consistent with $\mathcal{X}$.

Solving this problem can be done by dynamic programming by computing $d(v, X)$, for each $v \in V(\tilde{T})$ and each possible synteny $X$.

Let $v$ be an internal node of $\tilde{T}$ and $v_l$, $v_r$ be its two children. Let $X$, $X_l$, $X_r$ be valid assignments for respectively $v$, $v_l$ and $v_r$. Then $X_l$ and $X_r$ are subsequences of $X$. If $v$ is a speciation, then all missing genes in $X_l$ and $X_r$ are the result of losses. Otherwise, if $v$ is a duplication, then for at most one of $X_l$ and $X_r$, the missing prefix or suffix can be due to the partial duplication of a segment of $X$, and all other missing genes should be the result of losses. This motivates the following two variants of the loss distance between two syntenies.

Let $X$ and $Y$ be two syntenies with $Y$ being a subsequence of $X$. We let $D^T(X, Y)$ denote the minimum number of segmental losses required to transform $X$ to $Y$ and $D^P(X, Y)$ the minimum number of segmental losses required to transform a substring of $X$ to $Y$.

**Theorem 4.** *Let $v$ be a node of $\tilde{T}$, $X$ be a synteny and $\mathcal{S}(X)$ be the set of subsequences of $X$.*

- *If $v$ is a leaf, then $d(v, X) = 0$ if $X$ is the extant synteny corresponding to leaf $v$, and $+\infty$ otherwise;*
- *If $v$ is a speciation with children $v_l$ and $v_r$, then,*

$$d(v, X) = min_{(X_l \in \mathcal{S}(X))}(D^T(X, X_l) + d(v_l, X_l)) + \\ min_{(X_r \in \mathcal{S}(X))}(D^T(X, X_r) + d(v_r, X_r));$$

– *If $v$ is a duplication node with children $v_l$ and $v_r$, then*

$$d(v, X) = 1+$$

$$min \begin{cases} min_{(X_l \in S(X))}(D^T(X, X_l) + d(v_l, X_l))+ \\ \quad min_{(X_r \in S(X))}(D^T(X, X_r) + d(v_r, X_r)), \\ min_{(X_l \in S(X))}(D^T(X, X_l) + d(v_l, X_l))+ \\ \quad min_{(X_r \in S(X))}(D^P(X, X_r) + d(v_r, X_r)), \\ min_{(X_l \in S(X))}(D^P(X, X_l) + d(v_l, X_l))+ \\ \quad min_{(X_r \in S(X))}(D^T(X, X_r) + d(v_r, X_r)) \end{cases}$$

The above can be used to solve the SMALL-PHYLOGENY FOR SYNTENIES problem with dynamic programming. To do this, one can simply traverse $\tilde{T}$ in post-order, and apply the recurrences of Theorem 4 at each node encountered. We finish this section by analyzing the complexity of this algorithm. Let $n = |V(\tilde{T})|$ and let $t$ be the number of gene families involved in the SMALL-PHYLOGENY FOR SYNTENIES problem instance. For a node $v \in V(\tilde{T})$ and a synteny $X$, there are $O(2^t)$ possible subsequences of $X$. The value of $d(v, X)$ thus depends on the $O(2^t)$ values for its left child $v_l$ and the $O(2^t)$ values for its right child $v_r$. If these are known, then $d(v, X)$ can be computed in time $O(t2^t)$ (it is straightforward to check that $D^T$ and $D^P$ can be computed in time $O(t)$).

Let us now consider the number of possible entries in our dynamic programming table. The possible syntenies for $X$ correspond to the subsequences of a topological sorting of an acyclic directed graph with $t$ nodes (see supplementary material). In the worst case, there are $O(2^t \cdot t!) = O(2^{t \log t + t})$ such syntenies. It follows that there are at most $O(n2^{t \log t + t})$ entries in the dynamic programming table, and each entry takes time $O(t2^t)$. It is known that if there are $k$ possible topological sortings in a directed acyclic graph, then they can be enumerated in time $O(k)$ [24] (it is worth noting however that counting the number of such topological sortings in #P-complete [7]). Therefore, if $t$ is not too large, then the above recurrences can solve the small phylogeny problem relatively quickly, even if $n$ is large. Put differently, the SMALL-PHYLOGENY FOR SYNTENIES problem is *fixed-parameter tractable* with respect to parameter $t$.

**Corollary 1.** *The* SMALL-PHYLOGENY FOR SYNTENIES *problem can be solved in time $O(t2^{t \log t + 2t}n)$, where $t$ is the number of gene families present in the input and $n$ is the number of syntenies.*

## 6    Application

### 6.1    Simulated Datasets

The dynamic programming algorithm has been implemented in C++ [1] and tested on balanced trees obtained from simulated evolutionary histories. Simulations have been performed according to five parameters: $t$, the number of gene

---

[1] The program and simulations are available at:
https://github.com/UdeM-LBIT/SuperReconciliation.

families in the ancestral synteny; $d$, the maximum depth of the balanced tree; $p_{dupl}$, the probability for any given node to be a segmental duplication; $p_{loss}$, the probability for a loss to occur under any given node; and $p_{length}$, the probability to remove one gene in a segmental loss, defining the probability for a loss to remove $k$ genes (for $k \in \{1, 2, 3, ..\}$): $P(X = k) = (1 - p_{length})^{k-1}p_{length}$, following a shifted geometric distribution.

Simulations yield Super-Reconciliations leading to fully labelled trees. The input of the Super-Reconciliation algorithm is then obtained from those trees by removing loss nodes and synteny information on the internal, non-root nodes.

From an accuracy point of view (results not shown), as expected the larger the density of duplication and loss events, the further is the simulated history from a most parsimonious history, and thus from the inferred tree.

As for time-efficiency, values for inferring the Super-Reconciliation of a single tree, aggregated over 500 simulations per value of $t$, the size of the ancestral synteny (number of gene families), are given in Fig. 2. Computations have been done on the "Cedar" cluster of Compute Canada with 32 *Intel 8160* CPUs operating at 2.10 GHz. As expected, running time exponentially increases with respect to parameter $t$. This prevented us from extending the simulations beyond an ancestral synteny of size 14, for which the Super-Reconciliation of a single tree of depth 5 required around 15 min. However, if the synteny size remains fixed, running times increase polynomially with the size of the trees. As shown by the right diagram of Fig. 2, for an ancestral synteny of size 5, simulations exhibit a running time of no more than few seconds for trees with depth up to 15, representing balanced trees with up to $2^{15}$ leaves.

With real biological datasets, we are more likely to have to deal with large gene families rather than large sets of gene families evolving in concert. Thus, the increase in running time according to the size of the ancestral synteny is unlikely to be a bottleneck towards applying our Super-Reconciliation algorithm.

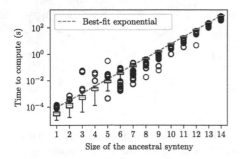 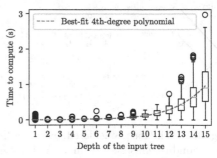

**Fig. 2.** Time-efficiency of the algorithm with respect to the size of the ancestral synteny (for $d = 5$) and the depth of the input tree (for $t = 5$), for $p_{dupl} = p_{loss} = p_{length} = 0.5$. Note that the leftmost graph uses a logarithmic scale.

## 6.2   The Opioid System

The opioid receptors, important regulators of neurotransmission and reward mechanisms in mammals, offer an interesting proof of concept, as these genes are present in clusters with conserved synteny in vertebrate genomes. Three genes for the opioid receptors (OPR) were identified and named OPRD1 (delta), OPRK1 (kappa) and OPRM1 (mu). A fourth gene was later found (OPRL1) in rodents and human. In human, they are located on the chromosomes 1, 6, 8 and 20.

Previous studies have considered the duplication scenario explaining the evolution of the opioid receptor genes [11,27,28]. The main question was whether observed paralogons arose from the two whole genome duplication events, often called 1R and 2R, known to have occurred early in vertebrate evolution. By exploring regions surrounding the OPR genes in human, four syntenic regions, containing genes from three other families (NKAIN, SRC-B and STMN) apparently sharing a common history, were identified. From the analysis of individual gene trees (Neighbor-joining and quartet-puzzling maximum likelihood trees), conclusions associating the evolution of the opioid system related genes to the 1R and 2R events were drawn.

Here, we consider the same four gene families OPR, NKAIN, STMN, and SRC-B, and further extend the OPR family with two neuropeptide NPBWR receptors, known to be closely related to the opioid receptors (Fig. 3.(i)). Protein sequences and gene orders were downloaded from the Ensembl database (Release

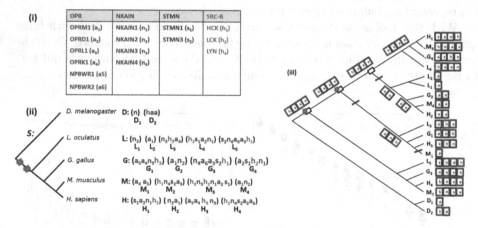

**Fig. 3.** (i) The four considered gene families. (ii) The considered species tree with the corresponding clusters: 19 in total involving 24 genes from the OPR family (genes named 'a'), 17 from the NKAIN family (named 'n'), 7 from the STMN family (named 's') and 13 from the SRC-B family (named 'h'). (iii) The Super-Reconciliation obtained form individual gene trees (not shown), and the induced duplication and loss history. Losses are indicated by red bars on the considered edges and duplications by rectangles. Yellow stars indicate the location of the 1R and 2R whole genome duplication events. Gene orders after removing duplicates (see text) are indicated on leaves, and chosen gene orders for internal nodes are shown.

$92)^2$ for the following five species: *Homo sapiens*, *Mus musculus*, *Gallus gallus*, *Lepisosteus oculatus* (spotted gar) and *Drosophila melanogaster*. Gene orders are given in Fig. 3.(ii).

For each gene family, we built a multiple sequence alignment with ClustalW [18] (Gonnet weight matrix and gap opening and extension penalties respectively set to 10 and 0.2). Maximum likelihood gene trees were subsequently constructed for each family using MEGA7 [19] (Jones-Taylor-Thornton substitution matrix and uniform rates among sites). As some syntenies contained paralogs (multiple copies from the same gene family, for example synteny $H_3$ contains two '$a$'), duplicates were removed in a way maximizing gene tree consistency. Although gene trees were still inconsistent, the overall clustering of gene copies was preserved among gene trees, and consistency could be attained after some local adjustments, using the species tree as reference.

The obtained Super-Reconciliation is given in Fig. 3.(iii). Notice however that gene orders are far from being consistent. In fact, all considered genomes are separated by a considerable evolutionary distance, and therefore, local rearrangements could have occurred along each lineage-specific branch. Choosing the $(h, s, a, n)$ order on every node of the tree and assuming rearrangements to occur at terminal edges, i.e. after duplication and loss events, leads to a history of three duplications and two losses before the speciation of bony fish and tetrapods, with two duplications correlating with the 1R and 2R tetraploidization events. This result is in agreement with previous studies on the opioid receptor genes [11].

Further analysis, using more genes and species, is required to provide a more detailed scenario for the evolution of the opioid receptor genes. Our objective here however, was not to verify a given hypothesis, but rather to provide a proof of concept and explore the applicability and limitations of the proposed reconciliation model on real data.

## 7   Conclusion

We have presented a natural extension of the DL reconciliation model, which is the first effort towards the development of a unifying automated method for reconciling a set of gene trees. It leads to a variety of problems requiring to be analysed from a complexity and algorithmic point of view.

In contrast with the inference of tandem duplications, where gene orders is a key information as created gene copies should be adjacent to the original ones, order is not a central information for the Super-Reconciliation problem. In fact, as chromosomal segments resulting from transposed duplications can be placed anywhere in the genome, gene order in syntenies is not a required information for the reconstruction of the supertree. However, labeling the supertree in a way minimizing the number of segmental duplications and losses still requires the knowledge of an ancestral gene order.

If, as we have considered in this paper, rearrangements are forbidden, then a duplication and loss history does not always exist for a set of syntenies, as the corresponding gene orders may be inconsistent. One solution would be to

---

[2] https://useast.ensembl.org/index.html.

minimally correct gene orders to ensure consistency, before applying the DL Super-Reconciliation model. Alternatively, an ancestral gene order can be inferred first, and all deviations from this order would be assumed to have occurred at terminal edges. As it clearly appears from the opioid receptor genes example, rearrangements could hardly be ignored.

A future extension of this work will be to minimize the segmental duplication and loss events explaining the evolution of a set of syntenies evolving through speciation and segmental duplication, loss, and rearrangements. In other words, we will infer gene orders leading to a most parsimonious history in terms of duplications and losses. The disagreement between the observed gene order at leaves and inferred orders can then simply be explained from rearrangements occurring after DL events. This is the approach we took to explain the supertree in Fig. 3. Other natural extensions of this work would be to account for the possibility of paralogous genes inside synteny blocks and expand the reconciliation model to horizontal gene transfers.

# References

1. Abbasi, A., Grzeschik, K.: An insight into the phylogenetic history of hox linked gene families in vertebrates. BMC Evol. Biol. **7**(1), 239 (2007)
2. Aho, A., Yehoshua, S., Szymanski, T., Ullman, J.: Inferring a tree from lowest common ancestors with an application to the optimization of relational expressions. SIAM J. Comput. **10**(3), 405–421 (1981)
3. Ajmal, W., Khan, H., Abbasi, A.: Phylogenetic investigation of human fgfr-bearing paralogons favors piecemeal duplication theory of vertebrate genome evolution. Mol. Phylogenet. Evol. **81**, 49–60 (2014)
4. Akerborg, O., Sennblad, B., Arvestad, L., Lagergren, J.: Simultaneous bayesian gene tree reconstruction and reconciliation analysis. Proc. Nat. Acad. Sci. USA **106**(14), 57145719 (2009)
5. Bansal, M., Alm, E., Kellis, M.: Efficient algorithms for the reconciliation problem with gene duplication, horizontal transfer and loss. Bioinformatics **28**(12), 283291 (2012). https://doi.org/10.1093/bioinformatics/bts225
6. Bérard, S., Gallien, C., Boussau, B., Szollosi, G., Daubin, V., Tannier, E.: Evolution of gene neighborhoods within reconciled phylogenies. Bioinformatics **28**(18), 382388 (2012)
7. Brightwell, G., Winkler, P.: Counting linear extensions. Order **8**(3), 225–242 (1991)
8. Constantinescu, M., Sankoff, D.: An efficient algorithm for supertrees. J. Classif. **12**, 101–112 (1995)
9. Dondi, R., Lafond, M., Scornavacca, C.: Reconciling multiple genes trees via segmental duplications and losses. WABI (2018, to appear)
10. Doyon, J., Ranwez, V., Daubin, V., Berry, V.: Models, algorithms and programs for phylogeny reconciliation. Briefings Bioinform. **12**(5), 392400 (2011)
11. Dreborg, S., Sundstrom, G., Larsson, T., Larhammar, D.: Evolution of vertebrate opioid receptors. Proc. Nat. Acad. Sci. USA **105**(40), 1548715492 (2008)
12. Duchemin, W.: Phylogeny of dependencies and dependencies of phylogenies in genes and genomes. Theses, Université de Lyon, Dec 2017. https://tel.archives-ouvertes.fr/tel-01779517
13. Ferrier, D.: Evolution of homeobox gene clusters in animals:the giga-cluster and primary vs. secondary clustering. Frontiers in Ecology and Evolution **4**(34) (2016)

14. Garcia-Fernàndez, J.: The genesis and evolution of homeobox gene clusters. Nat. Rev. Genet. **6**, 881892 (2005)
15. Goodman, M., Czelusniak, J., Moore, G., Romero-Herrera, A., Matsuda, G.: Fitting the gene lineage into its species lineage, a parsimony strategy illustrated by cladograms constructed from globin sequences. Syst. Zool. **28**, 132–163 (1979)
16. Hafeez, M., Shabbir, M., Altaf, F., Abbasi, A.: Phylogenomic analysis reveals ancient segmental duplications in the human genome. Mol. Phylogenet. Evol. **94**, 95–100 (2016)
17. Holyer, I.: The NP-completeness of edge-coloring. SIAM J. Comput. **10**(4), 718–720 (1981)
18. Thompson, J.D., Higgins, D., Gibson, T.: CLUSTAL W: improving the sensitivity of progressive multiple sequence alignment through sequence weighting, position-specific gap penalties and weight matrix choice. Nucleic Acids Res. **22**(22), 4673–4680 (1994)
19. Kumar, S., Stecher, G., Tamura, K.: Molecular evolutionary genetics analysis version 7.0 for bigger datasets. Mol. Biol. Evol. **33**(7), 18701874 (2016)
20. Lafond, M., Ouangraoua, A., El-Mabrouk, N.: Reconstructing a supergenetree minimizing reconciliation. BMC-Genomics 16, S4 (2015). special issue of RECOMB-CG 2015
21. Larsson, T., Olsson, F., Sundstrom, G., Lundin, L., Brenner, S., Venkatesh, B., Larhammar, D.: Early vertebrate chromosome duplications and the evolution of the neuropeptide y receptor gene regions. BMC Evol. Biol. **8**, 184 (2008)
22. Ng, M., Wormald, N.: Reconstruction of rooted trees from subtrees. Discrete Appl. Math **69**, 19–31 (1996)
23. Paszek, J., Gorecki, P.: Efficient algorithms for genomic duplication models. IEEE/ACM Trans. Comput. Biol. Bioinform. (2017)
24. Pruesse, G., Ruskey, F.: Generating linear extensions fast. SIAM J. Comput. **23**(2), 373–386 (1994)
25. Semple, C.: Reconstructing minimal rooted trees. Discrete Appl. Math. **127**(3), 489–503 (2003)
26. Sjöstrand, J., Tofigh, A., Daubin, V., Arvestad, L., Sennblad, B., Lagergren, J.: A bayesian method for analyzing lateral gene transfer. Syst. Biol. **63**(3), 409–420 (2014)
27. Stevens, C.: The evolution of vertebrate opioid receptors. Front. Biosci. J. Virtual Libr. **14**, 12471269 (2009)
28. Sundstrom, G., Dreborg, S., Larhammar, D.: Concomitant duplications of opioid peptide and receptor genes before the origin of jawed vertebrates. PLoS ONE **5**(5) (2010)
29. Szöllősi, G., Tannier, E., Daubin, V., Boussau, B.: The inference of gene trees with species trees. Syst. Biol. **64**(1), e42–e62 (2014)
30. Tofigh, A., Hallett, M., Lagergren, J.: Simultaneous identification of duplications and lateral gene transfers. IEEE/ACM Trans. Comput. BiolBioinform. **8**(2), 517–535 (2011). https://doi.org/10.1109/TCBB.2010.14
31. Duchemin, W., et al.: DeCoSTAR: Reconstructing the ancestral organization of genes or genomes using reconciled phylogenies. Genome Biol. Evol. **9**(5), 1312–1319 (2017)
32. Zhang, L.: On Mirkin-Muchnik-Smith conjecture for comparing molecular phylogenies. J. Comput. Biol. **4**, 177–188 (1997)
33. Zmasek, C.M., Eddy, S.R.: A simple algorithm to infer gene duplication and speciation events on a gene tree. Bioinformatics **17**, 821–828 (2001)

# On the Variance of Internode Distance Under the Multispecies Coalescent

Sébastien Roch(✉) [ID]

Department of Mathematics, University of Wisconsin–Madison,
Madison, WI 53706, USA
roch@math.wisc.edu

**Abstract.** We consider the problem of estimating species trees from
unrooted gene tree topologies in the presence of incomplete lineage sorting, a common phenomenon that creates gene tree heterogeneity in multilocus datasets. One popular class of reconstruction methods in this
setting is based on internode distances, i.e. the average graph distance
between pairs of species across gene trees. While statistical consistency in
the limit of large numbers of loci has been established in some cases, little
is known about the sample complexity of such methods. Here we make
progress on this question by deriving a lower bound on the worst-case
variance of internode distance which depends linearly on the corresponding graph distance in the species tree. We also discuss some algorithmic
implications.

**Keywords:** Species tree estimation · Incomplete lineage sorting
Internode distance · Sample complexity · ASTRID · NJst

## 1 Introduction

Species tree estimation is increasingly based on large numbers of loci or genes
across many species. Gene tree heterogeneity, i.e. the fact that different genomic
regions may be consistent with incongruent genealogical histories, is a common phenomenon in multilocus datasets that leads to significant challenges in
this type of estimation. One important source of incongruence is incomplete
lineage sorting (ILS), a population-genetic effect (see Fig. 1 below for an illustration), which is modeled mathematically using the multispecies coalescent
(MSC) process [14,19]. Many recent phylogenetic analyses of genome-scale biological datasets have indeed revealed substantial heterogeneity consistent with
ILS [3,6,27].

Standard methods for species tree estimation that do not take this heterogeneity into account, e.g. the concatenation of genes followed by a single-tree

This work was supported by funding from the U.S. National Science Foundation
DMS-1149312 (CAREER), DMS-1614242 and CCF-1740707 (TRIPODS). We thank
Tandy Warnow for suggesting the problem and for helpful discussions.

M. Blanchette and A. Ouangraoua (Eds.): RECOMB-CG 2018, LNBI 11183, pp. 196–206, 2018.
https://doi.org/10.1007/978-3-030-00834-5_11

maximum likelihood analysis, have been shown to suffer serious drawbacks under the MSC [20,23]. On the other hand, new methods have been developed for species tree estimation that specifically address gene tree heterogeneity. One popular class of methods, often referred to as summary methods, proceed in two steps: first reconstruct a gene tree for each locus; then infer a species tree from this collection of gene trees. Under the MSC, many of these methods have been proven to converge to the true species tree when the number of loci increases, i.e. the methods are said to be statistically consistent. Examples of summary methods that enable statistically consistent species tree estimation include MP-EST [12], NJst [11], ASTRID [26], ASTRAL [15,16], STEM [8], STEAC [13], STAR [13], and GLASS [17].

Here we focus on reconstruction methods, such as NJst and ASTRID, based on what is known as internode distances, i.e. the average of pairwise graph distances across genes. Beyond statistical consistency [1,7,11], little is known about the data requirement or sample complexity of such methods (unlike other methods such as ASTRAL [24] or GLASS [17] for instance). That is, how many genes or loci are needed to ensure that the true species tree is inferred with high probability under the MSC? Here we make progress on this question by deriving a lower bound on the worst-case variance of internode distance. Indeed the sample complexity of a reconstruction method depends closely on the variance of the quantities it estimates, in this case internode distances. Our bound depends linearly on the corresponding graph distance in the species tree which, as we explain below, has possible implications for the choice of an accurate reconstruction method.

The rest of the paper is structured as follows. In Sect. 2, we state our main results formally, after defining the MSC and the internode distance. In Sect. 3, we discuss algorithmic implications of our bound. Proofs can be found in Sect. 4.

## 2   Definitions and Results

In this section, we first introduce the multispecies coalescent. We also define the internode distance and state our results formally.

*Multilocus Evolution Under the Multispecies Coalescent.* Our analysis is based on the multispecies coalescent (MSC), a standard random gene tree model [14,19]. See Fig. 1 for an illustration. Consider a *species tree* $(\mathcal{S}, \Gamma)$ with $n$ leaves. Here $\mathcal{S} = (\mathcal{V}, \mathcal{E}, r)$ is a rooted binary tree with vertex and edge sets $\mathcal{V}$ and $\mathcal{E}$ and where each leaf is labeled by a species in $\{1, \ldots, n\}$. We refer to $\mathcal{S}$ as the *species tree topology*. The *branch lengths* $\Gamma = (\Gamma_e)_{e \in E}$ are expressed in so-called coalescent time units. We do not assume that $(\mathcal{S}, \Gamma)$ is ultrametric (see e.g. [25]). Each gene[1] $j = 1, \ldots, m$ has a genealogical history represented by its gene tree $\mathcal{T}_j$ distributed according to the following process: looking backwards in time, on each branch $e$ of

---

[1] In keeping with much of the literature on the MSC, we use the generic term *gene* to refer to any genomic region experiencing low rates of internal recombination, not necessarily a protein-coding region.

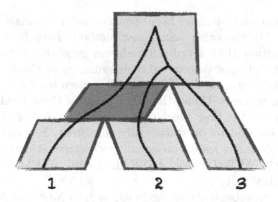

**1**      **2**      **3**

**Fig. 1.** An incomplete lineage sorting event (in the rooted setting). Although 1 and 2 are more closely related in the rooted species tree (fat tree), 2 and 3 are more closely related in the rooted gene tree (thin tree). This incongruence is caused by the failure of the lineages originating from 1 and 2 to coalesce within the shaded branch. The shorter this branch is, the more likely incongruence occurs.

the species tree, the coalescence of any two lineages is exponentially distributed with rate 1, independently from all other pairs; whenever two branches merge in the species tree, we also merge the lineages of the corresponding populations, that is, the coalescent proceeds on the union of the lineages; one individual is sampled at each leaf. The genes are assumed to be unlinked, i.e. the process above is run independently and identically for all $j = 1, \ldots, m$. More specifically, the probability density of a realization of this model for $m$ independent genes is

$$\prod_{j=1}^{m} \prod_{e \in E} \exp\left(-\binom{O_j^e}{2}\left[\gamma_j^{e,O_j^e+1} - \gamma_j^{e,O_j^e}\right]\right) \times \prod_{\ell=1}^{I_j^e - O_j^e} \exp\left(-\binom{\ell}{2}\left[\gamma_j^{e,\ell} - \gamma_j^{e,\ell-1}\right]\right),$$

where, for gene $j$ and branch $e$, $I_j^e$ is the number of lineages entering $e$, $O_j^e$ is the number of lineages exiting $e$, and $\gamma_j^{e,\ell}$ is the $\ell^{th}$ coalescence time in $e$; for convenience, we let $\gamma_j^{e,0}$ and $\gamma_j^{e,I_j^e - O_j^e + 1}$ be respectively the divergence times (expressed in coalescence time units) of $e$ and of its parent population (which depend on $\Gamma$). We write $\{\mathcal{T}_j\}_j \sim \mathcal{D}_s^m[\mathcal{S}, \Gamma]$ to indicate that the $m$ gene trees $\{\mathcal{T}_j\}_j$ are independently distributed according to the MSC on species tree $\mathcal{S}, \Gamma$. To be specific, $\mathcal{T}_j$ is the *unrooted gene tree topology*—without branch lengths— and we remark that, under the MSC, $\mathcal{T}_j$ is binary with probability 1. Throughout we assume that the $\mathcal{T}_j$'s are known and were reconstructed *without estimation error*.

*Internode Distance.* Assume we are given $m$ gene trees $\{\mathcal{T}_j\}_j$ over the $n$ species $\{1, \ldots, n\}$. For any pair of species $x, y$ and gene $j$, we let $d_g^j(x, y)$ be the *graph distance* between $x$ and $y$ on $\mathcal{T}_j$, i.e. the number of edges on the unique path between $x$ and $y$. The *internode distance* between $x$ and $y$ is defined as the average graph distance across genes, i.e.

$$\hat{\delta}_{\text{int}}^m(x, y) = \frac{1}{m} \sum_{j=1}^{m} d_{\text{g}}^{T_j}(x, y).$$

Under the MSC, the internode distances $(\hat{\delta}_{\text{int}}^m(x, y))_{x,y}$ are correlated random variables whose joint distribution depends in the a complex way on the species tree $(\mathcal{S}, \Gamma)$. Here follows a remarkable fact about internode distance [1,7,11]. Let $\bar{\delta}_{\text{int}}(x, y)$ be the expectation of $\hat{\delta}_{\text{int}}^m(x, y)$ under the MSC and let $\mathcal{S}_{\text{u}}$ be the unrooted version of the species tree $\mathcal{S}$. Then $(\bar{\delta}_{\text{int}}(x, y))_{x,y}$ is an additive metric associated[2] to $\mathcal{S}_{\text{u}}$ (see e.g. [25]). In particular, whenever $\mathcal{S}_{\text{u}}$ restricted to species $x, y, w, z$ has quartet topology $xy|wz$ (i.e. the middle edge of the restriction to $x, y, w, z$ splits $x, y$ from $w, z$), it holds that[3]

$$\bar{\delta}_{\text{int}}(x, w) + \bar{\delta}_{\text{int}}(y, z) = \bar{\delta}_{\text{int}}(x, z) + \bar{\delta}_{\text{int}}(y, w) \geq \bar{\delta}_{\text{int}}(x, y) + \bar{\delta}_{\text{int}}(w, z).$$

This result forms the basis for many popular multilocus reconstruction methods, including NJst [11] and ASTRID [26], which apply standard distance-based methods to the internode distances

$$(\hat{\delta}_{\text{int}}^m(x, y))_{x,y}.$$

*Main Results.* By the law of large numbers, for all pairs of species $x, y$

$$\hat{\delta}_{\text{int}}^m(x, y) \to \bar{\delta}_{\text{int}}(x, y),$$

with probability 1 as $m \to +\infty$, a fact that can be used to establish the statistical consistency (i.e. the guarantee that the true specie tree is recovered as long as $m$ is large enough) of internode distance-based methods such as NJst [11]. However, as far as we know, nothing is known about the sample complexity of internode distance-based methods, i.e. how many genes are needed to reconstruct the species tree with high probability—say 99%—as a function of some structural properties of the species tree—primarily the number of species $n$ and the shortest branch length $f$? We do not answer this important but technically difficult question here, but we make progress towards its resolution by providing a lower bound on the worst-case variance of internode distance. Let $d_{\text{g}}^{\mathcal{S}_{\text{u}}}(x, y)$ denote the graph distance between $x$ and $y$ on $\mathcal{S}_{\text{u}}$.

**Theorem 1 (Lower bound on the worst-case variance of internode distance).** *There exists a constant $C > 0$ such that, for any integer $n \geq 4$ and real $f > 0$, there is a species tree $(\mathcal{S}, \Gamma)$ with $n$ leaves and shortest branch length $f$*

---

[2] Note however that the associated branch lengths may differ from $\Gamma$.

[3] Note that it is trivial that $(d_{\text{g}}^{T_j}(x, y))_{x,y}$ is an additive metric associated to *gene tree* $T_j$. On the other hand it is far from trivial that averaging over the MSC leads to an additive metric associated to the *species tree*.

*such that the following holds: for all pairs of species $\ell, \ell'$ and all integers $m \geq 1$, if $\{T_j\}_j \sim \mathcal{D}_s^m[\mathcal{S}, \Gamma]$ then*

$$\mathbf{Var}\left[\hat{\delta}_{int}^m(\ell, \ell')\right] \geq C\frac{d_g^{\mathcal{S}_u}(\ell, \ell')}{m}, \tag{1}$$

*and, furthermore,*

$$\max_{\ell, \ell'} \mathbf{Var}\left[\hat{\delta}_{int}^m(\ell, \ell')\right] \geq C\frac{n}{m}, \tag{2}$$

In words, there are species trees for which the variance of internode distance scales as the graph distance—which can be of order $n$—divided by $m$. The proof of Theorem 1 is detailed in Sect. 4.

## 3    Discussion

How is Theorem 1 related to the sample complexity of species tree estimation methods? The natural approach for deriving bounds on the number of genes required for high-probability reconstruction in distance-based methods is to show that the estimated distances used are sufficiently concentrated around their expectations—provided that $m$ is large enough as a function of $n$ and $f$ (e.g. [2,9]; but see [22] for a more refined analysis). In particular, one needs to *control the variance* of distance estimates.

*Practical Implications.* Bound (2) in Theorem 1 implies that to make *all* variances negligible the number of genes $m$ is required to scale at least linearly in the number of species $n$. In contrast, certain quartet-based methods such as ASTRAL [15,16] have a sample complexity scaling only logarithmically in $n$ [24].

On the other hand, Bound (2) is only relevant for those reconstruction algorithms using *all* distances, for instance NJst which is based on Neighbor-Joining [2,10]. Many so-called fast-converging reconstruction methods purposely use only a *strict subset* of all distances, specifically those distances within a constant factor of the "depth" of the species tree. Refer to [9] for a formal definition of the depth, but for our purposes it will suffice to note that in the case of graph distance the depth is at most of the order of $\log n$. Hence Bound (1) suggests it may still possible to achieve a sample complexity comparable to that of ASTRAL—if one uses a fast-converging method (within ASTRID for instance).

*The Impact of Correlation.* Theorem 1 does not in fact lead to a bound on the sample complexity of internode distance-based reconstruction methods. For one, Theorem 1 only gives a lower bound on the variance. One may be able to construct examples where the variance is even larger. In general, analyzing the behavior of internode distance is quite challenging because it depends on the *full* multispecies coalescent process in a rather tangled manner.

Perhaps more importantly, the variance itself is not enough to obtain tight bounds on the sample complexity. One problem is *correlation*. Because $\hat{\delta}_{int}^m(x, y)$

and $\hat{\delta}_{\text{int}}^m(w, z)$ are obtained using the same gene trees, they are highly correlated random variables. One should expect this correlation to produce cancellations (e.g. in the four-point condition; see [25]) that could drastically lower the sample complexity. The importance of this effect remains to be studied.

*Gene Tree Estimation Error.* We pointed out above that quartet-based methods such as ASTRAL may be less sensitive to long distances than internode distance-based methods such as NJst. An important caveat is the assumption that gene trees are *perfectly reconstructed*. In reality, gene tree estimation errors are likely common and are also affected by long distances (see e.g. [9]). A more satisfactory approach would account for these errors or would consider simultaneously sequence-length and gene-number requirements. Few such analyses have so far been performed because of technical challenges [4,5,18,21].

# 4    Variance of Internode Distance

In this section, we prove Theorem 1. Our analysis of internode distance is based on the construction of a special species tree where its variance is easier to control. We begin with a high-level proof sketch:

- Our special example is a caterpillar tree with an alternation of *short* and *long* branches along the backbone.
- The *short* branches produce "local uncertainty" in the number of lineages that coalesce onto the path between two fixed leaves. The *long* branches make these contributions to the internode distance "roughly independent" along the backbone.
- As a result, the internode distance is, up to a small error, a sum of independent and identically distributed contributions. Hence, its variance grows linearly with graph distance.

*Setting for Analysis.* We fix the number of species $n$ and we assume for convenience that $n$ is even.[4] Recall also that $f$ will denote the length of the shortest branch in coalescent time units. We consider the species tree $(\mathcal{S}, \Gamma)$ depicted in Fig. 2. Specifically, $\mathcal{S}$ is a caterpillar tree: its *backbone* is an $n - 1$-edge path

$$(a, w_1), (w_1, z_1), (z_1, w_2), (w_2, z_2), \ldots, (w_{\frac{n-2}{2}}, z_{\frac{n-2}{2}}), (z_{\frac{n-2}{2}}, r)$$

connecting leaf $a$ to root $r = w_{n/2}$; each vertex $w_i$ on the backbone is incident with an edge $(w_i, x_i)$ to leaf $x_i$; each vertex $z_i$ on the backbone is incident with an edge $(z_i, y_i)$ to leaf $y_i$; root $r$ is incident with an edge $(r, b)$ to leaf $b$. Each edge of the form $e = (w_i, z_i)$ is a *short* edge of length $\Gamma_e = f$, while all other edges are *long* edges of length $g = 4 \log n$.

---

[4] A straightforward modification of the argument also works for odd $n$.

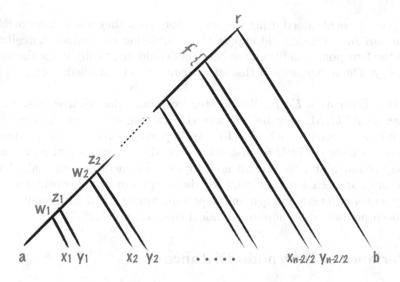

**Fig. 2.** The species tree used in the analysis.

*Proof of Theorem 1.* Recall that our goal is to prove that for all pairs of species $\ell, \ell'$ and all integers $m \geq 1$, if $\{T_j\}_j \sim \mathcal{D}_s^m[\mathcal{S}, \Gamma]$ then

$$\text{Var}\left[\hat{\delta}_{\text{int}}^m(\ell, \ell')\right] \geq C \frac{d_g^{\mathcal{S}_u}(\ell, \ell')}{m}.$$

To simplify the analysis, we detail the argument in the case $\ell = a$ and $\ell' = b$ only. The other cases follow similarly.

We first reduce the computation to a single gene. Recall that

$$\hat{\delta}_{\text{int}}^m(a, b) = \frac{1}{m} \sum_{j=1}^m d_g^{T_j}(a, b).$$

**Lemma 1 (Reduction to a single gene).** *For any $m$, it holds that*

$$\text{Var}\left[\hat{\delta}_{\text{int}}^m(\ell, \ell')\right] = \frac{1}{m} \text{Var}\left[d_g^{T_1}(a, b)\right].$$

*Proof.* Because the $T_j$'s are independent and identically distributed, it follows that

$$\text{Var}\left[\hat{\delta}_{\text{int}}^m(\ell, \ell')\right] = \text{Var}\left[\frac{1}{m} \sum_{j=1}^m d_g^{T_j}(a, b)\right] = \frac{1}{m^2} \sum_{j=1}^m \text{Var}\left[d_g^{T_j}(a, b)\right] = \frac{1}{m} \text{Var}\left[d_g^{T_1}(a, b)\right],$$

as claimed.

We refer to the 2-edge path $\{(w_i, z_i), (z_i, w_{i+1})\}$ as the *i-th block*. The purpose of the long backbone edges is to create independence between the contributions of the blocks. To make that explicit, let $\mathcal{F}_i$ be the event that, in $T_1$, all lineages entering the edge $(z_i, w_{i+1})$ have coalesced by the end of the edge (backwards in time). And let $\mathcal{F} = \cap_i \mathcal{F}_i$.

**Lemma 2 (Full coalescence on all blocks).** *It holds that*

$$\mathbf{P}[\mathcal{F}] \geq 1 - 1/n.$$

*Proof.* By the multiplication rule and the fact that $\mathcal{F}_i$ only depends on the number of lineages entering $(w_i, z_i)$, we have

$$\mathbf{P}[\mathcal{F}] = \prod_i \mathbf{P}[\mathcal{F}_i \mid \mathcal{F}_1 \cap \cdots \cap \mathcal{F}_{i-1}] = (\mathbf{P}[\mathcal{F}_1])^{n/2-1} \geq 1 - (n/2 - 1)(1 - \mathbf{P}[\mathcal{F}_1]).$$

It remains to upper bound $\mathbf{P}[\mathcal{F}_1^c]$. We have either 2 or 3 lineages entering $(z_1, w_2)$. In the former case, the failure to coalesce has probability $e^{-g}$, i.e. the probability that an exponential with rate 1 is greater than $g$. In the latter case, the failure to fully coalesce has probability at most $e^{-3(g/2)} + e^{-g/2}$, i.e. the probability that either the first coalescence (happening at rate 3) or the second one (happening at rate 1) takes more than $g/2$. Either way this gives at most $\mathbf{P}[\mathcal{F}_1^c] \leq 2e^{-g/2}$. With $g = 4 \log n = 2 \log n^2$ above, we get the claim.

We now control the contribution from each block. Let $X_i$ be the number of lineages coalescing into the path between $a$ and $b$ *on the $i$-th block*. Conditioning on $\mathcal{F}$, we have $X_i \in \{1, 2\}$ and we have further that all $X_i$'s are independent and identically distributed. This leads to the following bound.

**Lemma 3 (Linear variance).** *It holds that*

$$\mathbf{Var}\left[d_g^{T_1}(a, b)\right] \geq \frac{n-2}{2} \mathbf{Var}\left[X_1 \mid \mathcal{F}_1\right] \mathbf{P}[\mathcal{F}].$$

*Proof.* By the conditional variance formula, letting $\mathbb{1}_{\mathcal{F}}$ be the indicator of $\mathcal{F}$,

$$\mathbf{Var}\left[d_g^{T_1}(a, b)\right] \geq \mathbf{E}\left[\mathbf{Var}\left[d_g^{T_1}(a, b) \mid \mathbb{1}_{\mathcal{F}}\right]\right] \geq \mathbf{Var}\left[d_g^{T_1}(a, b) \mid \mathcal{F}\right] \mathbf{P}[\mathcal{F}].$$

On the event $\mathcal{F}$, it holds that

$$d_g^{T_1}(a, b) = \sum_i X_i.$$

Moreover, conditioning on $\mathcal{F}$ makes the $X_i$'s independent and identically distributed. Hence we have finally

$$\mathbf{Var}\left[d_g^{T_1}(a, b)\right] \geq \frac{n-2}{2} \mathbf{Var}\left[X_1 \mid \mathcal{F}\right] \mathbf{P}[\mathcal{F}] \geq \frac{n-2}{2} \mathbf{Var}\left[X_1 \mid \mathcal{F}_1\right] \mathbf{P}[\mathcal{F}],$$

where we used the fact that $X_1$ depends on $\mathcal{F}$ only through $\mathcal{F}_1$.

The final step is to bound the contribution to the variance from a single block.

**Lemma 4 (Contribution from a block).** *It holds that*

$$\mathbf{Var}\left[X_1 | \mathcal{F}_1\right] = \frac{1}{3}e^{-f}\left(1 - \frac{1}{3}e^{-f}\right) = \frac{2}{9}\left(1 - \Theta(f)\right),$$

*for f small, where we used the standard Big-Theta notation.*

*Proof.* As we pointed out earlier, conditioning on $\mathcal{F}_1$, we have $X_1 \in \{1, 2\}$. In particular $X_1 - 1$ is a Bernoulli random variable whose variance $\mathbf{P}\left[X_1 - 1 = 1 | \mathcal{F}_1\right](1 - \mathbf{P}\left[X_1 - 1 = 1 | \mathcal{F}_1\right])$ is the same as the variance of $X_1$ itself. So we need to compute the probability that $X_1 = 2$, conditioned on $\mathcal{F}_1$. There are four scenarios to consider (depending on whether or not there is coalescence in the short branch $(w_1, z_1)$ and which coalescence occurs first in the long branch $(z_1, w_2)$), only one of which produces $X_1 = 1$:

– No coalescence occurs in $(w_1, z_1)$ and the first coalescence in $(z_1, w_2)$ is between the lineages coming from $x_1$ and $y_1$. This event has probability $\frac{1}{3}e^{-f}$ by symmetry when conditioning on $\mathcal{F}_1$.

Hence $\mathbf{P}\left[X_1 = 2 | \mathcal{F}_1\right] = 1 - \frac{1}{3}e^{-f}$.

By combining Lemmas 1, 2, 3 and 4, we get that

$$\mathbf{Var}\left[\hat{\delta}_{\text{int}}^m(\ell, \ell')\right] \geq \frac{1}{m} \times \frac{n-2}{2} \times \frac{1}{3}e^{-f}\left(1 - \frac{1}{3}e^{-f}\right) \times \left(1 - \frac{1}{n}\right).$$

Choosing $C$ small enough concludes the proof of the theorem.

## 5    Conclusion

To summarize, we have derived a new lower bound on the worst-case variance of internode distance under the multispecies coalescent. No such bounds were previously known as far as we know. Our results suggest it may be preferable to use fast-converging methods when working with internode distances for species tree estimation. The problem of providing tight upper bounds on the sample complexity of internode distance-based methods remains however an important open question.

## References

1. Allman, E.S., Degnan, J.H., Rhodes, J.A.: Species tree inference from gene splits by unrooted star methods. IEEE/ACM Trans. Comput. Biol. Bioinform. **15**(1), 337–342 (2018)
2. Atteson, K.: The performance of neighbor-joining methods of phylogenetic reconstruction. Algorithmica **25**(2), 251–278 (1999)
3. Cannon, J.T., Vellutini, B.C., Smith, J., Ronquist, F., Jondelius, U., Hejnol, A.: Xenacoelomorpha is the sister group to nephrozoa. Nature **530**(7588), 89–93 (2016)

4. Dasarathy, G., Mossel, E., Nowak, R.D., Roch, S.: Coalescent-based species tree estimation: a stochastic farris transform. CoRR, abs/1707.04300 (2017)
5. Dasarathy, G., Nowak, R.D., Roch, S.: Data requirement for phylogenetic inference from multiple loci: a new distance method. IEEE/ACM Trans. Comput. Biology Bioinform. **12**(2), 422–432 (2015)
6. Jarvis, E., et al.: Whole-genome analyses resolve early branches in the tree of life of modern birds. Science **346**(6215), 1320–1331 (2014)
7. Kreidl, M.: Note on expected internode distances for gene trees in species trees (2011)
8. Kubatko, L.S., Carstens, B.C., Knowles, L.L.: STEM: species tree estimation using maximum likelihood for gene trees under coalescence. Bioinformatics **25**(7), 971–973 (2009)
9. Erdős, P.L., Steel, M.A., Székely, L.A., Warnow, T.J.: A few logs suffice to build (almost) all trees (i). Random Structures Algorithms **14**(2), 153–184 (1999)
10. Lacey, M.R., Chang, J.T.: A signal-to-noise analysis of phylogeny estimation by neighbor-joining: Insufficiency of polynomial length sequences. Math. Biosci. **199**(2), 188–215 (2006)
11. Liu, L., Yu, L.: Estimating species trees from unrooted gene trees. Syst. Biol. **60**(5), 661–667 (2011)
12. Liu, L., Yu, L., Edwards, S.V.: A maximum pseudo-likelihood approach for estimating species trees under the coalescent model. BMC Evol. Biol. **10**(1), 302 (2010)
13. Liu, L., Lili, Y., Pearl, D.K., Edwards, S.V.: Estimating species phylogenies using coalescence times among sequences. Syst. Biol. **58**(5), 468–477 (2009)
14. Maddison, W.P.: Gene trees in species trees. Syst. Biol. **46**(3), 523–536 (1997)
15. Mirarab, S., Reaz, R., Bayzid, M.S., Zimmermann, T., Swenson, M.S., Warnow, T.: ASTRAL: accurate species TRee ALgorithm. Bioinformatics **30**(17), i541–i548 (2014)
16. Mirarab, S., Warnow, T.: ASTRAL-II: coalescent-based species tree estimation with many hundreds of taxa and thousands of genes. Bioinformatics **31**(12), i44–i52 (2015)
17. Mossel, E., Roch, S.: Incomplete lineage sorting: consistent phylogeny estimation from multiple loci. IEEE/ACM Trans. Comput. Biol. Bioinform. **7**(1), 166–71 (2010)
18. Mossel, E., Roch, S.: Distance-based species tree estimation: Information-theoretic trade-off between number of loci and sequence length under the coalescent. In: Approximation, Randomization, and Combinatorial Optimization. Algorithms and Techniques, APPROX/RANDOM 2015, Princeton, NJ, USA, August 24–26, 2015, pp. 931–942 (2015)
19. Rannala, B., Yang, Z.: Bayes estimation of species divergence times and ancestral population sizes using DNA sequences from multiple loci. Genetics **164**(4), 1645–1656 (2003)
20. Roch, S., Steel, M.A.: Likelihood-based tree reconstruction on a concatenation of aligned sequence data sets can be statistically inconsistent. Theor. Popul. Biol. **100**, 56–62 (2015)
21. Roch, S., Warnow, T.: On the robustness to gene tree estimation error (or lack thereof) of coalescent-based species tree methods. Syst. Biol. **64**(4), 663–676 (2015)
22. Roch, S.: Toward extracting all phylogenetic information from matrices of evolutionary distances. Science **327**(5971), 1376–1379 (2010)
23. Roch, S., Nute, M., Warnow, T.J.: Long-branch attraction in species tree estimation: inconsistency of partitioned likelihood and topology-based summary methods. CoRR, abs/1803.02800 (2018)

24. Shekhar, S., Roch, S., Mirarab, S.: Species tree estimation using ASTRAL: how many genes are enough? IEEE/ACM Trans. Comput. Biol. Bioinform., 1 (2018)
25. Steel, M.: Phylogeny–discrete and random processes in evolution. In: CBMS-NSF Regional Conference Series in Applied Mathematics, vol. 89. Society for Industrial and Applied Mathematics (SIAM), Philadelphia (2016)
26. Vachaspati, P., Warnow, T.: ASTRID: accurate species TRees from internode distances. BMC Genomics **16**(Suppl 10), S3 (2015)
27. Wickett, N.J., et al.: Phylotranscriptomic analysis of the origin and early diversification of land plants. Proc. Natl. Acad. Sci. **111**(45), E4859–E4868 (2014)

# Phylogenetics

# Linear-Time Algorithms for Some Phylogenetic Tree Completion Problems Under Robinson-Foulds Distance

Mukul S. Bansal$^{(\boxtimes)}$

Department of Computer Science and Engineering and Institute for Systems
Genomics, University of Connecticut, Storrs, USA
mukul.bansal@uconn.edu

**Abstract.** We consider two fundamental computational problems that
arise when comparing phylogenetic trees, rooted or unrooted, with non-
identical leaf sets. The first problem arises when comparing two trees
where the leaf set of one tree is a proper subset of the other. The second
problem arises when the two trees to be compared have only partially
overlapping leaf sets. The traditional approach to handling these problems is to first restrict the two trees to their common leaf set. An alternative approach that has shown promise is to first *complete* the trees by
adding missing leaves, so that the resulting trees have identical leaf sets.
This requires the computation of an optimal completion that minimizes
the distance between the two resulting trees over all possible completions.

We provide optimal linear-time algorithms for both completion problems under the widely-used Robinson-Foulds (RF) distance measure. Our
algorithm for the first problem improves the time complexity of the current fastest algorithm from quadratic (in the size of the two trees) to
linear. No algorithms have yet been proposed for the more general second problem where both trees have missing leaves. We advance the study
of this general problem by proposing a biologically meaningful restricted
version of the general problem and providing optimal linear-time algorithms for the restricted version. Our experimental results on biological
data sets suggest that using completion-based RF distances can result in
different evolutionary inferences compared to traditional RF distances.

## 1 Introduction

A *phylogenetic tree*, or *phylogeny*, is a leaf-labeled tree that shows the evolutionary relationships between different biological entities, generally either species
or genes. Phylogenies may be either rooted or unrooted. The leaf nodes of a
phylogeny represent the extant set of entities on which the phylogeny is built,
while internal nodes represent hypothetical ancestors. The comparison of different phylogenetic trees is one of the most fundamental tasks in evolutionary
biology and computational phylogenetics. Many biologically relevant distance
or similarity measures have been defined in the literature for the case when

© Springer Nature Switzerland AG 2018
M. Blanchette and A. Ouangraoua (Eds.): RECOMB-CG 2018, LNBI 11183, pp. 209–226, 2018.
https://doi.org/10.1007/978-3-030-00834-5_12

210     M. S. Bansal

the two phylogenies to be compared have the same leaf set. These include
the widely used Robinson-Foulds distance [27], triplet and quartet distances
[13,19], nearest neighbor interchange (NNI) and subtree prune and regraft
(SPR) distances [20,30,33], maximum agreement subtrees [2,14,21], nodal distance [7], geodesic distance [23] and several others. Often, however, this comparison involves two trees that have non-identical leaf sets. The need to compare
trees that do not have identical leaf sets arises naturally in several situations: For
instance, algorithms for computing phylogenetic supertrees are typically based
on comparing input trees on partial leaf sets with candidate supertrees on the
complete leaf set [1,3,9,24,31]. Likewise, searching for phylogenies similar to a
query tree in a phylogenetic database [10,25,26,29], and clustering of phylogenetic trees [34] often involve comparisons between trees with only partially
overlapping leaf sets.

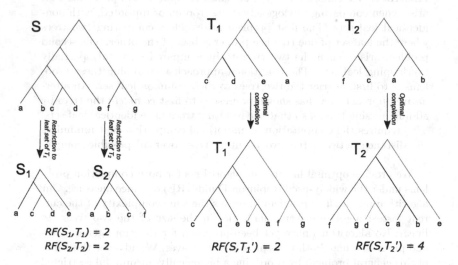

**Fig. 1. RF(-) and RF(+) distances.** This figure illustrates the difference between
the traditional (RF(-)) and RF(+) distance measures when applied to trees with partially overlapping leaf sets. In this example, the leaf sets of $T_1$ and $T_2$ are a subset
of the leaf set of $S$. To compute the RF(-) distance between $T_1$ and $S$, we must first
restrict $S$ to the leaf set of $T_1$, resulting in tree $S_1$. The RF(-) distance between $S$ and
$T_1$ is thus $RF(S_1, T_1)$, which is 2. Likewise, to compute the RF(-) distance between $T_2$
and $S$, we must first restrict $S$ to the leaf set of $T_2$, resulting in tree $S_2$. The RF(-)
distance between $S$ and $T_2$ is thus $RF(S_2, T_2)$, which is also 2. In contrast, to compute
the RF(+) distance between $T_1$ and $S$, we must first compute an optimal completion
of $T_1$ on the leaf set of $S$ (denoted by the dashed red lines), resulting in tree $T_1'$. The
RF(+) distance between $S$ and $T_1$ is thus $RF(S, T_1')$, which is 2. Likewise, to compute
the RF(+) distance between $T_2$ and $S$, we must first compute an optimal completion
of $T_2$ on the leaf set of $S$, resulting in tree $T_2'$. The RF(+) distance between $S$ and $T_2$
is thus $RF(S, T_2')$, which is 4. Observe that while both $T_1$ and $T_2$ are equidistant from
$S$ under RF(-), computing the RF(+) distances reveals that $T_1$ is more similar to $S$
than is $T_2$.

The traditional approach to comparing two phylogenies on non-identical leaf sets is to first restrict the two phylogenies to their common leaf set and then apply one of the distance or similarity measures that compare two trees on the same leaf set. However, an alternative, and perhaps more useful, approach to comparing trees with non-identical taxa is to *fill-in* or *complete* the two trees to be compared with the leaves missing from each, resulting in two trees on the same leaf set, and then apply the distance or similarity measure. This completion based approach is especially desirable when used with the Robinson-Foulds (RF) distance measure [27], the most commonly used distance measure in evolutionary biology. Indeed, several important biological applications would directly bene- fit from the use of this completion-based RF distance, such as the construction of majority-rule(+) supertrees [12,17,18,22], construction of Robinson-Foulds supertrees [3,9,28], phylogenetic database search [10,25,26,29], and clustering of phylogenetic trees [34]. To distinguish between the two methods for com- puting RF distance between two trees with non-identical leaf sets, we refer to the completion-based RF distance as RF(+) distance and to the traditional pruning-based RF distance as RF(−). Figure 1 shows an example of two trees with partially overlapping leaf sets and these two ways of computing the RF distance between them.

**Previous Work.** The idea of a completion-based RF(+) distance was proposed at least a decade ago. Cotton and Wilkinson were among the first to propose such a distance measure in their seminal paper describing majority-rule supertrees [12]. Specifically, they defined two types of majority-rule supertrees: majority-rule(− ) and majority-rule(+) supertrees. The majority-rule(-) supertrees were based on traditional RF(−) distances between trees, while majority-rule(+) supertrees were based on completion-based RF(+) distances. Majority-rule(+) supertrees and its variants have been shown to have many desirable properties [16] and there have been efforts to develop exact (ILP based) and heuristic methods for computing majority-rule(+) supertrees [17,22]. Though these methods only work for small datasets, they have been shown to result in biologically meaningful supertrees [17]. The paper by Kupczok [22] characterizes the RF(+) distance in the case when the leaf set of one tree is a subset of the leaf set of the other in terms of incompatible splits between the two trees, but does not provide an efficient algorithm for computing this distance or for computing an actual completion. More recently, Christensen et al. [11] provided an $O(n^2)$ time algorithm for the case when the leaf set of one tree is a subset of the leaf set of the other and applied the algorithm to compute optimal completions for gene trees with respect to a species tree. To the best of our knowledge, no algorithms (polynomial time or otherwise) currently exist for the general problem where the two trees have only partially overlapping leaf sets, or for any of its variants.

**Our Contribution.** In this work, we address an important gap in the algorith- mics of phylogenetic tree comparison. Specifically, we provide the first optimal, linear-time algorithms for two fundamental computational problems that arise when comparing phylogenetic trees with non-identical leaf sets. For the first problem, which arises when computing the RF(+) distance between two binary trees where the leaf set of one tree is a proper subset of the other, we improve upon the time complexity of the previous fastest algorithm for this problem by a factor of $n$, where $n$ is the number of leaves in the larger of the two trees. For the second problem, which is a generalization of the first and arises when computing the RF(+) distance between two binary trees that have only partially overlapping leaf sets, we show that the default problem formulation can result in biologically meaningless results, propose a modification of the problem formulation that corrects this deficiency, and provide optimal linear-time algorithms for the modified problem. Crucially, no polynomial time algorithms currently exist for the default formulation of the second problem, and our modified problem formulation can be viewed as a biologically meaningful restricted version of the general problem. Our algorithms are easy to understand and implement, work for both rooted and unrooted trees, and are scalable to the entire tree of life. These algorithms can be applied wherever phylogenetic distances must be computed between trees with non-identical leaf sets and enable new kinds of phylogenetic and comparative analyses that have been computationally infeasible.

We implemented our algorithm for the first problem and applied it to three published biological supertree data sets to study how RF(+) distances differ from RF(-) distances in practice. For each data set, we ordered the input trees according to their RF(+) and RF(-) distances to a precomputed supertree and measured how often the relative pairwise ranking between any pair of input trees differs between the two rankings. We found a large number of such pairs for each data set, demonstrating, for the first time, that using the RF(+) distance could result in different evolutionary inferences compared to inferences using the RF(-) distance.

RF(+) distances have several desirable properties compared to RF(-) distances. For instance, the range of possible values RF(+) distance can take ranges from 0 to about twice the size of the *union* of the leaf sets of the two trees, while for RF(-) distance this range is only from 0 to about twice the size of the *intersection* of the two leaf sets. Thus, RF(+) distances have significantly more discriminatory power than RF(-) distances. In applications such as median supertree construction, RF(+) distance has the distinct advantage that each input tree gets an equal "vote" in the supertree construction since all input trees contribute an RF distance within the same range. With RF(-) distances, larger trees can contribute much more to the total distance than smaller trees. Finally, in computing RF(-) distances we ignore the additional topological information provided by leaves that are present in only one tree, while RF(+) distance makes complete use of the information in the topologies of the two trees. RF(+) distances thus make more efficient use of the available information. Despite these advantages, RF(+) distances have not been applied in practice due to unavailability of efficient algorithms. In contrast, RF(-) distances can be computed in time linear in the sizes of the input trees. Our new algorithms address this discrepancy by making it equally computationally efficient to compute RF(+) distances.

The remainder of this manuscript is organized as follows. The next section includes basic definitions, notation, and problem formulations. Sections 3, 4, and 5 describe our algorithms for the problems considered in this work. Experimental results appear in Sect. 6 and concluding remarks appear in Sect. 7. For brevity, some proofs and certain details are deferred to the full version of this manuscript.

## 2  Preliminaries and Problem Definitions

Given a tree $T$, we denote its node set, edge set, and leaf set by $V(T)$, $E(T)$, and $Le(T)$, respectively. The set of all non-leaf (i.e., internal) nodes of $T$ is denoted by $I(T)$.

If $T$ is rooted, the root node of $T$ is denoted by $rt(T)$, the parent of a node $v \in V(T)$ by $pa_T(v)$, its set of children by $Ch_T(v)$, and the (maximal) subtree of $T$ rooted at $v$ by $T(v)$. If two nodes in $T$ have the same parent, they are called *siblings* of each other. The *least common ancestor*, denoted $lca_T(L)$, of a set $L \subseteq Le(T)$ in $T$ is defined to be the node $v \in V(T)$ such that $L \subseteq Le(T(v))$ and $L \not\subseteq Le(T(u))$ for any child $u$ of $v$. A rooted tree is *binary* if all of its internal nodes have exactly two children, while an unrooted tree is *binary* if all its nodes have degree either 1 or 3. Throughout this work, the term *tree* refers to binary trees with uniquely labeled leaves.

Let $T$ be a rooted or unrooted tree. Given a set $L \subseteq Le(T)$, let $T'$ be the subtree of $T$ with leaf set $L$. We define the *leaf induced subtree* $T[L]$ of $T$ on leaf set $L$ to be the tree obtained from $T'$ by successively removing each non-root node of degree two and adjoining its two neighbors.

**Definition 1 (Completion of a tree).** *Given a tree $T$ and a set $L'$ such that $Le(T) \subseteq L'$, a completion of $T$ on $L'$ is a tree $T'$ such that $Le(T') = L'$ and $T'[Le(T)] = T$.*

If $T$ is a rooted tree, for each node $v \in V(T)$, the *clade* $C_T(v)$ is defined to be the set of all leaf nodes in $T(v)$; i.e. $C_T(v) = Le(T(v))$. We denote the set of all clades of a rooted tree $T$ by $Clade(T)$. This concept can be extended to unrooted trees as follows. If $T$ is an unrooted tree, each edge $(u, v) \in E(T)$ defines a partition of the leaf set of $T$ into two disjoint subsets $Le(T_u)$ and $Le(T_v)$, where $T_u$ is the subtree containing node $u$ and $T_v$ is the subtree containing node $v$, obtained when edge $(u, v)$ is removed from $T$. The partition induced by any edge $(u, v) \in E(T)$ is called a *split* and is represented by the set $\{Le(T_u), Le(T_v)\}$. The set of all splits in an unrooted tree $T$ is denoted by $Split(T)$.

The *symmetric difference* of two sets $A$ and $B$, denoted by $A \triangle B$, is the set $(A \setminus B) \cup (B \setminus A)$.

**Definition 2 (Robinson-Foulds distance).** *The Robinson-Foulds (RF) distance, $RF(S, T)$, between two trees $S$ and $T$ is defined to be $|Clade(S) \triangle Clade(T)|$ if $S$ and $T$ are rooted trees, and $|Split(S) \triangle Split(T)|$ if $S$ and $T$ are unrooted trees.*

Let $S$ and $T$ be two trees. Without loss of generality, we will assume that $|Le(T)| \leq |Le(S)|$. When $Le(S) \neq Le(T)$, there are two possible scenarios: (1) $Le(T) \subsetneq Le(S)$, i.e., the leaf set of $T$ is a proper subset of the leaf set of $S$, and (2) $Le(S) \cap Le(T) \subsetneq Le(T)$, i.e., each of $S$ and $T$ contains leaves not found in the other. Based on these two scenarios, and depending on whether the two trees are rooted or unrooted, we define the following four problems.

**Problem 1 (Rooted One-Tree RF(+) (ROT-RF(+))).** *Given two rooted trees $S$ and $T$, such that $Le(T) \subseteq Le(S)$, compute a completion $T'$ of $T$ on $Le(S)$ such that $RF(S, T')$ is minimized.*

**Problem 2 (Unrooted One-Tree RF(+) (UOT-RF(+))).** *Given two unrooted trees $S$ and $T$, such that $Le(T) \subseteq Le(S)$, compute a completion $T'$ of $T$ on $Le(S)$ such that $RF(S, T')$ is minimized.*

**Problem 3 (Rooted RF(+) (R-RF(+))).** *Given two rooted trees $S$ and $T$, compute a completion $S'$ of $S$ on $Le(S) \cup Le(T)$ and a completion $T'$ of $T$ on $Le(S) \cup Le(T)$ such that $RF(S', T')$ is minimized.*

**Problem 4 (Unrooted RF(+) (U-RF(+))).** *Given two unrooted trees $S$ and $T$, compute a completion $S'$ of $S$ on $Le(S) \cup Le(T)$ and a completion $T'$ of $T$ on $Le(S) \cup Le(T)$ such that $RF(S', T')$ is minimized.*

We show how to solve Problems 1 and 2 in $O(|V(S)|)$ time. As we will see later, Problems 3 and 4 can actually lead to biologically meaningless completions. We will therefore define biologically meaningful variants of Problems 3 and 4 (requiring only a slight variation on the original problems) and show how to solve them in $O(|V(S)| + |V(T)|)$ time. Throughout this work, we assume that the leaves of $S$ and $T$ are labeled by integers from the set $\{1, \ldots, |Le(S) \cup Le(T)|\}$. However, our algorithms work even if the leaf labels are arbitrary, and universal hashing [8] or perfect hashing [15] can be used to guarantee expected $O(|V(S)| + |V(T)|)$ time complexity.

## 3    A Linear-Time Algorithm for ROT-RF(+)

To solve the ROT-RF(+) problem, our algorithm starts with the trees $S$ and $T$ and modifies $T$ by adding to it, according to a particular scheme, the leaves from $Le(S) \setminus Le(T)$. The completed tree thus produced, denoted by $T'$, will be such that $RF(S, T')$ is minimized.

We define $Tree\text{-}Add(T, v, X)$ to be the tree obtained from $T$ by attaching to it a tree $X$, where $Le(X) \cap Le(T) = \emptyset$, as follows: If $v$ is not the root of $T$, then attach $X$ onto the edge $(pa(v), v)$ (by subdividing $(pa(v), v)$ into two edges) such that $rt(X)$ becomes the sibling of the node $v \in V(T)$. If $v$ is the root of $T$, then $Tree\text{-}Add(T, v, X)$ is the tree obtained by creating a new root node and setting $v$ and $rt(X)$ as its two children.

The main idea behind our algorithm can be illustrated by the following simple example. Suppose the given trees $S$ and $T$ are such that $Le(S) = Le(T) \cup \{l\}$.

The goal is to add to $T$ this leaf $l$, so as to minimize the RF distance. Let $v$ denote the sibling of $l$ in $S$. Let $u$ denote the node $lca_T(Le(S(v)))$. As we will prove later, $T' = Tree\text{-}Add(T, u, l)$ must be an optimal completion for $T$. Our algorithm extends this idea to the case when $T$ has multiple missing leaves. A description of the algorithm follows:

**Algorithm** *OneTreeCompletion*$(S, T)$
1: **for** each $v \in V(S)$ in post-order **do**
2:     Initialize the mapping $\mathcal{M}_S(v)$ to be NULL.
3:     **if** $v \in Le(S)$ **then**
4:         **if** leaf $v$ is also present in tree $T$ **then**
5:             Color $v$ green.
6:         **else**
7:             Color $v$ red.
8:     **else**
9:         **if** $v$ has two green children **then**
10:         Color $v$ green.
11:         **else if** $v$ has two red children **then**
12:         Color $v$ red.
13:         **else if** $v$ has exactly one red child **then**
14:         Color $v$ blue and label $v$ as "marked".
15:         **else**
16:         Color $v$ blue.
17: **for** each green or blue node $v$ from $V(S)$ in post-order **do**
18:     Assign $\mathcal{M}_S(v) = lca_T(X)$, where $X = \{g | g \in Le(S(v))$ and $g$ is green$\}$.
19: **for** each marked node $v \in V(S)$ in pre-order **do**
20:     $Tree\text{-}Add(T, \mathcal{M}_S(v), R)$, where $R$ is the subtree rooted at the red child of $v$.
21: Return the completed tree $T$.

Figure 2 illustrates the algorithm through an example. Next, we prove the correctness and analyze the time complexity of this algorithm. We need the following additional definitions:

**Definition 3 (Matched clade).** *Given any two rooted trees $A$ and $B$ on the same leaf set, and $v \in V(A)$, we say that clade $C_A(v)$ has a* match *in $B$ if $Clade(B)$ contains $C_A(v)$.*

**Definition 4 (Matchable clade of $S$).** *Given any $v \in I(S)$, we call the clade $C_S(v)$* matchable *if there exists some completion of $T$ on $Le(S)$ that contains the clade $C_S(v)$.*

The correctness of Algorithm *OneTreeCompletion* follows from the following lemma.

**Lemma 1.** *Let $T'$ denote the completion of $T$ returned by Algorithm One-TreeCompletion on trees $S$ and $T$. Let $T^*$ denote an optimal completion of $T$ on*

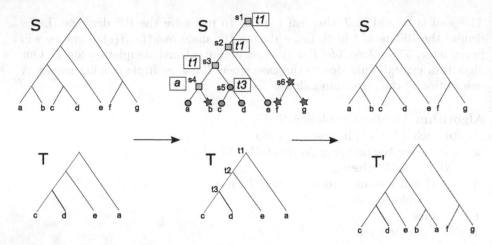

**Fig. 2. Algorithm for ROT-RF(+).** Given $S$ and $T$ as shown in the left column of the figure, Algorithm *OneTreeCompletion* first colors each node of $S$ either green (circles), red (stars), or blue (squares) as shown in the middle column of the figure. A node is colored green if all leaves in the subtree rooted at that node are present in both $S$ and $T$, red if all leaves in that subtree are present only in $S$, and blue if that subtree has both green and red descendants. If a blue node $v$ has exactly one red child, then it is "marked". In this example, $s_1$ and $s_4$ are marked nodes. The algorithm then computes the LCA mapping, defined to be $lca_T(Le(S(v)) \cap Le(T))$, for each green or blue node $v$ of $S$. These LCA mappings appear in the square boxes on $S$ in the middle column. The algorithm then performs a pre-order traversal of $S$, grafting copies of the red subtrees at each marked node onto the appropriate edges of $T$. The grafted subtrees are shown using dashed red lines on $T'$ in the right column. Tree $T'$ is an optimal completion of $T$ on $Le(S)$.

$Le(S)$ that minimizes $RF(S, T^*)$. Then, $RF(S, T') = RF(S, T^*)$, implying that $T'$ is a solution for the ROT-RF(+) problem.

*Proof.* It suffices to show that $T'$ maximizes the number of matched clades $C_S(v)$, for $v \in V(S)$.

Observe that Algorithm *OneTreeCompletion* partitions $V(S)$ into three sets according to the color assigned to each node: red, green, or blue. We will consider these three sets of nodes separately.

*Case 1: Red nodes.* All maximal subtrees in $S$ that contain only red nodes are included as-is in the completed tree $T'$. Thus, if $v$ is a red node then $C_S(v)$ has a match in $T'$. Thus, $T'$ maximizes the number of matched clades $C_S(v)$ over all red $v$.

*Case 2: Green nodes.* We claim that if $v$ is green and $C_S(v)$ does not have a match in $T'$ then it must be unmatchable. Suppose $C_S(v)$ has a match in $T$, and let $u \in V(T)$ be such that $C_S(v) = C_T(u)$. Observe that the clade $C_T(u)$ must also appear in $T'$ since no blue node $x \in V(S)$ will be such that $M_S(x) \in V(T(u))$. This implies that if $C_S(v)$ has a match in $T$ then $C_S(v)$ must also have a match in $T'$. In other words, if $C_S(v)$ does not have a match in $T'$ then $C_S(v)$ can not have a match in $T$. Now, since $C_S(v)$ only contains leaves that are already present in $T$, no completion of $T$ on $Le(S)$ can create clade $C_S(v)$ if $C_S(v)$ is not already present in $Clade(T)$. Thus, if $C_S(v)$ has no match in $T$, then $C_S(v)$ must be unmatchable. This proves our claim, and so $T'$ must maximize the number of matched clades $C_S(v)$ for green $v$.

*Case 3: Blue nodes.* We claim that if $v$ is blue and $C_S(v)$ does not have a match in $T'$ then it must be unmatchable. Let $C'_S(v)$ denote the set containing only the green nodes from $C_S(v)$. We will say that clade $C_S(v)$ has a partial-match in $T$ if and only if $C'_S(v) \in Clade(T)$. Suppose $C_S(v)$ has a partial-match in $T$, and let $u$ be the node from $T$ for which $C_T(u) = C'_S(v)$ (note that, in fact, $u = \mathcal{M}_S(v)$). Observe that any marked node $x \in V(S(v))$ must be such that $\mathcal{M}_S(x) \in V(T(u))$. This implies that Algorithm *OneTreeCompletion* adds all the maximal red subtrees within $S(v)$ (i.e., subtrees rooted at a red child of a marked node in $S(v)$) to one or more of the edges in the set $\{(pa(t), t) | t \in T(u)\}$. Moreover, since $C_T(u) = C'_S(v)$, none of the other marked nodes $y \in V(S) \setminus V(S(v))$ can be such that $\mathcal{M}_S(y) \in V(T(u))$. Thus, there must be a node $u' \in T'$ for which $C_{T'}(u') = C_T(u) \cup \{r | r \text{ is a red leaf from } S(v)\}$, and so $C_S(v)$ must have a match in $T'$. Consequently, if $C_S(v)$ has a partial-match in $T$ then $C_S(v)$ must have match in $T'$. In other words, if $C_S(v)$ does not have a match in $T'$ then $C_S(v)$ can not have a partial-match in $T$.

Now, suppose $v \in V(S)$ is such that $C_S(v)$ has no partial-match in $T$. Since, $C'_S(v)$ only contains leaves that are already present in $T$, and there exists no node $u \in V(T)$ for which $C_T(u) = C'_S(v)$, no completion of $T$ on $Le(S)$ can create clade $C_S(v)$. Thus, if $C_S(v)$ has no partial-match in $T$, then $C_S(v)$ must be unmatchable. This proves our claim, and so $T'$ must maximize the number of matched clades $C_S(v)$ for blue $v$.

In summary, the tree $T'$ maximizes the number of matched clades for each of the three sets into which $V(S)$ is partitioned, thereby maximizing the number of matched clades over all of $V(S)$. Hence, $T'$ must be a solution for the ROT-RF(+) problem. $\qquad\square$

**Theorem 1.** *Algorithm OneTreeCompletion solves the ROT-RF(+) problem in $O(|V(S)|)$ time.*

*Proof.* Lemma 1 establishes that Algorithm *OneTreeCompletion* solves the ROT-RF(+) problem. It therefore suffices to show that this algorithm can be implemented in $O(|V(S)|)$ time. We consider the complexity of each of the three 'for' loops separately.

The 'for' loop of Step 1 executes a single post-order traversal of the tree $S$, and so Steps 2 through 16 are executed a total of $O(|V(S)|)$ times. Each of the Steps 2 through 16, except for Step 16, clearly requires only $O(1)$ time per iteration. Step 16 can also be executed in $O(1)$ time after an $O(|S|)$ preprocessing step to construct a lookup table that enables $O(1)$ time lookup of whether a given leaf label from $S$ occurs in tree $T$ as well. This lookup table can be easily implemented using an array since the leaves of $S$ (and $T$) are uniquely labeled by integers from the set $\{1, \ldots, |Le(S)|\}$. The indices of the array correspond to the leaf labels, and the entries correspond to whether the corresponding leaf appears only in $S$ or in both $T$ and $S$. Such an array can be constructed using a single traversal through the leaf sets of $S$ and $T$. Even if the leaves have arbitrary labels, $O(|S|)$ preprocessing time and expected $O(1)$ lookup time can be achieved through hashing [8].

Step 18 is executed a total of $O(|V(S)|)$ times through the 'for' loop of Step 17. After an $O(|V(T)|)$ preprocessing step on $T$, the least common ancestor of any pair of nodes from $V(T)$ can be computed in constant time [5]. For any node $v$ considered in the 'for' loop of Step 17, computing the least common ancestor mapping for that node (in Step 18) is equivalent to computing the least common ancestor of the mappings of its (up to two) blue or green children. Thus, after an $O(|Le(T)|)$ preprocessing step on $T$ to enable fast least common ancestor computation [5], each execution of Step 18 requires only $O(1)$ time. This gives a total time complexity of $O(|V(S)|)$ for Steps 17 and 18.

The 'for' loop of Step 19 executes Step 20 a total of $O(|V(S)|)$ times. For a marked node $v$, Step 20 requires $O(|V(R)|)$ time, where $R$ is the subtree rooted at the red child of $v$, to copy over the subtree $R$ to $T$. Since each such $R$ is disjoint from the others, over all possible marked nodes $v$, the total number of nodes in all the corresponding $R$s is bounded by $O(|V(S)|)$. Thus, the total time complexity of Steps 19 and 20 is $O(|V(S)|)$.

Finally, Step 21 requires $O(|V(S)|)$ time to write the completed version of $T$. The total time complexity is thus $O(|V(S)|)$.                                    □

Note that Algorithm *OneTreeCompletion* computes a single optimal completion, and that optimal completions need not be unique.

## 4   The R-RF(+) Problem

Observe how an optimal completion of $T$ in the ROT-RF(+) problem maximizes the number of clades that have a match in $S$. This ensures a biologically meaningful completion of $T$. However, in the R-RF(+) problem, where both trees may have missing leaves, it is possible that optimal completions of the two trees contain "extraneous" clades that contain leaves from both $S$ and $T$ but do not contain any leaves common to $S$ and $T$. Extraneous clades are created by pairing a subtree containing only missing leaves from one tree with a subtree containing only missing leaves from the other tree. Such clades can help to lower the RF distance between the two completed trees, but are not biologically meaningful since they are completely unsupported by the topologies of $S$ and $T$. This phenomenon is illustrated through an example in Fig. 3. We therefore define a biologically meaningful variant of the R-RF(+) problem that only allows completions that do not result in extraneous clades. Crucially, this restriction to only non-extraneous clades also makes the underlying completion problem easier to solve.

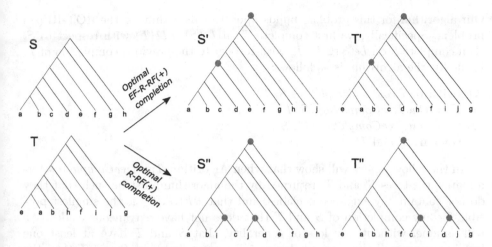

**Fig. 3. Extraneous clades and R-RF(+) and EF-R-RF(+) completions.** This figure shows two trees $S$ and $T$ with partial leaf set overlap whose optimal completions under the R-RF(+) problem result in extraneous clades. The tree $S$ contains two leaves $c$ and $d$ that are absent from $T$, and the tree $T$ contains two leaves $i$ and $j$ absent from $S$. The lower-right part of the figure shows optimal completions of $S$ and $T$, labeled $S''$ and $T''$, respectively, that minimize the RF distance over all possible completions. The nodes marked in red denote (non-leaf) clades common to both $S''$ and $T''$. Observe that of the three nodes that $S''$ and $T''$ have in common, the lower two, i.e., $\{c, i\}$ and $\{d, j\}$ are extraneous clades that have no support in either $S$ or $T$ and do not contain any of the leaves shared by both $S$ and $T$. Optimal completions under EF-R-RF(+) disallow such extraneous clades. The upper-right part of the figure shows optimal completions of $S$ and $T$ that minimize the RF distance over all completions without any extraneous clades. The completions $S'$ and $T'$ are more biologically meaningful since they only contain clades that have at least one leaf shared by both trees.

## Definition 5 (Extraneous clade). *Suppose $S$ and $T$ are rooted trees. Given completions $S'$ and $T'$ of $S$ and $T$, respectively, on $Le(S) \cup Le(T)$, we define a clade of $S'$ or $T'$ to be an extraneous clade if it contains leaves from both $S$ and $T$ but no leaves from $Le(S) \cap Le(T)$.*

## Problem 5 (Extraneous-Clade-Free R-RF(+) (EF-R-RF(+))). *Given two rooted trees $S$ and $T$, compute a completion $S'$ of $S$ on $Le(S) \cup Le(T)$ and a completion $T'$ of $T$ on $Le(S) \cup Le(T)$ such that $S'$ and $T'$ do not contain any extraneous clades and $RF(S', T')$ is minimized.*

An example of an optimal EF-R-RF(+) completion appears in Fig. 3. Next, we show how to solve the EF-R-RF(+) problem in linear time.

### 4.1 A Linear-Time Algorithm for EF-R-RF(+)

For the EF-R-RF(+) problem, $Le(S)$ and $Le(T)$ are both proper subsets of $Le(S) \cup Le(T)$, i.e., both $S$ and $T$ must be completed on the leaf set $Le(S) \cup Le(T)$.

Our algorithm for this problem builds upon the algorithm for the ROT-RF($+$) problem. Specifically, we first complete $T$ on $Le(S) \cup Le(T)$ with respect to $S$, then complete $S$ on $Le(S) \cup Le(T)$ with respect to the previous completion of $T$. Formally, the algorithm is as follows:

**Algorithm** *TwoTreeCompletion($S$, $T$)*
1: $T' = OneTreeCompletion(S, T)$.
2: $S' = OneTreeCompletion(T', S)$.
3: **return** $S'$ and $T'$.

In the following, we will show that when Algorithm *TwoTreeCompletion* terminates, the trees $S'$ and $T'$ returned by the algorithm must be such that they do not contain any extraneous clades, and that $RF(S', T')$ is the smallest possible for any completion of $S$ and $T$ that does not have extraneous clades. We will assume, without any loss of generality, that $S$ and $T$ have at least one leaf in common; if there are no leaves in common between $S$ and $T$ then the EF-R-RF($+$) problem has no solution since any completion of $S$ and $T$ would necessarily contain extraneous clades.

For brevity, in the remainder of this section, we will implicitly assume that all completions of $S$ and $T$ are on the leaf set $Le(S) \cup Le(T)$. Next, we define the notions of *original nodes*, *grafted nodes*, and *grafted subtrees* in tree completions.

**Definition 6 (Original nodes).** *Let $S'$ and $T'$ denote any completions of $S$ and $T$. Observe that completing a tree creates new internal nodes in the tree but preserves all original internal nodes (though not necessarily the clades rooted at those nodes). Thus, we have $I(S) \subset I(S')$ and $I(T) \subset I(T')$. The set of nodes in $I(S')$ that are also present in $I(S)$ are called the* original nodes *of $S'$, denoted $\mathcal{O}(S')$. Analogously, the set of nodes in $I(T')$ that are also present in $I(T)$ are called the* original nodes *of $T'$, denoted $\mathcal{O}(T')$.*

**Definition 7 (Grafted nodes).** *Let $S'$ and $T'$ denote any completions of $S$ and $T$. Observe that any node $u \in I(S') \setminus \mathcal{O}(S')$ is either a node that was already present in a subtree from $T$ (consisting of leaves missing from $S$) as that subtree was grafted into $S$, or a new node that was created as a subtree from $T$ (consisting of leaves missing from $S$) was grafted into $S$. We refer to the new nodes created by the grafting of a subtree from $T$ into $S'$ as the* grafted nodes *of $S'$, denoted $\mathcal{G}(S')$. Analogously, the set of nodes in $I(T') \setminus \mathcal{O}(T')$ that were newly created through the process of grafting a subtree from $S$ into $T$ are called the* grafted nodes *of $T'$, denoted $\mathcal{G}(T')$.*

**Definition 8 (Grafted subtrees).** *If $S'$ denotes any completion of $S$ and $u \in \mathcal{G}(S')$, then $u$ is created by the grafting of a subtree of $T$ (consisting of leaves missing from $S$) at that node $u$ in $S'$. We denote the grafted subtree of $T$ at $u$ by graft($u$). Similarly, if $T'$ denotes any completion of $T$ and $v \in \mathcal{G}(T')$, then $v$ is created by the grafting of a subtree of $S$ at that node $v$ in $T'$. We denote the grafted subtree of $S$ at $v$ by graft($v$).*

**Node Colorings.** For convenience, we will color the nodes of $S$ and $T$ according to the coloring scheme used in Algorithm *One Tree Completion*. Thus, each node of $S$ and $T$ is colored either red, or green, or blue. We will assume that these colored nodes maintain their original colors in the completed trees $S'$ and $T'$, and thus both $S'$ and $T'$ contain nodes that are red, green, and blue, as well as nodes that are uncolored.

We now show that the completed trees $S'$ and $T'$ returned by Algorithm *Two Tree Completion* must be free of extraneous clades.

**Lemma 2.** *The trees $S'$ and $T'$ returned by Algorithm Two Tree Completion do not have any extraneous clades.*

*Proof.* Let us first consider the tree $T'$. Any non-original node in $T'$ is either a node from a maximal red subtree of $S$ or is a grafted node created by grafting a maximal red subtree of $S$ into $T'$ using the *Tree-Add* operation. Based on Algorithm *One Tree Completion*, each grafted node created through the *Tree-Add* operation has at least one green descendant, and so it cannot be extraneous. Moreover, any node inside a maximal red subtree of $S$ only has descendants from $S$, not from $T$. Thus, since $T$ did not contain any extraneous clades to begin with, neither can $T'$. An analogous argument applies to $S'$.    □

The next lemma identifies an important property of optimal completions.

**Lemma 3.** *Let $S^*$ and $T^*$ be any optimal completions of $S$ and $T$, respectively, under the EF-R-RF(+) problem. Then, for any $u \in \mathcal{G}(S^*)$, graft(u) must be a maximal red subtree of $T$ and, for any $v \in \mathcal{G}(T^*)$, graft(v) must be a maximal red subtree of $S$.*

*Proof.* Observe that any maximal red subtree of $T$ must appear as-is in the tree $T^*$, since grafting a red leaf or subtree from $S$ into any of the red subtrees of $T$ would result in an extraneous clade. We will show that if there exists a node $u \in \mathcal{G}(S^*)$ for which $graft(u)$ is not a maximal red subtree of $T$, it is possible to modify the tree $S^*$ so that the modified tree has more matched clades than $S^*$, a contradiction. An analogous argument applies to $T^*$. Suppose there exists such a node $u$. Then, there must exist a red internal node $r$ of $T$ such that the two subtrees, denoted $R'$ and $R''$, rooted at the two children of $r$ appear as-is in the tree $S^*$ but not as siblings of each other (i.e., their roots do not have the same parent in $S^*$). Let $r'$ and $r''$ denote the root nodes of $R'$ and $R''$, respectively, and $s'$ and $s''$ denote the parents of $r'$ and $r''$ in $S^*$. Thus, $R' = graft(s')$ and $R'' = graft(s'')$. Now, observe that all clades of $S^*$ rooted either at a node on the path from $lca_{S^*}(s', s'')$ to $s'$ or on the path from $lca_{S^*}(s', s'')$ to $s''$, except for the node $lca_{S^*}(s', s'')$ itself, must be mismatched clades (since all maximal red subtrees of $T$ appear as-is in the tree $T^*$). Also, note that if $S^*$ is modified by pruning out the subtree $R'$ and regrafting it on the edge $(s'', r'')$, then the only matched clades that can become mismatched are the ones whose roots lie on the path from $lca_{S^*}(s', s'')$ to $s'$ or from $lca_{S^*}(s', s'')$ to $s''$, except for node $lca_{S^*}(s', s'')$. Thus, modifying the tree $S^*$ in this fashion does not result in any additional mismatched clades, but results in a new matched clade rooted at the node where $R'$ is regrafted. Thus, the modified tree has a larger number of matched clades than $S^*$, which is a contradiction. □

We also have the following simple observation about optimal completions.

**Observation 1.** *Let $S^*$ and $T^*$ be optimal completions of $S$ and $T$, respectively, that satisfy the property described in Lemma 3. Then any $u \in \mathcal{G}(S^*)$ and any $v \in \mathcal{G}(T^*)$ must have at least one green leaf as a descendant.*

*Proof.* This follows immediately from the fact that, under EF-R-RF(+), each clade must contain at least one green leaf (otherwise it would be an extraneous clade). □

Finally, the following lemma proves the correctness of Algorithm *TwoTreeCompletion*. For brevity, its proof is deferred to the full version of this paper.

**Lemma 4.** *Let $S'$ and $T'$ denote the completions of $S$ and $T$, respectively, returned by Algorithm TwoTreeCompletion. Let $S^*$ and $T^*$ denote optimal completions of $S$ and $T$, respectively, under the EF-R-RF(+) problem. Then, $RF(S', T') = RF(S^*, T^*)$.*

The next theorem now follows immediately based on Algorithm *TwoTreeCompletion*, Theorem 1, and Lemma 4.

**Theorem 2.** *Algorithm TwoTreeCompletion solves the EF-R-RF(+) problem in $O(|V(S)| + |V(T)|)$ time.*

## 5 Extension to Unrooted Trees

The linear-time algorithms for the ROT-RF(+) and EF-R-RF(+) problems described in the previous two sections can be easily extended to unrooted trees without any increase in time complexity. The idea is to first root the two unrooted trees at any leaf-edge that is common to both trees, and then apply the algorithm for ROT-RF(+) or EF-R-RF(+) on the resulting rooted trees. It can be shown that this is guaranteed to result in optimal solutions for UOT-RF(+) and EF-U-RF(+). Further details and proofs are deferred to the full version of this paper.

## 6 Experimental Evaluation

We implemented our algorithm for the ROT-RF(+) problem and applied it to three large biological supertree data sets with the goal of assessing the impact of using RF(+) distance instead of the traditional RF(-) distance in practice. Specifically, we computed a supertree (using a standard supertree method; RFS [3] in this case) for each of the supertree data sets, and computed the RF(+) and RF(-) distances between the supertree and the input trees for each data set.

Let the RF(+) distance between a supertree $S$ and an input tree $I$ be denoted by $RF^+(S, I)$, and the RF(-) distance those two trees by $RF^-(S, I)$. For each data set, we ordered the input trees according to their RF(+) and RF(-) distances to the supertree and measured how often the relative ranking between any pair of input trees differs between the two rankings. More precisely, given a supertree $S$ and its set of input trees $\mathcal{I}$, we computed $RF^-(S, I)$ and $RF^+(S, I)$ for each $I \in \mathcal{I}$, and counted the number of *Type-1*, *Type-2*, and *Type-3* pairs $\{I', I''\}$, where $I', I'' \in \mathcal{I}$, as follows:

**Type-1 pairs.** Pair $\{I', I''\}$ is Type-1 if either $RF^-(S, I') < RF^-(S, I'')$ but $RF^+(S, I') > RF^+(S, I'')$, or $RF^-(S, I') > RF^-(S, I'')$ but $RF^+(S, I') < RF^+(S, I'')$. These are pairs for which the RF(+) and RF(-) distances impose completely opposite orderings relative to the supertree.

**Type-2 pairs.** Pair $\{I', I''\}$ is Type-2 if $RF^-(S, I') = RF^-(S, I'')$ but $RF^+(S, I') \neq RF^+(S, I'')$. For these pairs, RF(-) distances are identical but RF(+) distances are not.

**Type-3 pairs.** Pair $\{I', I''\}$ is Type-3 if $RF^-(S, I') \neq RF^-(S, I'')$ but $RF^+(S, I') = RF^+(S, I'')$. For these pairs, RF(+) distances are identical but RF(-) distances are not.

The three data sets, marsupials [6], placental mammals [4], and legumes [32], contain 272, 116, and 571 species, and 158, 726, and 22 input trees, respectively. We observed that for the 158 input trees of the marsupial data set, there were 521 Type-1 pairs, 619 Type-2 pairs, and 376 Type-3 pairs. For the 726 input trees of the placental mammals data set, there were 5,816 Type-1 pairs, 14,344 Type-2 pairs, and 6,238 Type-3 pairs. Likewise, for the 22 input trees in the legumes data set, we observed 8 Type-1 pairs, 3 Type-2 pairs, and no Type-3 pairs. These results show that there can be substantial difference between RF(-) and RF(+) distances and suggest that using RF(+) distances can result in different evolutionary inferences compared to inferences using RF(-).

Our current implementation is available from the author upon request. An improved open-source version, currently under development, will be released with the full version of this paper.

# 7 Conclusion

In this work, we provide the first optimal, linear-time algorithms for two fundamental computational problems that arise when comparing phylogenetic trees with non-identical leaf sets. For the first problem, which arises when computing the RF(+) distance between two trees where the leaf set of one tree is a proper subset of the other, we improved upon the time complexity of the previous fastest algorithm by a factor of $n$, where $n$ is the size of the larger of the two trees. For the second problem, which arises when computing the RF(+) distance between two trees that have only partially overlapping leaf sets, and for which there are no existing algorithms, we defined a biologically meaningful restriction of the problem and provided an optimal linear-time algorithm for it. Our algorithms are easy to implement and should be scalable even to trees with millions of taxa.

The algorithms work for both rooted and unrooted trees, and can be directly applied wherever phylogenetic distances must be computed between trees with non-identical leaf sets. Furthermore, our experiments with three large biological supertree data sets suggest that using the RF(+) distance can result in different evolutionary inferences compared to using the RF(-) distance.

The algorithms presented here have several important, well-established applications, including construction of majority-rule(+) supertrees and supertree construction in general, phylogenetic database search, and clustering of phylogenetic trees, and these applications should be studied and developed further. A more detailed experimental study is needed to properly assess the impact of using RF(+) distances and to systematically study the effect of factors such as fraction of leaf set overlap and degree of discordance between trees. This work also motivates several theoretical questions for future investigation. For instance, our algorithms for the EF-R-RF(+) and EF-U-RF(+) problems cannot be easily extended to solve the R-RF(+) and U-RF(+) problems. In particular, if optimal completions are allowed to contain extraneous clades, then inferring the number and composition of these extraneous clades (to attain overall optimality) appears to be computationally challenging. It would be interesting to determine if linear or near-linear time algorithms exist for R-RF(+) and U-RF(+).

**Funding.** This work was supported in part by NSF awards IIS 1553421 and MCB 1616514 to MSB.

# References

1. Akanni, W.A., Wilkinson, M., Creevey, C.J., Foster, P.G., Pisani, D.: Implementing and testing Bayesian and maximum-likelihood supertree methods in phylogenetics. R. Soc. Open Sci. **2**(8), 140436 (2015)
2. Amir, A., Keselman, D.: Maximum agreement subtree in a set of evolutionary trees: metrics and efficient algorithms. SIAM J. Comput. **26**(6), 1656–1669 (1997)
3. Bansal, M.S., Burleigh, J.G., Eulenstein, O., Fernández-Baca, D.: Robinson-foulds supertrees. Algorithms Mol. Biol. **5**(1), 18 (2010)
4. Beck, R., Bininda-Emonds, O., Cardillo, M., Liu, F.-G., Purvis, A.: A higher-level MRP supertree of placental mammals. BMC Evol. Biol. **6**(1), 93 (2006)
5. Bender, M.A., Farach-Colton, M., Pemmasani, G., Skiena, S., Sumazin, P.: Lowest common ancestors in trees and directed acyclic graphs. J. Algorithms **57**(2), 75–94 (2005)
6. Cardillo, M., Bininda-Emonds, O.R.P., Boakes, E., Purvis, A.: A species-level phylogenetic supertree of marsupials. J. Zool. **264**, 11–31 (2004)
7. Cardona, G., Llabrés, M., Rosselló, F., Valiente, G.: Nodal distances for rooted phylogenetic trees. J. Math. Biol. **61**(2), 253–276 (2010)
8. Carter, J., Wegman, M.N.: Universal classes of hash functions. J. Comput. Syst. Sci. **18**(2), 143–154 (1979)
9. Chaudhary, R., Burleigh, J.G., Fernandez-Baca, D.: Fast local search for unrooted robinson-foulds supertrees. IEEE/ACM Trans. Comput. Biol. Bioinform. (TCBB) **9**(4), 1004–1013 (2012)

10. Chen, D., Burleigh, J.G., Bansal, M.S., Fernández-Baca, D.: Phylofinder: an intelligent search engine for phylogenetic tree databases. BMC Evol. Biol. **8**(1), 90 (2008)
11. Christensen, S., Molloy, E.K., Vachaspati, P., Warnow, T.: Optimal completion of incomplete gene trees in polynomial time using OCTAL. In: Schwartz, R., Reinert, K. (eds.) 17th International Workshop on Algorithms in Bioinformatics (WABI 2017), Leibniz International Proceedings in Informatics (LIPIcs), vol. 88, pp. 27:1–27:14. Schloss Dagstuhl-Leibniz-Zentrum fuer Informatik, Dagstuhl (2017)
12. Cotton, J.A., Wilkinson, M., Steel, M.: Majority-rule supertrees. Syst. Biol. **56**(3), 445–452 (2007)
13. Critchlow, D.E., Pearl, D.K., Qian, C., Faith, D.: The triples distance for rooted bifurcating phylogenetic trees. Syst. Biol. **45**(3), 323–334 (1996)
14. de Vienne, D.M., Giraud, T., Martin, O.C.: A congruence index for testing topological similarity between trees. Bioinformatics **23**(23), 3119–3124 (2007)
15. Dietzfelbinger, M., Karlin, A., Mehlhorn, K., Meyer auf der Heide, F., Rohnert, H., Tarjan, R.E.: Dynamic perfect hashing: upper and lower bounds. SIAM J. Comput. **23**(4), 738–761 (1994)
16. Dong, J., Fernandez-Baca, D.: Properties of majority-rule supertrees. Syst. Biol. **58**(3), 360–367 (2009)
17. Dong, J., Fernández-Baca, D., McMorris, F.: Constructing majority-rule supertrees. Algorithms Mol. Biol. **5**(1), 2 (2010)
18. Dong, J., Fernández-Baca, D., McMorris, F., Powers, R.C.: An axiomatic study of majority-rule (+ ) and associated consensus functions on hierarchies. Discret. Appl. Math. **159**(17), 2038–2044 (2011)
19. Estabrook, G.F., McMorris, F.R., Meacham, C.A.: Comparison of undirected phylogenetic trees based on subtrees of four evolutionary units. Syst. Zool. **34**(2), 193–200 (1985)
20. Felsenstein, J.: Inferring Phylogenies. Sinauer Associates, Sunderland (2003)
21. Finden, C.R., Gordon, A.D.: Obtaining common pruned trees. J. Classif. **2**(1), 255–276 (1985)
22. Kupczok, A.: Split-based computation of majority-rule supertrees. BMC Evol. Biol. **11**(1), 205 (2011)
23. Kupczok, A., Haeseler, A.V., Klaere, S.: An exact algorithm for the geodesic distance between phylogenetic trees. J. Comput. Biol. **15**(6), 577–591 (2008)
24. Lin, H.T., Burleigh, J.G., Eulenstein, O.: Triplet supertree heuristics for the tree of life. BMC Bioinform. **10**(1), S8 (2009)
25. McMahon, M.M., Deepak, A., Fernndez-Baca, D., Boss, D., Sanderson, M.J.: STBase: one million species trees for comparative biology. PLOS ONE **10**(2), 1–17 (2015)
26. Piel, W.H., Donoghue, M., Sanderson, M., Netherlands, L.: TreeBASE: a database of phylogenetic information. In: Proceedings of the 2nd International Workshop of Species 2000 (2000)
27. Robinson, D., Foulds, L.: Comparison of phylogenetic trees. Math. Biosci. **53**(1), 131–147 (1981)
28. Vachaspati, P., Warnow, T.: FastRFS: fast and accurate robinson-foulds supertrees using constrained exact optimization. Bioinformatics **33**(5), 631–639 (2017)
29. Wang, J.T., Shan, H., Shasha, D., Piel, W.H.: Fast structural search in phylogenetic databases. Evol. Bioinform. **1**, 37–46 (2005). 2007
30. Waterman, M., Smith, T.: On the similarity of dendrograms. J. Theor. Biol. **73**(4), 789–800 (1978)

31. Whidden, C., Zeh, N., Beiko, R.G.: Supertrees based on the subtree prune-and-regraft distance. Syst. Biol. **63**(4), 566–581 (2014)
32. Wojciechowski, M., Sanderson, M., Steele, K., Liston, A.: Molecular phylogeny of the "temperate herbaceous tribes" of papilionoid legumes: a supertree approach. In: Herendeen, P., Bruneau, A. (eds.) Advances in Legume Systematics, vol. 9, pp. 277–298. Royal Botanic Gardens, Kew (2000)
33. Wu, Y.: A practical method for exact computation of subtree prune and regraft distance. Bioinformatics **25**(2), 190–196 (2009)
34. Yoshida, R., Fukumizu, K., Vogiatzis, C.: Multilocus phylogenetic analysis with gene tree clustering. Ann. Oper. Res. (2017). https://doi.org/10.1007/s10479-017-2456-9

# Multi-SpaM: A Maximum-Likelihood Approach to Phylogeny Reconstruction Using Multiple Spaced-Word Matches and Quartet Trees

Thomas Dencker[1], Chris-André Leimeister[1], Michael Gerth[2],
Christoph Bleidorn[3,4], Sagi Snir[5], and Burkhard Morgenstern[1,6(✉)]

[1] Department of Bioinformatics, Institute of Microbiology and Genetics,
University of Goettingen, Goldschmidtstr. 1, 37077 Goettingen, Germany
bmorgen@gwdg.de
[2] Institute for Integrative Biology, University of Liverpool, Biosciences Building,
Crown Street, Liverpool L69 7ZB, UK
[3] Department of Animal Evolution and Biodiversity, University of Goettingen,
Untere Karspüle 2, 37073 Goettingen, Germany
[4] Museo Nacional de Ciencias Naturales, Spanish National Research Council (CSIC),
28006 Madrid, Spain
[5] Institute of Evolution, Department of Evolutionary and Environmental Biology,
University of Haifa, 199 Aba Khoushy Ave. Mount Carmel, Haifa, Israel
[6] Goettingen Center of Molecular Biosciences (GZMB), Justus-von-Liebig-Weg 11,
37077 Goettingen, Germany

**Abstract.** Word-based or 'alignment-free' methods for phylogeny reconstruction are much faster than traditional, alignment-based approaches, but they are generally less accurate. Most alignment-free methods calculate *pairwise* distances for a set of input sequences, for example from *word frequencies*, from so-called *spaced-word matches* or from the average *length of common substrings*. In this paper, we propose the first word-based phylogeny approach that is based on *multiple* sequence comparison and *Maximum Likelihood*. Our algorithm first samples small, gap-free alignments involving four taxa each. For each of these alignments, it then calculates a quartet tree and, finally, the program *Quartet MaxCut* is used to infer a super tree for the full set of input taxa from the calculated quartet trees. Experimental results show that trees calculated with our approach are of high quality.

**Keywords:** Alignment-free · Phylogeny · Likelihood · Spaced words

## 1 Introduction

Sequence-based phylogeny reconstruction is a fundamental task in computational biology. Standard phylogeny methods rely on *sequence alignments* of either entire genomes or of sets of orthologous genes or proteins. *Character based* methods

© Springer Nature Switzerland AG 2018
M. Blanchette and A. Ouangraoua (Eds.): RECOMB-CG 2018, LNBI 11183, pp. 227–241, 2018.
https://doi.org/10.1007/978-3-030-00834-5_13

such as *Maximum Parsimony* [14,17] or *Maximum Likelihood* [15] infer trees based on evolutionary substitution events that may have happened since the species evolved from their last common ancestor. These methods are generally considered to be accurate as long as the underlying alignment is of high quality and as long as suitable substitution models are used. However, for the task of multiple alignment no exact polynomial-time algorithm exists, and even heuristic approaches are relatively time consuming [45]. Moreover, exact algorithms for character-based approaches are themselves *NP hard* [10,18].

*Distance* methods, by contrast, infer phylogenies by estimating evolutionary distances for all pairs of input taxa. Here, pairwise alignments are sufficient which can be faster calculated than multiple alignments, but still require runtime proportional to the product of the lengths of the aligned sequences. However, there is a loss in accuracy compared to character-based approaches, as all of the information about evolutionary events is reduced to a single number for each pair of taxa, and not more than two sequences are considered simultaneously, as opposed to character-based approaches, where all sequences are examined simultaneously. The final trees are obtained by clustering based on the distance matrices, most commonly with *Neighbor Joining* [44]. Since both pairwise and multiple sequence alignments are computationally expensive, they are ill-suited for the increasingly large datasets that are available today due to the next generation sequencing techniques.

In recent years, *alignment free* approaches to genome-based phylogeny reconstruction have been published which are very fast in comparison to alignment-based methods [5,7,39,41,48,57]. Some of these approaches – despite being called 'alignment-free' – are using pairwise gap-free 'mini-alignments'; recently, methods have been proposed that estimate phylogenetic distances based on the relative frequency of mismatches in such 'mini-alignments'. Another advantage of the so-called 'alignment-free' methods for genome comparison is that they can circumvent common problems of alignment-based approaches such as genome rearrangements and duplications. Moreover, alignment-free methods can be applied not only to entire genomes, but also to partially sequenced genomes or even to unassembled reads [11,43,49,56]. A disadvantage of these methods is that they are considerably less accurate than slower, alignment-based methods.

A recently proposed 'alignment-free' method is *co-phylog* [56]. This approach finds short, gap-free alignments of a fixed length, consisting of matching nucleotide pairs only, except for the middle position in each alignment, where mismatches are allowed. Phylogenetic distances are estimated from the fraction of such alignments for which the middle position is a mismatch. As a generalization of this approach, *andi* [23] uses pairs of maximal exact word matches that have the same distance to each other in both sequences and uses the frequency of mismatches in the segments between those matches to estimate the number of substitutions per position between two input sequences. Since *co-phylog* and *andi* require a minimum length of the flanking word matches in order to reduce the number of matches that are mere random background matches, they tend not to perform well on distantly related sequences where long exact matches are less frequent. Moreover, the number of random segment matches grows quadratically with the length of the input sequences while the expected number of homologous matches grows only linearly. Thus, longer exact matches are necessary in these approaches to limit the number of background matches if longer sequences are compared. This, in turn, reduces the number of homologies that are found, and therefore the amount of information that can be used to calculate accurate distances. Other alignment-free approaches are based on the length of maximal common substrings between sequences that can be rapidly found using suffix trees or related data structures [24,55]. As a generalization of this approach, some methods use longest common substrings with a certain number of mismatches [3,29,32,53,54].

In previous publications, we proposed to use words with *wildcard characters* – so-called *spaced words* – for alignment-free sequence comparison [25,28]. Here, a binary pattern of *match* and *don't-care* positions specifies the positions of the *wildcard* characters, see also [20,35,37]. In *Filtered Spaced-Word Matches (FSWM)* [31] and *Proteome-based Spaced-Word Matches (Prot-SpaM)* [30], alignments of such spaced words are used where sequence positions must match at the *match* positions while mismatches are allowed at the *don't care positions*. A score is calculated for every such spaced-word match in order to remove – or *filter out* – *background* spaced-word matches; the mismatch frequency of the remaining *homologous* spaced-word matches is then used to estimate the number of substitutions per position that happened since two sequences evolved from their last common ancestor. The filtering step allows us to use patterns with fewer match positions in comparison to above mentioned methods *co-phylog* and *andi*, since the vast majority of the background noise can be eliminated reliably. Thus, the phylogenetic distances calculated by *FSWM* or *Prot-SpaM* are generally rather accurate, even for large and distantly related sequences.

In this paper, we introduce a novel approach to phylogeny reconstruction called <u>Multiple</u> *Spaced-Word* <u>M</u>atches *(Multi-SpaM)* that combines the *speed* of the so-called 'alignment free' methods with the *accuracy* of the *Maximum-Likelihood* approach. While other alignment free methods are limited to *pairwise* sequence comparison, we generalize the spaced-word approach to *multiple* sequence comparison. For a binary pattern of *match* and *don't care* positions, *Multi-SpaM* identifies *quartet blocks* of four matching spaced words each, *i.e.* gap-free four-way alignments with matching nucleotides at the *match* positions of the underlying binary pattern and possible mismatches at the *don't care* positions (see Fig. 2 for an example). For each such quartet block, an optimal *Maximum-Likelihood* tree topology is calculated with the software *RAxML* [50]. We then use the *Quartet MaxCut* algorithm [47] to combine the calculated quartet tree topologies into a super tree. We show that on both simulated and real data, *Multi-SpaM* produces phylogenetic trees of high quality and often outperforms other alignment-free methods. An earlier version of the present paper has been uploaded to the preprint server *arXiv* [13].

## 2    Method

### 2.1    Spaced Words and P-blocks

To describe our method, we first need to introduce some formal definitions. We want to compare sequences over an alphabet $\mathcal{A}$; since our approach is dealing with DNA sequences, our alphabet is $\mathcal{A} = \{A, C, G, T\}$. For a given binary pattern $P \in \{0,1\}^\ell$, a *spaced word* with respect to $P$ is a word $W$ of length $\ell$ over $\mathcal{A} \cup \{*\}$, such that $W(i) = *$ if and only if $P(i) = 0$. A spaced word $W$ can be considered as a regular expression where '$*$' is a *wildcard character*. A position $i \in \{1, \ldots, \ell\}$ is called a *match position* if $P(i) = 1$ and a *don't care position* otherwise. The number of match positions in $P$ is called the *weight* of the $P$. For a DNA Sequence $S$ of length $n$ and a position $1 \le i \le n - \ell + 1$, we say that a *spaced word* $W$ with respect to $P$ occurs in $S$ at position $i$, or that $[S, i]$ is an *occurrence* of $W$ – if $S(i + j - 1) = W(j)$ for all match positions $j$. This corresponds to the definition previously used in [28,33].

A pair $([S, i], [S', i'])$ of occurrences of the same spaced word $W$ is called a *spaced-word match*. For a substitution matrix assigning a *score* $s(X, Y)$ to $X, Y \in \mathcal{A}$, we define the *score* of a spaced word match $([S, i], [S', i'])$ of length $\ell$ as

$$\sum_{1 \le k \le \ell} s(S(i + k - 1), S'(i' + k - 1))$$

That is, if we align the two occurrences of $W$ to each other, the score of the spaced-word match is the sum of the scores of the nucleotides aligned to each other. In *Multi-SpaM*, we are using the following nucleotide substitution matrix that has been proposed in [9]:

$$
\begin{array}{ccccc}
 & A & C & G & T \\
A & 91 & -114 & -31 & -123 \\
C & & 100 & -125 & -31 \\
G & & & 100 & -114 \\
T & & & & 91
\end{array}
\tag{1}
$$

*Multi-SpaM* starts with generating a binary pattern $P$ with user-defined length $\ell$ and weight $w$; by default, we use values $\ell = 110$ and $w = 10$, *i.e.* by default the pattern has 10 *match positions* and 100 *don't-care* positions. We are using a low *weight* to obtain a large number of spaced-word matches when comparing two sequences. This includes necessarily a high proportion of random spaced-word matches. The high number of *don't-care* positions, on the other hand, allows us to accurately distinguish between *homologous* and *background* spaced-word matches.

Given these parameters, a pattern $P$ is calculated by running our previously developed software tool *rasbhari* [21] minimizing the *overlap complexity* [26,27]. As a basis for phylogeny reconstruction, we are using four-way alignments consisting of occurrences of the same spaced word with respect to $P$ in four different sequences or their reverse complements. We call such an alignment a *quartet P-block* or a *P-block*, for short. A P-block is thus a gap-free alignment of length $\ell$

where in the $k$-th column identical nucleotides are aligned if $k$ is a *match* position in $P$, while mismatches are possible if $k$ is a *don't-care* position (see Fig. 2). Note that a $P$-block can involve spaced words from both strands of the input sequences. The number of such $P$-blocks can be very large: if there are $n$ occurrences of a spaced-word $W$ in $n$ different sequences, then this gives rise to $\binom{n}{4}$ different $P$-blocks. Thus, instead of using all possible $P$-blocks, *Multi-SpaM* randomly samples a limited number of $P$-blocks to keep the program run time under control (Fig. 1).

$$S_1 : T\,A\,C\,T\,A\,G\,C\,G\,T\,C\,G$$
$$S_2 : A\,C\,T\,C\,C\,T\,A\,G\,T\,G\,T\,T\,G$$

**Fig. 1.** Spaced-word match with respect to a pattern $P = 1101001$.

Moreover, for phylogeny reconstruction, we want to use $P$-blocks that are likely to represent true homologies. Therefore, we introduce the following definition: a $P$-block – *i.e.* a set of four occurrences of the same spaced word $W$ – is called a *homologous P-block* if it contains at least *one* occurrence $[S_i, p]$ of $W$ such that all remaining three occurrences of $W$ have positive scores when compared to $[S_i, p]$. To sample a list of homologous $P$-blocks, we randomly select spaced-word occurrences with respect to $P$ from the input sequences and their reverse complements. For each selected $[S_i, p]$, we then randomly select occurrences of the same spaced word from sequences $S_j \neq S_i$, until we have found three occurrences of $W$ from three different sequences that all have positive scores with $[S_i, p]$.

$$S_1 : C\,C\,C\,A\,A\,G\,G\,A\,C$$
$$S_2 : A\,A\,C\,T\,A\,C\,G\,T\,A\,C\,C\,T$$
$$S_3 : A\,A\,C\,T\,A\,C\,G\,T\,A\,C\,C$$
$$S_4 : C\,C\,A\,C\,G\,T\,C\,C\,G\,C\,G$$
$$S_5 : A\,G\,A\,C\,T\,C\,C\,C\,A\,A\,G\,G\,A$$
$$S_6 : T\,C\,C\,C\,A\,T\,G\,G\,A\,C\,C$$
$$S_7 : A\,A\,C\,T\,A\,C\,G\,T\,A\,C\,C\,A$$
$$1\ 2\ 3\ 4\ 5\ 6\ 7\ 8\ 9\ 10\ 11\ 12\ 13$$

**Fig. 2.** $P$-block for a pattern $P = 11001$: the spaced word $W = CC * *G$ occurs at $[S_1, 2], [S_4, 1], [S_5, 7]$ and $[S_6, 3]$.

To find spaced-word matches efficiently, we first sort the list of all occurrences of spaced words with respect to $P$ in lexicographic order. This way, we obtain a list of spaced-word occurrences where all occurrences of the same spaced word $W$ are appearing as a contiguous block. Once we have sampled a homologous $P$-block as described, we remove the four occurrences of $W$ from our list of spaced-word occurrences, so no two of the sampled $P$-blocks can contain the same occurrence of a spaced word. The algorithm continues to sample $P$-blocks until no further $P$-blocks can be found, or until a maximal number of $M$ $P$-blocks is reached. By default, *Multi-SpaM* uses a maximal number of $M = 1,000,000$ $P$-blocks, but this parameter can be adjusted by the user.

## 2.2  Quartet Trees

For each of the sampled quartet $P$-blocks, we infer an unrooted tree topology. This most basic *unrooted* phylogenetic unit is called a *quartet* topology; there are three possible different quartet topologies for a set of four taxa. To identify the best of these three topologies, we use the *Maximum-Likelihood* program *RAxML* [50]. We integrated parts of the *RAxML* code into our software and used the program with the *GTR* model [52]. This corresponds to using the command-line version of *RAxML* with the option "-m GTRGAMMA -f q -p 12345". We note that *RAxML* is a general *Maximum-Likelihood* software, its use in our context is fairly degenerated, as we only use it to infer optimal quartet topologies.

After having the optimal tree topology for each of the sampled quartet P-blocks, we need to amalgamate them into a single tree spanning the entire taxa set. This task is denoted the *Supertree Task* [6] and is known to be *NP hard*, even for the special case where the input is limited to quartets topologies, as in our case [51]. Nevertheless there are several heuristics for this task, with *MRP* [4,40] the most popular. Here we chose to use *Quartet MaxCut* [46,47] that proved to be faster and more accurate for this kind of input [2]. In brief, *Quartet MaxCut* partitions recursively the taxa set where each such partition corresponds to a split in the final tree. In each such recursive step, a graph over the taxa set is built where the set of quartets induces the edge set in that graph. The idea is to partition the vertex set (the taxa) such that the minimum quartets are violated. This is achieved by a *semidefinite-programming*-like algorithm that embeds the graph on the unit sphere and applies a random hyperplane through the sphere.

## 2.3  Implementation

To keep the runtime of our software manageable, we integrated the *RAxML* code directly into our program code. We parallelized our program with *openmp* [36].

# 3  Test Results

To evaluate *Multi-SpaM* and to compare it to other fast, alignment-free methods, we applied these approaches to both simulated and real sequence data and compared the resulting trees to reference trees. In phylogeny reconstruction, artificial benchmark data are often used since here, 'correct' reference trees are known. For the real-world sequence data that we used in our study, we had to rely on reference trees that are believed to reflect the true evolutionary history, or on trees calculated using traditional, alignment-based methods that can be considered to be reasonably accurate. In our test runs, we used standard parameters for all methods, if such parameters were suggested by the respective program authors. The program *kmacs* [29] that was one of the programs that we evaluated, has no default value for its only parameter, the number $k$ of allowed mismatches in common substrings. Here, we chose a value of $k = 4$. While *Multi-SpaM* produces tree topologies without branch lengths, all other methods that we compared produce distance matrices. To generate trees with these methods, we applied *Neighbor-joining* [44] to the distances produced by these methods.

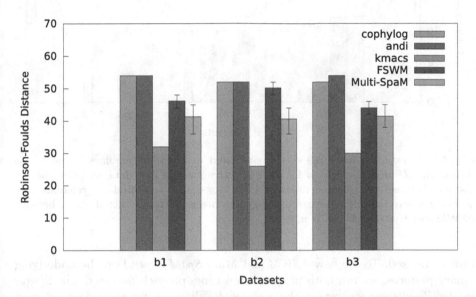

**Fig. 3.** Average *Robinson-Foulds (RF)* distances between trees calculated with alignment-free methods and reference trees for three datasets of simulated bacterial genomes. *FSWM* and *Multi-SpaM* were run 10 times, with different patterns $P$ generated (see main text). Error bars indicate the lowest and highest *RF* distances, respectively.

To compare the trees produced by the different alignment-free methods to the respective benchmark trees, we used the *Robinson-Foulds (RF)* metric [42], a standard measure to compare how different two tree topologies are. Thus, the smaller the *RF* distances between reconstructed trees and the corresponding reference trees are, the better a method is. To calculate *Neighbor-joining* trees and to calculate *RF* distances between the obtained trees and the respective reference trees, we used the *PHYLIP* package [16]. As explained above, both *FSWM* and *Multi-SpaM* rely on binary patterns of *match* and *don't care* positions; the results of these programs therefore depend on the underlying patterns. Both programs use the software *rasbhari* [21] to calculate binary patterns. *rasbhari* uses a probabilistic algorithm, so different program runs usually return different patterns and, as a result, different program runs with *FSWM* and *Multi-SpaM* may produce slightly different distance estimates, even if the same parameter

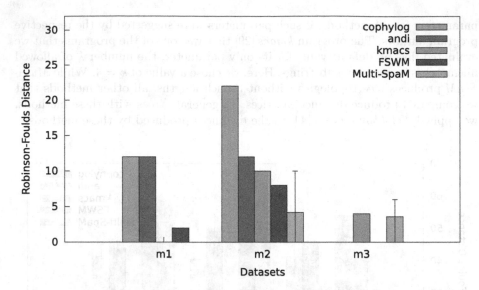

**Fig. 4.** *RF* distances for three sets of simulated mammalian genomes. If no bar is shown, the *RF* distance is zero for the respective method and data set. *E.g.* the *RF* distance between the tree generated by *kmacs* for data set *m1* and the reference tree is zero, *i.e.* here the reference tree topology was precisely reconstructed. Error bars for *FSWM* and *Multi-SpaM* are as in Fig. 3.

values are used. To see how *FSWM* and *Multi-SpaM* depend on the underlying binary patterns, we ran both programs ten times on each data set. The figures in the *Results* section report the *average RF*-distance for each data set over the ten program runs. Error bars indicate the highest and lowest *RF*-distances, respectively, for the 10 program runs.

## 3.1   Simulated Sequences

At first, we evaluated *Multi-SpaM* on datasets generated with the *Artificial Life Framework (ALF)* [12]. *ALF* starts by simulating an ancestral genome that includes a number of genes. According to a guide tree which is either provided by the user or randomly generated, *ALF* simulates speciation events and other evolutionary events such as substitutions, insertions and deletions for nucleotides, as well as duplications, deletions and horizontal transfer of entire genes. A large number of parameters can be specified by the user for these events. We used parameter files that were used in a study by the authors of *ALF* [12]. This way, we generated six datasets, three with simulated $\gamma$-proteobacterial genomes $(b1, b2, b3)$, and three with simulated mammalian genomes $(m1, m2, m3)$. We used the base parameter sets for each dataset and only slightly modified them to generate *DNA* sequences for roughly 1,000 genes per taxon which we then concatenated to full genomes. As in [12], we used parameter values 7.2057 and 401.4189 for the length distribution of the simulated bacterial sequences and

1.7781 and 274.1061, respectively, for the length distribution of the simulated mammalian sequences. Within each data set, we used the same rate for gene duplication, gene loss and horizontal gene transfer, but we used different rates for different data sets. For the six data sets, the corresponding rates were set to 0.0025 ($b1$), 0.0018 ($b2$), 0.0017 ($b3$), 0.0058 ($m1$), 0.0068 ($m2$), 0.011 ($m3$), respectively.

Each data generated in this way set contains 30 genomes (taxa) and has a size of around 10 mb. As shown in Fig. 3, none of the tools that we evaluated were able to exactly reconstruct the reference tree topologies for the simulated bacterial genomes. In contrast, reference topologies for the simulated mammalian genomes could be reconstructed by some tools, although no method could reconstruct all three reference topologies exactly, see Fig. 4.

## 3.2    Real Genomes

We also applied the programs that we evaluated to real genomes to see if the results are similar to our results on simulated genomes. Here, our first dataset were 29 *E. coli* and *Shigella* genomes which are commonly used as a benchmark dataset to evaluate alignment-free methods [23]. As a reference, we used a tree calculated with *Maximum Likelihood*, based on a *mugsy* alignment [1]. The dataset is 144 mb large and the average distance between two sequences in this set is about 0.0166 substitutions per sequence position. Next, we used a set of 32 *Roseobacter* genomes of 132 mb with a reference tree published by [34]; here the distance between sequence pairs was 0.233 substitutions per position on average. As a third benchmark set, we used 19 *Wolbachia* genomes that have been analyzed by [19]; we used the phylogeny published in their paper as a reference. The total size of this sequence set is 25 mb, the average pairwise distance is 0.06 substitutions per position. The results of these three series of test runs are summarized in Fig. 5.

Finally, we applied our dataset to a much larger dataset of eukaryotic genomes. It consists of 14 plant genomes totalling 4.8 gb. Figure 6 shows the resulting trees. For this data set, we used a pattern with a weight of $w = 12$ instead of the default value $w = 10$, to keep the number of background spaced-word matches manageable. For a dataset of this size, the number of additional score calculations would increase the runtime unnecessarily if one would use the default weight of $w = 10$. As can be seen in Fig. 6, *Multi-SpaM* and *FSWM* produced fairly accurate trees for this data set, with only minor differences to the reference tree: *Multi-SpaM* misclassified *Carica papaya*, whereas *FSWM* failed to classify *Brassica rapa* correctly. None of the other alignment-free tools that we evaluated could produce a reasonable tree for this data set: *andi* returned a tree that is rather different to the reference tree, while *kmacs* and *co-phylog* could not finish the program runs in a reasonable timeframe.

As explained in the *Method* section, *Multi-SpaM* calculates an optimal tree topology for each of the sampled *quartet P-blocks*. Here, it can happen that no single best topology is found. In particular for closely related sequences, this happens for a large fraction of the sampled quartet P-blocks. For the *E. coli/Shigella*

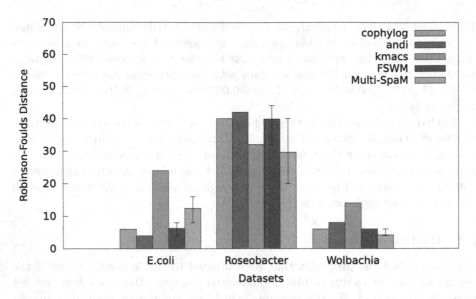

**Fig. 5.** *RF* distances for three sets of benchmark genomes: 29 *E. coli/Shigella* genomes, 32 *Roseobacter* genomes and 19 *Wolbachia* genomes. Error bars for *FSWM* and *Multi-SpaM* as in Fig. 3.

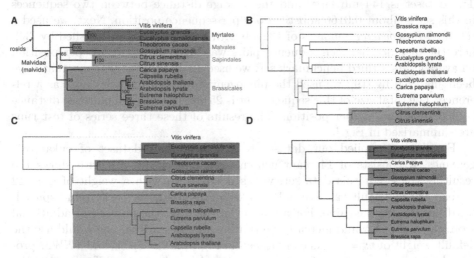

**Fig. 6.** Reference tree (**A**) from [22] and trees reconstructed by *andi* (**B**), *FSWM* (**C**) and *Multi-SpaM* (**D**) for a set of 14 plant genomes.

data set, for example, around 50% of the quartet blocks were inconclusive, *i.e.* *RAxML* could find no single best tree topology. We observed a similar result for a dataset of 13 *Brucella* genomes where the pairwise phylogenetic distances are even smaller than for the *E. coli/Shigella* data set, namely 0.0019 substitutions per site, on average. Here, roughly 80% of the blocks were inconclusive. For all other datasets, the fraction of inconclusive quartet $P$-blocks was negligible. For example, for the set of 14 plant genomes, only ~250 out of the 1,000,000 sampled $P$-blocks were inconclusive.

### 3.3   Program Runtime and Memory Usage

Table 1 shows the program run time for *Multi-SpaM, FSWM, kmacs, andi* and *co-phylog* on the three real-world data sets in our program comparison. The test runs were done on a 5 x Intel(R) Xeon(R) CPU E7-4850 with 2.00 GHz, a total of 40 threads (20 cores). For the largest data set in our study, the set of 14 plant genomes, the peak $RAM$ usage was 76 GB for *FSWM*, 110 GB for *andi* and 142 GB for *Multi-SpaM*. In memory saving mode, the peak $RAM$ usage of *Mult-SpaM* could be reduced to 10.5 GB, but this roughly doubles the program run time.

**Table 1.** Runtime in seconds for different alignment-free approaches on the four sets of real-world genomes that we used in this study. On the largest data set, the 14 plant genomes, *kmacs* and *co-phylog* did not terminate the program run. On this data set, we increased the pattern weight for *Multi-SpaM* from the default value of $w = 10$ to $w = 12$, in order to reduce the run time. Note that *Multi-SpaM, FSWM* and *andi* are parallelized, so we could run them on multiple processors, while *kmacs* and *co-phylog* had to be run on single processors. The reported run times are *wall-clock* times.

|            | M.-SpaM | FSWM      | kmacs  | andi  | co-phylog |
|------------|---------|-----------|--------|-------|-----------|
| E. coli/Shig. | 683  | 906       | 41,336 | 11    | 443       |
| Roseobac.  | 7,991   | 746       | 13,163 | 17    | 615       |
| Wolbachia  | 382     | 70        | 17,581 | 2.6   | 58        |
| Plants     | 27,072  | 1,107,720 | -      | 1,808 | -         |

## 4   Discussion

While standard methods for phylogeny reconstruction are time consuming because they rely on multiple sequence alignments and on time-consuming probabilistic calculations, recent so-called 'alignment-free' methods are orders of magnitudes faster. Existing alignment-free phylogeny methods are *distance based* approaches which are generally regarded to be less accurate than *character-based* approaches. In this paper, we introduced a novel approach to phylogeny reconstruction called *Multi-SpaM* to combine the speed of alignment-free methods with the accuracy of *Maximum Likelihood*. To our knowledge, this is the first alignment-free approach that uses multiple sequence comparison and Likelihood.

Our test runs show that *Multi-SpaM* produces phylogenetic trees of high quality. It outperforms other alignment-free methods, in particular on sequences with large evolutionary distances. For closely related input sequences, such as different strains of the same bacterial species, however, our approach was sometimes outperformed by other alignment-free methods. As shown in Fig. 5, the programs *andi, co-phylog* and *FSWM* produce better results on a set of *E. coli/Shigella* genomes than *Multi-SpaM*. This may be due to our above mentioned observation that for many *quartet P-blocks* no single best tree topology can be found if the compared sequences are very similar to each other.

Calculating optimal tree topologies for the sampled *quartet P-blocks* is a relatively time-consuming step in *Multi-SpaM*. In fact, we observed that the program run time is roughly proportional to the number of *quartet blocks* for which topologies are calculated. However, the maximal number of *quartet blocks* that are sampled is a user-defined parameter. By default we sample up to $M = 1,000,000$ quartet blocks; in our test runs, the quality of the resulting trees could not be significantly improved by further increasing $M$. Consequently, our method is relatively fast on large data sets, where only a small fraction of the possible quartet-blocks is sampled. By contrast, on small data sets, *Multi-SpaM* is slower than other alignment-free methods. We have parallelized our software to run on multiple cores; the run times in Table 1 are wall-clock run times. It should be straight-forward to adapt our software to be run on distributed systems, as has been done for other alignment-free approaches [8, 38].

Apart from the maximum number of sampled quartet blocks, the only relevant parameters of our approach are the *length* and the *weight* – *i.e.* number of *match positions* – of the underlying binary pattern. For *Multi-SpaM*, we used similar default parameter values as in *Filtered Spaced Word Matches (FSWM)* [31], namely a weight of $w = 10$ and a pattern length of $\ell = 110$, *i.e.* our default patterns have 10 *match* positions and 100 *don't-care* positions. As mentioned in Sect. 2.1, a large number of *don't care* positions is important in *FSWM* and *Multi-SpaM* as this makes it easier to distinguish homologous from random background spaced-word matches. Also, with a large number of *don't-care* positions, the number of *inconclusive* quartet *P*-blocks is reduced for sequences with a high degree of similarity. On the other hand, we found that the number of *match* positions has less impact on the performance of *Multi-SpaM*. On large data sets it is advisable to increase the *weight* of $P$ since this reduces the fraction of background spaced-word matches, and therefore the number of spaced words for which their scores need to be calculated by the program. A higher weight, thus, reduces the program run time. For the largest data set our study, the set of plant genomes, we increased the pattern weight from the default value of 10 to a value of 12 to keep the run time of *Multi-SpaM* low.

Our current implementation uses the previously developed software *Quartet MaxCut* by [46, 47] to calculate supertrees from quartet tree topologies. We are using this program since it is faster than other supertree approaches. As a result, the current version of *Multi-SpaM* generates tree *topologies* only, *i.e.* trees without branch lengths. We will investigate in the future, if our approach can be extended to calculate full phylogenetic trees with branch lengths, based on the same *quartet P*-blocks that we have used in the present study.

**Funding.** The project was funded by *VW Foundation*, project *VWZN3157*. We acknowledge support by the *Open Access Publication Funds of Göttingen University*.

# References

1. Angiuoli, S.V., Salzberg, S.L.: Mugsy: fast multiple alignment of closely related whole genomes. Bioinformatics **27**, 334–342 (2011)
2. Avni, E., Yona, Z., Cohen, R., Snir, S.: The performance of two supertree schemes compared using synthetic and real data quartet input. J. Mol. Evol. **86**, 150–165 (2018)
3. Ayad, L.A., Charalampopoulos, P., Iliopoulos, C.S., Pissis, S.P.: Longest common prefixes with $k$-errors and applications. arXiv:1801.04425 [cs.DS] (2018)
4. Baum, B.: Combining trees as a way of combining data sets for phylogenetic inference. Taxon **41**, 3–10 (1992)
5. Bernard, G., Chan, C.X., Ragan, M.A.: Alignment-free microbial phylogenomics under scenarios of sequence divergence, genome rearrangement and lateral genetic transfer. Sci. Rep. **6**, 28970 (2016)
6. Bininda-Emonds, O.R.P.: Phylogenetic Supertrees: Combining Information to Reveal the Tree of Life. Computational Biology. Springer, Netherlands (2004). https://doi.org/10.1007/978-1-4020-2330-9
7. Bromberg, R., Grishin, N.V., Otwinowski, Z.: Phylogeny reconstruction with alignment-free method that corrects for horizontal gene transfer. PLoS Comput. Biol. **12**, e1004985 (2016)
8. Cattaneo, G., Ferraro Petrillo, U., Giancarlo, R., Roscigno, G.: An effective extension of the applicability of alignment-free biological sequence comparison algorithms with Hadoop. J. Supercomput. **73**, 1467–1483 (2017)
9. Chiaromonte, F., Yap, V.B., Miller, W.: Scoring pairwise genomic sequence alignments. In: Altman, R.B., Dunker, A.K., Hunter, L., Klein, T.E. (eds.) Pacific Symposium on Biocomputing, Lihue, Hawaii, pp. 115–126 (2002)
10. Chor, B., Tuller, T.: Maximum likelihood of evolutionary trees is hard. In: Miyano, S., Mesirov, J., Kasif, S., Istrail, S., Pevzner, P.A., Waterman, M. (eds.) RECOMB 2005. LNCS, vol. 3500, pp. 296–310. Springer, Heidelberg (2005). https://doi.org/10.1007/11415770_23
11. Comin, M., Schimd, M.: Assembly-free genome comparison based on next-generation sequencing reads and variable length patterns. BMC Bioinform. **15**, S1 (2014)
12. Dalquen, D.A., Anisimova, M., Gonnet, G.H., Dessimoz, C.: ALF - a simulation framework for genome evolution. Mol. Biol. Evol. **29**, 1115–1123 (2012)
13. Dencker, T., Leimeister, C.A., Morgenstern, B.: *Multi-SpaM*: a maximum-likelihood approach to phylogeny reconstruction based on multiple spaced-word matches. arxiv.org/abs/1803.09222 [q-bio.PE] (2018). http://arxiv.org/abs/1703.08792
14. Farris, J.S.: Methods for computing wagner trees. Syst. Biol. **19**, 83–92 (1970)
15. Felsenstein, J.: Evolutionary trees from DNA sequences:a maximum likelihood approach. J. Mol. Evol. **17**, 368–376 (1981)
16. Felsenstein, J.: PHYLIP - phylogeny inference package (version 3.2). Cladistics **5**, 164–166 (1989)
17. Fitch, W.: Toward defining the course of evolution: minimum change for a specific tree topology. Syst. Zool. **20**, 406–416 (1971)
18. Foulds, L., Graham, R.: The steiner problem in phylogeny is NP-complete. Adv. Appl. Math. **3**, 43–49 (1982)
19. Gerth, M., Bleidorn, C.: Comparative genomics provides a timeframe for *Wolbachia* evolution and exposes a recent biotin synthesis operon transfer. Nat. Microbiol. **2**, 16241 (2016)

20. Girotto, S., Comin, M., Pizzi, C.: FSH: fast spaced seed hashing exploiting adjacent hashes. Algorithms Mol. Biol. **13**, 8 (2018)
21. Hahn, L., Leimeister, C.A., Ounit, R., Lonardi, S., Morgenstern, B.: *rasbhari*: optimizing spaced seeds for database searching, read mapping and alignment-free sequence comparison. PLOS Comput. Biol. **12**(10), e1005107 (2016)
22. Hatje, K., Kollmar, M.: A phylogenetic analysis of the brassicales clade based on an alignment-free sequence comparison method. Front. Plant Sci. **3**, 192 (2012)
23. Haubold, B., Klötzl, F., Pfaffelhuber, P.: andi: fast and accurate estimation of evolutionary distances between closely related genomes. Bioinformatics **31**, 1169–1175 (2015)
24. Haubold, B., Pfaffelhuber, P., Domazet-Loso, M., Wiehe, T.: Estimating mutation distances from unaligned genomes. J. Comput. Biol. **16**, 1487–1500 (2009)
25. Horwege, S., et al.: *Spaced words and kmacs*: fast alignment-free sequence comparison based on inexact word matches. Nucl. Acids Res. **42**, W7–W11 (2014)
26. Ilie, L., Ilie, S., Bigvand, A.M.: SpEED: fast computation of sensitive spaced seeds. Bioinformatics **27**, 2433–2434 (2011)
27. Ilie, S.: Efficient Computation of Spaced Seeds. BMC Res. Notes **5**, 123 (2012)
28. Leimeister, C.A., Boden, M., Horwege, S., Lindner, S., Morgenstern, B.: Fast alignment-free sequence comparison using spaced-word frequencies. Bioinformatics **30**, 1991–1999 (2014)
29. Leimeister, C.A., Morgenstern, B.: kmacs: the $k$-mismatch average common substring approach to alignment-free sequence comparison. Bioinformatics **30**, 2000–2008 (2014)
30. Leimeister, C.A., Schellhorn, J., Schöbel, S., Gerth, M., Bleidorn, C., Morgenstern, B.: Prot-SpaM: Fast alignment-free phylogeny reconstruction based on whole-proteome sequences. bioRxiv (2018). https://doi.org/10.1101/306142
31. Leimeister, C.A., Sohrabi-Jahromi, S., Morgenstern, B.: Fast and accurate phylogeny reconstruction using filtered spaced-word matches. Bioinformatics **33**, 971–979 (2017)
32. Morgenstern, B., Schöbel, S., Leimeister, C.A.: Phylogeny reconstruction based on the length distribution of k-mismatch common substrings. Algorithms Mol. Biol. **12**, 27 (2017)
33. Morgenstern, B., Zhu, B., Horwege, S., Leimeister, C.A.: Estimating evolutionary distances between genomic sequences from spaced-word matches. Algorithms Mol. Biol. **10**, 5 (2015)
34. Newton, R., et al.: Genome characteristics of a generalist marine bacterial lineage. ISME J. **4**, 784–798 (2010)
35. Noé, L.: Best hits of 11110110111: model-free selection and parameter-free sensitivity calculation of spaced seeds. Algorithms Mol. Biol. **12**, 1 (2017)
36. OpenMP Forum: OpenMP C and C++ Application Program Interface, Version 2.0. Technical report (2002). http://www.openmp.org
37. Ounit, R., Lonardi, S.: Higher classification accuracy of short metagenomic reads by discriminative spaced $k$-mers. In: Pop, M., Touzet, H. (eds.) WABI 2015. LNCS, vol. 9289, pp. 286–295. Springer, Heidelberg (2015). https://doi.org/10.1007/978-3-662-48221-6_21
38. Petrillo, U.F., Guerra, C., Pizzi, C.: A new distributed alignment-free approach to compare whole proteomes. Theor. Comput. Sci. **698**, 100–112 (2017)
39. Pizzi, C.: MissMax: alignment-free sequence comparison with mismatches through filtering and heuristics. Algorithms Mol. Biol. **11**, 6 (2016)
40. Ragan, M.: Matrix representation in reconstructing phylogenetic-relationships among the eukaryotes. Biosystems **28**, 47–55 (1992)

41. Ren, J., Bai, X., Lu, Y.Y., Tang, K., Wang, Y., Reinert, G., Sun, F.: Alignment-free sequence analysis and applications. Annu. Rev. Biomed. Data Sci. **1**, 93–114 (2018)

42. Robinson, D.F., Foulds, L.: Comparison of phylogenetic trees. Math. Biosci. **53**, 131–147 (1981)

43. Roychowdhury, T., Vishnoi, A., Bhattacharya, A.: Next-generation anchor based phylogeny (NexABP): constructing phylogeny from next-generation sequencing data. Sci. Rep. **3**, 2634 (2013)

44. Saitou, N., Nei, M.: The neighbor-joining method: a new method for reconstructing phylogenetic trees. Mol. Biol. Evol. **4**, 406–425 (1987)

45. Sievers, F., et al.: Fast, scalable generation of high-quality protein multiple sequence alignments using Clustal Omega. Mol. Syst. Biol. **7**, 539 (2011)

46. Snir, S., Rao, S.: Quartets MaxCut: a divide and conquer quartets algorithm. IEEE/ACM Trans. Comput. Biology Bioinform. **7**, 704–718 (2010)

47. Snir, S., Rao, S.: Quartet MaxCut: a fast algorithm for amalgamating quartet trees. Mol. Phylogenetics Evol. **62**, 1–8 (2012)

48. Song, K., Ren, J., Reinert, G., Deng, M., Waterman, M.S., Sun, F.: New developments of alignment-free sequence comparison: measures, statistics and next-generation sequencing. Brief. Bioinform. **15**, 343–353 (2014)

49. Song, K., Ren, J., Zhai, Z., Liu, X., Deng, M., Sun, F.: Alignment-free sequence comparison based on next-generation sequencing reads. J. Comput. Biol. **20**, 64–79 (2013)

50. Stamatakis, A.: RAxML version 8: a tool for phylogenetic analysis and post-analysis of large phylogenies. Bioinformatics **30**, 1312–1313 (2014)

51. Steel, M.: The complexity of reconstructing trees from qualitative characters and subtress. J. Classif. **9**, 91–116 (1992)

52. Tavaré, S.: Some probabilistic and statistical problems on the analysis of DNA sequences. Lect. Math. Life Sci. **17**, 57–86 (1986)

53. Thankachan, S.V., Apostolico, A., Aluru, S.: A provably efficient algorithm for the $k$-mismatch average common substring problem. J. Comput. Biol. **23**, 472–482 (2016)

54. Thankachan, S.V., Chockalingam, S.P., Liu, Y., Aluru, A.K.S.: A greedy alignment-free distance estimator for phylogenetic inference. BMC Bioinform. **18**, 238 (2017)

55. Ulitsky, I., Burstein, D., Tuller, T., Chor, B.: The average common substring approach to phylogenomic reconstruction. J. Comput. Biol. **13**, 336–350 (2006)

56. Yi, H., Jin, L.: Co-phylog: an assembly-free phylogenomic approach for closely related organisms. Nucl. Acids Res. **41**, e75 (2013)

57. Zielezinski, A., Vinga, S., Almeida, J., Karlowski, W.M.: Alignment-free sequence comparison: benefits, applications, and tools. Genome Biol. **18**, 186 (2017)

# FastNet: Fast and Accurate Statistical Inference of Phylogenetic Networks Using Large-Scale Genomic Sequence Data

Hussein A. Hejase[1], Natalie VandePol[2], Gregory M. Bonito[2], and Kevin J. Liu[3(✉)]

[1] Simons Center for Quantitative Biology, Cold Spring Harbor Laboratory, Cold Spring Harbor, NY 11724, USA
[2] Department of Plant, Soil and Microbial Sciences, Michigan State University, East Lansing, MI 48824, USA
[3] Department of Computer Science and Engineering, Michigan State University, East Lansing, MI 48824, USA
kjl@msu.edu

**Abstract.** An emerging discovery in phylogenomics is that interspecific gene flow has played a major role in the evolution of many different organisms. To what extent is the Tree of Life not truly a tree reflecting strict "vertical" divergence, but rather a more general graph structure known as a phylogenetic network which also captures "horizontal" gene flow? The answer to this fundamental question not only depends upon densely sampled and divergent genomic sequence data, but also computational methods which are capable of accurately and efficiently inferring phylogenetic networks from large-scale genomic sequence datasets. Recent methodological advances have attempted to address this gap. However, in the 2016 performance study of Hejase and Liu, state-of-the-art methods fell well short of the scalability requirements of existing phylogenomic studies.

The methodological gap remains: how can phylogenetic networks be accurately and efficiently inferred using genomic sequence data involving many dozens or hundreds of taxa? In this study, we address this gap by proposing a new phylogenetic divide-and-conquer method which we call FastNet. We conduct a performance study involving a range of evolutionary scenarios, and we demonstrate that FastNet outperforms state-of-the-art methods in terms of computational efficiency and topological accuracy.

## 1 Introduction

Recent advances in biomolecular sequencing [30] and evolutionary modeling and inference [10,34] set the stage for a new era of phylogenomics. One major outcome is the discovery that interspecific gene flow has played a major role in the

**Electronic supplementary material** The online version of this chapter (https://doi.org/10.1007/978-3-030-00834-5_14) contains supplementary material, which is available to authorized users.

© Springer Nature Switzerland AG 2018
M. Blanchette and A. Ouangraoua (Eds.): RECOMB-CG 2018, LNBI 11183, pp. 242–259, 2018.
https://doi.org/10.1007/978-3-030-00834-5_14

evolution of many different organisms across the Tree of Life [1,23,29], including humans and ancient hominins [15,39], butterflies [44], mice [28], and fungi [14]. These findings point to new directions for phylogenetics and phylogenomics: to what extent is the Tree of Life not truly a tree reflecting strict vertical divergence, but rather a more general graph structure known as a phylogenetic network where reticulation edges and nodes capture gene flow? And what is the evolutionary role of gene flow? In addition to densely sampled and divergent genomic sequence data, one additional ingredient is needed to make progress on these questions: computational methods which are capable of accurately and efficiently inferring phylogenetic networks on large-scale genomic sequence datasets.

Recent methodological advances have attempted to address this gap. Solís-Lemus and Ané proposed SNaQ [42], a new statistical method which seeks to address the computational efficiency of species network inference using a pseudo-likelihood approximation. The method of Yu and Nakhleh [45] (referred to here as MPL, which stands for maximum pseudo-likelihood) substitutes pseudo-likelihoods in place of the full model likelihoods used by the methods of Yu et al. [48] (referred to here as MLE, which stands for maximum likelihood estimation, and MLE-length, which differ based upon whether or not gene tree branch lengths contribute to model likelihood). Two of us recently conducted a performance study which demonstrated the scalability limits of SNaQ, MPL, MLE, MLE-length, and other state-of-the-art phylogenetic methods in the context of phylogenetic network inference [17]. The scalability of the state of the art falls well short of that required by current phylogenetic studies, where many dozens or hundreds of divergent genomic sequences are common [34]. The most accurate phylogenetic network inference methods performed statistical inference under phylogenomic models [42,47,48] that extended the multi-species coalescent model [16,24]. MPL and SNaQ were among the fastest of these methods while MLE and MLE-length were the most accurate. None of the statistical phylogenomic inference methods completed analyses of datasets with 30 taxa or more after many weeks of CPU runtime – not even the pseudo-likelihood-based methods which were devised to address the scalability limitations of other statistical approaches. The remaining methods fell into two categories: split-based methods [4,7] and the parsimony-based inference method of Yu et al. [46] (which we refer to as MP in this study). Both categories of methods were faster than the statistical phylogenomic inference methods but less accurate.

The methodological gap remains: how can species networks be accurately and efficiently inferred using large-scale genomic sequence datasets? In this study, we address this question and propose a new method for this problem. We investigate this question in the context of two constraints. We focus on dataset size in terms of the number of taxa and the number of reticulations in the species phylogeny. We note that scalability issues arise due to other dataset features as well, including population-scale allele sampling for each taxon in a study.

## 2  Methods

One path forward is through the use of divide-and-conquer. The general idea behind divide-and-conquer is to split the full problem into smaller and more closely related subproblems, analyze the subproblems using state-of-the-art phylogenetic network inference methods, and then merge solutions on the subproblems into a solution on the full problem. Viewed this way, divide-and-conquer can be seen as a computational framework that "boosts" the scalability of existing methods (and which is distinct from boosting in the context of machine learning). The advantages of analyzing smaller and more closely related subproblems are two-fold. First, smaller subproblems present more reasonable computational requirements compared to the full problem. Second, the evolutionary divergence of taxa in a subproblem is reduced compared to the full set of taxa, which has been shown to improve accuracy for phylogenetic tree inference [11,19,26]. We and others have successfully applied divide-and-conquer approaches to enable scalable inference in the context of species tree estimation [26,27,33].

Here, we consider the more general problem of inferring species phylogenies that are directed phylogenetic networks. A directed phylogenetic network $N = (V, E)$ consists of a set of nodes $V$ and a set of directed edges $E$. The set of nodes $V$ consists of a root node $r(N)$ with in-degree 0 and out-degree 2, leaves $\mathcal{L}(N)$ with in-degree 1 and out-degree 0, tree nodes with in-degree 1 and out-degree 2, and reticulation nodes with in-degree 2 and out-degree 1. A directed edge $(u, v) \in E$ is a tree edge if and only if $v$ is a tree node, and is otherwise a reticulation edge. Following the instantaneous admixture model used by Durand et al. [9], each reticulation node contributes a parameter $\gamma$, where one incoming edge has admixture frequency $\gamma$ and the other has admixture frequency $1 - \gamma$. The edges in a network $N$ can be labeled by a set of branch lengths $\ell$. A directed phylogenetic tree is a special case of a directed phylogenetic network which contains no reticulation nodes (and edges). An unrooted tree can be obtained from a directed tree by ignoring edge directionality.

The phylogenetic network inference problem consists of the following. One input is a partitioned multiple sequence alignment $A$ containing data partitions $a_i$ for $1 \leq i \leq k$, where each partition corresponds to the sequence data for one of $k$ genomic loci. Each of the $n$ rows in the alignment $A$ is a sample representing taxon $x \in X$, and each taxon is represented by one or more samples. Similar to other approaches [42,48], we also require an input parameter $C_r$ which specifies a hypothesized number of reticulations. We note that increasing $C_r$ for a given input alignment $A$ results in a solution with either better or equal likelihood under the evolutionary models used in our study and others [42,48]. As is common practice for this and many other statistical inference/learning problems, inference can be coupled with standard model selection techniques (e.g., information criteria [2,3,20,41], cross-validation, etc.) to balance model fit to the observed data against model complexity, thereby determining a suitable choice for parameter $C_r$ in an automated manner. The output consists of a directed phylogenetic network $N$ where each leaf in $\mathcal{L}(N)$ corresponds to a taxon $x \in X$.

## 2.1    The FastNet Algorithm

We now describe our new divide-and-conquer algorithm, which we refer to as FastNet. A flowchart of the algorithm is shown in Fig. 1. (Detailed pseudocode can be found in the Appendix's Supplementary Methods section.)

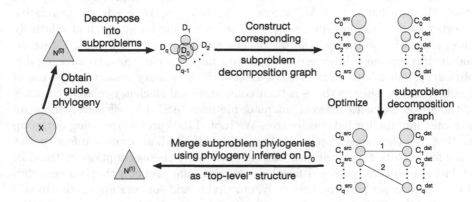

**Fig. 1. A high-level illustration of the FastNet algorithm.** First, a guide phylogeny $N^{(0)}$ is inferred on the full set of taxa $X$. Next, the guide phylogeny $N^{(0)}$ is used to decompose $X$ into subproblems $\{D_0, D_1, D_2, \ldots, D_{q-1}, D_q\} = D$. Then, the subproblem decomposition $D$ is used to construct a bipartite graph $G_D = (V_D, E_D)$, which is referred to as the subproblem decomposition graph. The set of vertices $V_D$ consist of two partitions: source vertices $V_D^{src} = \{C_0^{src}, C_1^{src}, \ldots, C_q^{src}\}$ where each subproblem $D_i$ has a corresponding source vertex $C_i^{src}$, and destination vertices $V_D^{dst} = \{C_0^{dst}, C_1^{dst}, \ldots, C_q^{dst}\}$ similarly. The subproblem decomposition graph $G_D$ is optimized to infer subproblem phylogenies and reticulations, where the latter are inferred based on the placement of weighted edges $e \in E_D$. Finally, the subproblem phylogenies are merged using the phylogeny inferred on $D_0$ as the "top-level" structure.

**Step Zero: Obtaining Local Gene Trees.** FastNet is a summary-based method for inferring phylogenetic networks. Subsequent steps of the FastNet algorithm (i.e., steps one and three) therefore utilize a list of gene trees $G$ as input, where the $i$th gene tree $g_i$ in list $G$ represents the evolutionary history of data partition $a_i$. The experiments in our study utilized either true or inferred gene trees as input to summary-based inference methods, including FastNet (see below for details). We used FastTree [37] to perform maximum likelihood estimation of local gene trees. Our study made use of an outgroup, and the unrooted gene trees inferred by FastTree were rooted on the leaf edge corresponding to the outgroup.

**Step One: Obtaining a Guide Phylogeny.** The subsequent subproblem decomposition step requires a rooted guide phylogeny $N^{(0)}$. The phylogenetic relationships need not be completely accurate. Rather, the guide phylogeny needs to be sufficiently accurate to inform subsequent divide-and-conquer steps.

Another requirement is that the method used for inferring the guide phylogeny must have reasonable computational requirements.

A range of different methods for obtaining guide phylogenies can satisfy these criteria. One option is the parsimony-based algorithm proposed by Yu et al. [46] to infer a rooted species network. The algorithm is implemented in the PhyloNet software package [43]. We refer to this method as MP. In a previous simulation study [17], we found that MP offers a significant runtime advantage relative to other state-of-the-art species network inference methods, but had relatively lower topological accuracy. Another option is using ASTRAL [31,32], a state-of-the-art phylogenomic inference method that infers species trees, to infer a guide phylogeny that is a tree rather than a network. A primary reason for the use of species tree inference methods is their computational efficiency relative to state-of-the-art phylogenetic network inference methods. ASTRAL effectively infers an unrooted and undirected species tree. We rooted the species tree using outgroup rooting. Another consideration is that, while ASTRAL accurately infers species trees for evolutionary scenarios lacking gene flow, the assumption of tree-like evolution is generally invalid for the computational problem that we consider. As we show in our performance study, our divide-and-conquer approach can still be applied despite this limitation, suggesting that FastNet is robust to guide phylogeny error. For this reason, the FastNet experiments in our study exclusively use ASTRAL to infer guide phylogenies.

**Step Two: Subproblem Decomposition.** The rooted and directed guide phylogeny $N^{(0)}$ is then used to produce a subproblem decomposition $D$. The decomposition $D$ consists of a "bottom-level" component and a "top-level" component, which refers to the subproblem decomposition technique. The bottom-level component is comprised of disjoint subsets $D_i$ for $1 \leq i \leq q$ which partition the set of taxa $X$ such that $\bigcup_{1 \leq i \leq q} D_i = X$. We refer to each subset $D_i$ as a bottom-level subproblem. The top-level component consists of a top-level subproblem $D_0$ which overlaps each bottom-level subproblem $D_i$ where $1 \leq i \leq q$.

The bottom-level component of the subproblem decomposition is obtained using the following steps. First, for each reticulation node in $N^{(0)}$, we delete the incoming edge with lower admixture frequency. Let $T^{(0)}$ be the resulting phylogeny, which contains no reticulation edges and is therefore a tree. Removal of any single edge in $T^{(0)}$ disconnects the tree into two subtrees; the leaves of the two subtrees will form two subproblems. We extend this observation to obtain decompositions with two or more subproblems. The decomposition is defined by $S$, a set of nodes in $T^{(0)}$. Each node $s \in S$ induces a corresponding subproblem $D_i$ for $1 \leq i \leq q$ which consists of the taxa corresponding to the leaves that are reachable from $s$ in $T^{(0)}$. Of course, not all decompositions are created equal. In this study, decompositions are constrained by the maximum subproblem size $c_m$; we also required a minimum of two subproblems in a decomposition. We obtained a decomposition using a greedy algorithm which is similar to the Center-Tree-$i$ decomposition used by Liu et al. [26] in the context of species tree inference. The two methods differ primarily due to their decomposition criteria. Initially the set $S$ consists of the root node $r(T^{(0)})$. The set $S$ is iteratively updated as follows: each iteration greedily selects a node $s \in S$ with maximal corresponding subproblem size, the node $s$ is removed from the set $S$ and replaced by its children. Iteration terminates when both decomposition criteria (the maximum subproblem size criterion and the minimum number of subproblems) are satisfied. If no decomposition satisfies the criteria, then the search is restarted using a maximum subproblem size of $c_m - 1$. In practice, the parameter $c_m$ is set to an empirically determined value which is based upon the largest datasets that state-of-the-art methods can analyze accurately within a reasonable timeframe [17]. The output of the search algorithm is effectively a search tree $T_{top}^{(0)}$ with a root corresponding to $r(T^{(0)})$, leaves corresponding to $s \in S$, and the subset of edges in $T^{(0)}$ which connect the root $r(T^{(0)})$ to the nodes $s \in S$ in $T^{(0)}$. The decomposition is obtained by deleting $T_{top}^{(0)}$'s corresponding structure in $T^{(0)}$, resulting in $q$ sub-trees which induce bottom-level subproblems as before.

The top-level component augments the subproblem decomposition with a single top-level subproblem $D_0$ which overlaps each bottom-level subproblem. Phylogenetic structure inferred on $D_0$ represents ancestral evolutionary relationships among bottom-level subproblems. Furthermore, overlap between the top-level subproblem $D_0$ and bottom-level subproblems is necessary for the subsequent merge procedure (see "Step four" below). The top-level subproblem $D_0$ contains representative taxa taken from each bottom-level subproblem $D_i$ for $1 \le i \le q$: for each bottom-level subproblem $D_i$, we choose the leaf in $T^{(0)}$ that is closest to the corresponding node $s \in S$ to represent $D_i$, and the corresponding taxon is included in the top-level subproblem $D_0$.

**Step Three: Subproblem Decomposition Graph Optimization.** Tree-based divide-and-conquer approaches reduce evolutionary divergence within subproblems by effectively partitioning the inference problem based on phylogenetic relationships. Within each part of the true phylogeny corresponding to a subproblem, the space of possible unrooted sub-tree topologies contributes a smaller set of distinct bipartitions (each corresponding to a possible tree edge) that need to be evaluated during search as compared to the full inference problem. The same insight can be applied to reticulation edges as well, except that a given reticulation is not necessarily restricted to a single subproblem.

We address the issue of "inter-subproblem" reticulations through the use of an abstraction which we refer to as a subproblem decomposition graph. A subproblem decomposition graph $G_D = (V_D, E_D)$ is a bipartite graph where the vertices $V_D$ can be partitioned into two sets: a set of source vertices $V_D^{src}$ and a set of destination vertices $V_D^{dst}$. There is a source vertex $C_i^{src} \in V_D^{src}$ for each distinct subproblem $D_i \in D$ where $0 \le i \le q$, and similarly for destination vertices $C_i^{dst} \in V_D^{dst}$. An undirected edge $e_{ij} \in E_D$ connects a source vertex $C_i^{src}$ to a destination vertex $C_j^{dst}$ where $i \le j$ and has a weight $w(e_{ij}) \in \mathcal{N}^+$. If an edge $e_{ii}$ connects nodes $C_i^{src}$ and $C_i^{dst}$ that correspond to the same subproblem $D_i \in D$, then the edge weight $w(e_{ii}) > 0$ specifies the number of reticulations in the phylogenetic network to be inferred on subproblem $D_i$; otherwise, a phylogenetic tree is to be inferred on subproblem $D_i$. If an edge $e_{ij}$ connects nodes $C_i^{src}$ and

$C_j^{\text{dst}}$ where $i < j$, then the edge weight $w(e_{ij}) > 0$ specifies the number of "inter-subproblem" reticulations between the subproblems $D_i$ and $D_j$ (where an inter-subproblem reticulation is a reticulation with one incoming edge which is incident from the phylogeny to be inferred on subproblem $D_i$ and the other incoming edge which is incident from the phylogeny to be inferred on $D_j$); otherwise, no reticulations are to be inferred between the two subproblems. A subproblem decomposition graph is constrained to have a total number of reticulations such that $\sum_{e \in E_D} w(e) = C_r$.

Given a subproblem decomposition $D$, FastNet's search routines make use of the correspondence between a subproblem decomposition graph $G_D$ and a multiset with cardinality $C_r$ that is chosen from $\binom{q+1}{2} + (q+1)$ elements, where $q$ is the number of bottom-level subproblems and there are $(q+1)$ subproblems in $D$. Enumeration over corresponding multisets is feasible when the number of subproblems and $C_r$ are sufficiently small; otherwise, perturbations of a corresponding multiset can be used as part of a local search heuristic. See Algorithm 1 in the Appendix's Supplementary Methods section for detailed pseudocode.

A subproblem decomposition graph $G_D$ facilitates phylogenetic inference given a subproblem decomposition $D$. The resulting inference is evaluated with respect to a pseudo-likelihood-based criterion. Pseudocode for the pseudo-likelihood calculation is shown in Algorithm 2 in the Appendix's Supplementary Methods section.

The first step is to analyze each individual subproblem $D_i \in D$ where $0 \leq i \leq q$. If an edge $e_{ii}$ exists, then a phylogenetic network with $w(e_{ii})$ reticulations is inferred on the corresponding subproblem $D_i$; otherwise, a phylogenetic tree is inferred. We used one of three different summary-based methods to perform phylogenetic inference on subproblems, which we refer to as a base method: two likelihood-based methods – MLE and MLE-length – as well as MPL, a pseudo-likelihood-based method. Due to the modular design of FastNet's divide-and-conquer algorithm, topological constraints on a base method's inference will also apply to FastNet. To simplify discussion, the remainder of the algorithm description will assume the use of MLE as a base method.

Next, reticulations are inferred "between" pairs of subproblems as follows. Let $N_i$ and $N_j$ where $i \neq j$ be the networks inferred on subproblems $D_i$ and $D_j$, respectively, using the above procedure. Construct the cherry given by the Newick-formatted [12] string "$(N_i: b_i, N_j: b_j)$ANC;", which consists of a new root node ANC with children $r(N_i)$ and $r(N_j)$ where $N_i$ and $N_j$ are respectively retained as sub-phylogenies. Then, infer branch lengths $b_i$ and $b_j$ and add $w(e_{ij})$ reticulations under the maximum likelihood criterion used by the base method. For pairs of subproblems not involving the top-level subproblem $D_0$, we used the base method to perform constrained optimization. For pairs of subproblems involving the top-level subproblem $D_0$, we used a greedy heuristic: initial placements were chosen arbitrarily for each reticulation, the source node for each reticulation edge was exhaustively optimized, and then the destination node for each reticulation edge was exhaustively optimized.

Inferred phylogenies and likelihoods were cached to ensure consistency among individual and pairwise subproblem analyses, which is necessary for the subsequent merge procedure. Caching also aids computational efficiency.

Finally, the subproblem decomposition graph and associated phylogenetic inferences are evaluated using a pseudo-likelihood criterion:

$$\prod_{0 \leq i \leq q} \delta[i, w(G_D, i, i)] \prod_{\substack{0 \leq i \leq q \\ i < j \leq q}} \psi[i, j, w(G_D, i, j), w(G_D, i, i), w(G_D, j, j)] \quad (1)$$

where $w(G_D, i, j)$ is the weight of edge $e_{ij}$ if it exists in $E(G_D)$ or 0 otherwise, $\delta[i, w(G_D, i, i)]$ is the cached likelihood for an individual subproblem $D_i$, and $\psi[i, j, w(G_D, i, j), w(G_D, i, i), w(G_D, j, j)]$ is the cached likelihood for a pair of subproblems $D_i$ and $D_j$ where $i < j$. The pseudo-likelihood calculation effectively assumes that subproblems are independent, although they are correlated through connecting edges in the model phylogeny. The choice of optimization criterion in this context represents a tradeoff between efficiency and accuracy, and several other state-of-the-art phylogenetic inference methods also use pseudo-likelihoods to analyze subsets of taxa (e.g., MPL and SNaQ). Other choices are possible. For example, an alternative would be to merge subproblem inferences into a single network hypothesis and calculate its likelihood under the multi-species network coalescent (MSNC) model.

We optimize subproblem decomposition graphs under the pseudo-likelihood criterion. Exhaustive enumeration of subproblem decomposition graphs is possible for the datasets in our study. Pseudocode to obtain a global optimum is shown in Algorithm 3 in the Appendix's Supplementary Methods section. For larger datasets with more reticulations, heuristic search techniques can be used to obtain local optima as a more efficient alternative.

**Step Four: Merge Subproblem Phylogenies into a Phylogeny on the Full Set of Taxa.** Given an optimal subproblem decomposition graph $G'_D$ returned by the previous step, the final step of the FastNet algorithm merges the "top-level" phylogenetic structure inferred on $D_0$ and "bottom-level" subproblem phylogenies $D_i$ for $1 \leq i \leq q$ (Algorithm 4 in the Appendix's Supplementary Methods section). First, the phylogeny inferred on the top-level subproblem $D_0$ serves as the top-level of the output phylogeny $N'$. Next, the $i$th taxon in $N'$ is replaced with the phylogeny inferred on bottom-level subproblem $D_i$, which was cached during the evaluation of $G'_D$. Finally, each "inter-subproblem" reticulation that was inferred for a pair of subproblems $D_i$ and $D_j$ where $i < j$ is added to the output phylogeny $N'$, which is compatible by construction of the decomposition $D$ and the optimal subproblem decomposition graph $G'_D$. The result of the merge procedure is an output phylogeny $N'$ on the full set of taxa $X$.

## 2.2    Performance Study

We conducted a simulation study to evaluate the performance of FastNet and existing state-of-the-art methods for phylogenetic network inference. The performance study utilized the following procedures. Detailed commands and software options are given in the Supplementary Material.

We also conducted an empirical study to evaluate FastNet's performance. Details about the empirical study are provided in the Appendix.

**Simulation of Model Networks.** For each model condition, random model networks were generated using the following procedure. First, r8s version 1.7 [40] was used to simulate random birth-death trees with $n$ taxa where $n \in \{15, 20, 25, 30\}$, which served as in-group taxa during subsequent analysis. The height of each tree was scaled to 5.0 coalescent units. Next, a time-consistent level-$r$ rooted tree-based network [13,21,49] was obtained by adding $r$ reticulations to each tree, where $r \in [1, 4]$. The procedure for adding a reticulation consists of the following steps: based on a consistent timing of events in the tree,

(1) choose a time $t_M$ uniformly at random between 0 and the tree height, (2) randomly select two tree edges for which corresponding ancestral populations existed during time interval $[t_A, t_B]$ such that $t_M \in [t_A, t_B]$, and (3) add a reticulation to connect the pair of tree edges. Finally, an outgroup was added to the resulting network at time 15.0.

Reticulations in our study have the same interpretation as in the study of Leaché et al. [25]. Gene flow is modeled using an isolation-with-migration model, where each reticulation is modeled as a unidirectional migration event with rate 5.0 during the time interval $[t_A, t_B]$. We focus on paraphyletic gene flow as described by Leaché et al.; their study also investigated two other classes of gene flow – both of which involve gene flow between two sister species after divergence. Our simulation study omits these two classes since several existing methods (i.e., MLE and MPL) have issues with identifiability in this context; thus, the model networks in our study are a proper subset of the class of level-$r$ rooted tree-based networks. We note that FastNet makes no assumptions about the type of gene flow to be inferred, and identifiability depends on the model used for inference by FastNet's base method.

As in the study of Leaché et al., we further classify simulation conditions based on whether gene flow is "non-deep" or "deep" based on topological constraints. Non-deep reticulations involve leaf edges only, and all other reticulations are considered to be deep. Similarly, model conditions with non-deep gene flow have model networks with non-deep reticulations only; all other model conditions include deep reticulations and are referred to as deep.

**Simulation of Local Genealogies and DNA Sequences.** We used ms [18] to simulate local gene trees for independent and identically distributed (i.i.d.) loci under an extended multi-species coalescent model, where reticulations correspond to migration events as described above. Each coalescent simulation sampled one allele per taxon. The primary experiments in our study simulated 1000 gene trees for each random model network. Our study also investigated data requirements of different methods by including additional datasets where either 200 or 100 gene trees were simulated for each random model network.

Sequence evolution was simulated using seq-gen [38], which takes the local genealogies generated by ms as input and simulates sequence evolution along each genealogy under a finite-sites substitution model. Our simulations utilized the Jukes-Cantor substitution model [22]. We simulated 1000 bp per locus, and the resulting multi-locus sequence alignment had a total length of 1000 kb.

**Replicate Datasets.** A model condition in our study consisted of fixed values for each of the above model parameters. For each model condition, the simulation procedure was repeated twenty times to generate twenty replicate datasets.

**Species Network Inference Methods.** Our simulation study compared the performance of FastNet against existing methods which were among the fastest and most accurate in our previous performance study of state-of-the-art species network inference methods [17]. Like FastNet, these methods perform summary-based inference – i.e., the input consists of gene trees inferred from sequence alignments for multiple loci, rather than the sequence alignments themselves. The methods are broadly characterized by their statistical optimization criteria: either maximum likelihood or maximum pseudo-likelihood under the multi-species network coalescent (MSNC) model [47]. The maximum likelihood estimation methods consisted of two methods proposed by Yu et al. [48] which are implemented in PhyloNet [43]. One method utilizes gene trees with branch lengths as input observations, whereas the other method considers gene tree topologies only; we refer to the methods as MLE-length and MLE, respectively. Our study also included the pseudo-likelihood-based method of [46], which we refer to as MPL. For each analysis in our study, all species network inference methods – MLE, MLE-length, MPL, and FastNet – were provided with identical inputs.

Our study included two categories of experiments. The "boosting" experiments in our simulation study compared the performance of FastNet against its base method; we refer to all other experiments in our study as "non-boosting". To make boosting comparisons explicit, each boosting experiment will refer to "FastNet(BaseMethod)" which is FastNet run with a specific base method "BaseMethod" – either MLE-length, MLE, or MPL. The input for each boosting experiment consisted of either true or inferred gene trees for all loci. The inferred gene trees were obtained using FastTree [37] with default settings to perform maximum likelihood estimation under the Jukes-Cantor substitution model [22]. The inferred gene trees were rooted using the outgroup. The non-boosting experiments focused on the performance of FastNet using MLE as a base method and inferred gene trees as input, where gene trees were inferred using the same procedure as in the boosting experiments.

**Performance Measures.** The species network inference methods in our study were evaluated using two different criteria.

The first criterion was topological accuracy. For each method, we compared the inferred species phylogeny to the model phylogeny using the tripartition fraction [35], which counts the proportion of tripartitions that are not shared between the inferred and model network. It has been shown that the tripartition fraction is not a metric on rooted phylogenetic networks in general [8]. However, the model networks in our study satisfy the tree-child condition (i.e., every internal node has at least one child that is a tree node) since the simulation procedure stipulates that reticulation placements can only connect tree edges; the reticulation placement procedure also naturally gives a temporal representation [5] and ensures that the parents of a reticulation node cannot be connected by a path. Cardona et al. [8] showed that the tripartition fraction is a metric for the subset of rooted phylogenetic networks that satisfy these constraints.

The second criterion was computational runtime. All computational analyses were run on computing facilities in Michigan State University's High Performance Computing Center. We used compute nodes in the intel16 cluster, each of which had a 2.5 GHz Intel Xeon E5-2670v2 processor with 64 GiB of main memory. All replicates completed with memory usage less than 32 GiB.

## 3  Results

FastNet's use of phylogenetic divide-and-conquer is compatible with a range of different methods for inferring rooted species networks on subproblems, which we refer to as "base" methods. From a computational perspective, FastNet can be seen as a general-purpose framework for boosting the performance of base methods. We began by assessing the relative performance boost provided by FastNet when used with two different state-of-the-art network inference methods. We evaluated two different aspects of performance: topological error as measured by the tripartition fraction [35] between an inferred species network and the model network, and computational runtime. The initial set of boosting experiments focused on species network inference in isolation of upstream inference accuracy by providing true gene trees as input to all of the summary-based inference methods.

In the performance study of Hejase and Liu [17], the probabilistic network inference methods were found to be the most accurate among state-of-the-art methods, and MPL was among the fastest methods in this class. MPL utilized a pseudo-likelihood-based approximation for increased computational efficiency compared with full likelihood methods [45]. However, the tradeoff netted efficiency that was well short of current phylogenomic dataset sizes [17].

Table 1 shows the performance of FastNet(MPL) relative to MPL on model conditions with increasing numbers of taxa and non-deep reticulations. On model conditions with dataset sizes ranging from 15 to 30 taxa and from 1 to 4 reticulations, FastNet(MPL)'s improvement in topological error relative to its base method was statistically significant (one-sided pairwise t-test with Benjamini-Hochberg correction for multiple tests [6]; $\alpha = 0.05$ and $n = 20$) and substantial in magnitude – an absolute improvement that amounted to as much as 41%. Furthermore, the improvement in topological error grew as datasets became larger and involved more reticulations: the largest improvements were seen on the 30-taxon 4-reticulation model condition. Runtime improvements were also statistically significant and represented speedups which amounted to as much as a day and a half of runtime.

Next, we evaluated FastNet's performance when boosting MLE-length, the most accurate state-of-the-art method from the performance study of

**Table 1. FastNet(MPL) "boosts" MPL's runtime and topological accuracy, where a greater performance boost occurs as dataset sizes increase.** The relative performance of FastNet(MPL) and MPL is compared on model conditions with 15–30 taxa and 1–4 non-deep reticulations. The performance measures consisted of topological error as measured by the tripartition fraction between an inferred species network and the model network and computational runtime in hours. Average ("Avg") and standard error ("SE") of FastNet(MPL)'s performance improvement over MPL is reported ($n = 20$). All methods were provided with true gene trees as input. The statistical significance of FastNet(MPL)'s performance improvement over MPL was assessed using a one-sided t-test. Corrected q-values are reported where multiple test correction was performed using the Benjamini-Hochberg method [6].

| Number of taxa | Number of reticulations | Improvement in topological error | | | Improvement in runtime (h) | | |
|---|---|---|---|---|---|---|---|
| | | Avg | SE | Corrected q-value | Avg | SE | Corrected q-value |
| 15 | 1 | 0.087 | 0.036 | $3.3 \times 10^{-2}$ | 2.8 | 0.3 | $7.2 \times 10^{-5}$ |
| 20 | 2 | 0.346 | 0.036 | $1.1 \times 10^{-5}$ | 9.6 | 0.1 | $1.1 \times 10^{-2}$ |
| 25 | 3 | 0.281 | 0.024 | $7.9 \times 10^{-5}$ | 35.6 | 5.6 | $8.5 \times 10^{-4}$ |
| 30 | 4 | 0.413 | 0.001 | $8.8 \times 10^{-12}$ | 30.3 | 6.5 | $2.8 \times 10^{-2}$ |

Hejase and Liu [17]. On model conditions with non-deep reticulations, Fast-Net (MLE-length) had a similar boosting effect as compared to FastNet (MPL) (Table 2). On the 15-taxon single-reticulation model condition, FastNet's average improvement in topological error was greater when MLE-length was used as a base method rather than MPL. An even greater improvement in computational runtime was seen: FastNet(MLE-length)'s runtime improvement over MLE-length was over an order of magnitude greater than FastNet(MPL)'s improvement over MPL. As the number of taxa increased from 15 to 20 (but the number of reticulations was fixed to one), FastNet(MLE-length)'s advantage in topological error and runtime relative to its base method nearly doubled. In all cases, FastNet(MLE-length)'s performance improvements were statistically significant (Benjamini-Hochberg-corrected one-sided pairwise t-test; $\alpha = 0.05$ and $n = 20$). Although FastNet(MLE-length) successfully completed analysis of larger datasets (i.e., model conditions with more than 20 taxa and/or more than one reticulation), we were unable to quantify FastNet(MLE-length)'s performance relative to its base method due to MLE-length's scalability limitations.

We further evaluated FastNet's performance in the context of additional experimental and methodological considerations. On model conditions with deep gene flow (Table 3), FastNet returned significant improvements in topological accuracy and runtime relative to its base method – either MPL or MLE-length – with one exception: on the 15-taxon single-reticulation model condition, Fast-Net(MPL) returned a small and statistically insignificant improvement in topological error over MPL. Otherwise, FastNet's performance boost was robust to the choice of base method. As dataset sizes increased, the average performance boost increased when MPL was the base method; a similar finding applied to runtime improvements when MLE-length was the base method, whereas

**Table 2. FastNet(MLE-length) "boosts" MLE-length's runtime and topological accuracy, where a greater performance boost occurs as dataset sizes increase.** The relative performance of FastNet(MLE-length) and MLE-length is compared on model conditions with 15–20 taxa and 1–2 non-deep reticulations. Note that, for the model condition with 20 taxa and 2 reticulations, MLE-length did not finish analysis of any replicates after a week of runtime. Otherwise, table layout and description are identical to Table 1.

| Number of taxa | Number of reticulations | Improvement in topological error | | | Improvement in runtime (h) | | |
|---|---|---|---|---|---|---|---|
| | | Avg | SE | Corrected q-value | Avg | SE | Corrected q-value |
| 15 | 1 | 0.103 | 0.021 | $8.8 \times 10^{-4}$ | 49.4 | 6.9 | $9.1 \times 10^{-7}$ |
| 20 | 1 | 0.195 | 0.024 | $6.1 \times 10^{-5}$ | 114.3 | 14.7 | $3.3 \times 10^{-7}$ |
| 20 | 2 | Base method DNF | | | | | |

topological error improvements were largely unchanged. We note that Fast-Net's performance boost was somewhat smaller on model conditions involving deep gene flow as opposed to non-deep gene flow. When maximum-likelihood-estimated gene trees were used as input to summary-based inference in lieu of true gene trees (Table 4), FastNet boosted the topological accuracy and runtime of its base method in all cases and the improvements were statistically significant. As dataset sizes increased, FastNet's improvement in topological accuracy and runtime grew when MPL was its base method; runtime improvements grew and topological error improvements were largely unchanged when MLE-length was the base method. Finally, we conducted an additional experiment to evaluate FastNet's statistical efficiency when given a finite number of observations in terms of the number of loci (Table 5). As the number of loci ranged from genome-scale (i.e., on the order of 1000 loci) to sizes that were smaller by up to an order of magnitude, FastNet's average topological error increased by less than 0.02.

**Table 3. Boosting experiments on model conditions with deep gene flow.** The performance improvement of FastNet over its base method (either MPL or MLE-length) is reported for two different performance measures: topological error as measured by tripartition fraction and computational runtime in hours. The simulation conditions involved either 15 or 20 taxa and a single deep reticulation. Otherwise, table layout and description are identical to Table 1.

| Number of taxa | Boosted method | Improvement in topological error | | | Improvement in runtime (h) | | |
|---|---|---|---|---|---|---|---|
| | | Avg | SE | q-value | Avg | SE | q-value |
| 15 | MPL | 0.015 | 0.017 | $3.8 \times 10^{-1}$ | 2.3 | 0.2 | $5.1 \times 10^{-4}$ |
| 20 | MPL | 0.166 | 0.035 | $3.2 \times 10^{-3}$ | 8.0 | 1.5 | $3.2 \times 10^{-3}$ |
| 15 | MLE-length | 0.066 | 0.001 | $1.5 \times 10^{-2}$ | 35.0 | 4.1 | $1.3 \times 10^{-7}$ |
| 20 | MLE-length | 0.070 | 0.014 | $1.1 \times 10^{-2}$ | 71.1 | 7.7 | $8.7 \times 10^{-8}$ |

**Table 4. Boosting experiments using inferred gene trees.** The performance improvement of FastNet over its base method (either MPL or MLE-length) is reported for two different performance measures: topological error as measured by tripartition fraction and computational runtime in hours. For each replicate dataset, all summary-based methods were provided with the same input: a set of rooted gene trees that was inferred using FastTree and outgroup rooting (see Methods section for more details). The simulation conditions involved either 15 or 20 taxa and 1–2 non-deep reticulations. Otherwise, table layout and description are identical to Table 1.

| Number of taxa | Number of reticulations | Boosted method | Improvement in topological error | | | Improvement in runtime (h) | | |
|---|---|---|---|---|---|---|---|---|
| | | | Avg | SE | q-value | Avg | SE | q-value |
| 15 | 1 | MPL | 0.071 | 0.021 | $1.2 \times 10^{-2}$ | 3.8 | 0.5 | $7.7 \times 10^{-5}$ |
| 20 | 2 | MPL | 0.134 | 0.017 | $1.4 \times 10^{-2}$ | 15.1 | 1.7 | $6.9 \times 10^{-6}$ |
| 15 | 1 | MLE-length | 0.231 | 0.002 | $1.3 \times 10^{-4}$ | 15.4 | 2.0 | $6.7 \times 10^{-7}$ |
| 20 | 1 | MLE-length | 0.195 | 0.005 | $5.8 \times 10^{-5}$ | 43.2 | 7.3 | $1.7 \times 10^{-5}$ |

**Table 5. The impact of the number of observed loci on FastNet(MLE)'s topological error.** The inputs to FastNet(MLE) consisted of gene trees that were inferred using FastTree and outgroup rooting (see Methods section for more details). The simulations sampled between 100 and 1000 loci for a single 20-taxon 1-reticulation model condition involving non-deep gene flow. Topological error was evaluated based upon the tripartition fraction between the model phylogeny and the species phylogeny inferred by FastNet(MLE); average ("Avg") and standard error ("SE") are shown ($n = 20$).

| Number of loci | Topological error | |
|---|---|---|
| | Avg | SE |
| 100 | 0.094 | 0.028 |
| 200 | 0.078 | 0.024 |
| 1000 | 0.075 | 0.027 |

# 4   Discussion

Relative to the state-of-the-art methods that served as base methods, FastNet consistently returned sizeable and statistically significant improvements in topological error and computational runtime across a range of dataset scales and gene flow scenarios. There was only a single experimental condition where comparable error without statistically significant improvements was seen. This exception occurred when FastNet was used to boost a relatively inaccurate base method (MPL) on the smallest dataset sizes in our study and with deep gene flow; even still, large and statistically significant runtime improvements were seen in this case. In contrast, with a more accurate base method (i.e., MLE-length), large and statistically significant performance improvements were seen throughout our simulation study.

FastNet's boosting effect on topological error and runtime were robust to several different experimental and design factors. The boosting performance obtained using different base methods – one with lower computational requirements but higher topological error relative to a more computationally intensive alternative – suggests that, while accuracy improvements can be obtained even using less accurate subproblem inference, even greater accuracy improvements can be obtained when reasonably accurate subproblem phylogenies can be inferred. We note that the base methods were run in default mode. More intensive search settings for each base method's optimization procedures may allow a tradeoff between topological accuracy and computational runtime. We stress that our goal was *not* to make specific recommendations about the nuances of running the base methods. Rather, FastNet's divide-and-conquer framework can be viewed as orthogonal to the specific algorithmic approaches utilized by a base method. In this sense, improvements to the latter accrue to the former in a straightforward and modular manner. Furthermore, FastNet's performance effect was robust to gene tree error and varying numbers of observed loci.

The biggest performance gains were observed on the largest, most challenging datasets. The findings in our earlier performance study [17] suggest that, given weeks of computational runtime, even the fastest statistical methods (including MPL) would not complete analysis of datasets with more than 50 taxa or so and several reticulations. In comparison to MPL, FastNet(MPL) was faster by more than an order of magnitude on the largest datasets in our study, and we predict that FastNet(MPL) would readily scale to datasets with many dozens of taxa and multiple reticulations.

# 5   Conclusions

In this study, we introduced FastNet, a new computational method for inferring phylogenetic networks from large-scale genomic sequence datasets. Fast-Net utilizes a divide-and-conquer algorithm to constrain two different aspects of scale: the number of taxa and evolutionary divergence. We evaluated the performance of FastNet in comparison to state-of-the-art phylogenetic network inference methods. We found that FastNet improves upon existing methods in terms of computational efficiency and topological accuracy. On the largest datasets explored in our study, the use of the FastNet algorithm as a boosting framework enabled runtime speedups that were over an order of magnitude faster than standalone analysis using a state-of-the-art method. Furthermore, FastNet returned comparable or typically improved topological accuracy compared to the state-of-the-art-methods that were used as its base method.

**Acknowledgments.** We gratefully acknowledge the following support: NSF grants no. CCF-1565719 (to KJL), CCF-1714417 (to KJL), and DEB-1737898 (to GMB and KJL), BEACON grants (NSF STC Cooperative Agreement DBI-093954) to GMB and KJL, and computing resources provided by MSU HPCC. We would also like to acknowledge Daniel Neafsey for kindly sending us a processed version of the genomic sequence dataset from [36].

# References

1. Abbott, R.J., Rieseberg, L.H.: Hybrid speciation. In: Seligman, E.R.A., Johnson, A. (eds.) Encyclopaedia of Life Sciences. Wiley, Hoboken (2012)
2. Akaike, H.: Information theory and an extension of the maximum likelihood principle. In: Parzen, E., Tanabe, K., Kitagawa, G. (eds.) Selected Papers of Hirotugu Akaike. Springer Series in Statistics (Perspectives in Statistics). Springer, New York (1998). https://doi.org/10.1007/978-1-4612-1694-0_15
3. Akaike, H.: A new look at the statistical model identification. IEEE Trans. Autom. Control **19**(6), 716–723 (1974)
4. Bandelt, H.-J., Dress, A.W.M.: A canonical decomposition theory for metrics on a finite set. Adv. Math. **92**(1), 47–105 (1992)
5. Baroni, M., Semple, C., Steel, M.: Hybrids in real time. Syst. Biol. **55**(1), 46–56 (2006)
6. Benjamini, Y., Hochberg, Y.: Controlling the false discovery rate: a practical and powerful approach to multiple testing. J. R. Stat. Soc. Ser. B (Methodological) **57**(1), 289–300 (1995)
7. Bryant, D., Moulton, V.: Neighbor-Net: an agglomerative method for the construction of phylogenetic networks. Mol. Biol. Evol. **21**(2), 255–265 (2004)
8. Cardona, G., Rosselló, F., Valiente, G.: Tripartitions do not always discriminate phylogenetic networks. Math. Biosci. **211**(2), 356–370 (2008)
9. Durand, E.Y., Patterson, N., Reich, D., Slatkin, M.: Testing for ancient admixture between closely related populations. Mol. Biol. Evol. **28**(8), 2239–2252 (2011)
10. Edwards, S.V.: Is a new and general theory of molecular systematics emerging? Evolution **63**(1), 1–19 (2009)
11. Felsenstein, J.: Cases in which parsimony or compatibility methods will be positively misleading. Syst. Biol. **27**(4), 401–410 (1978)
12. Felsenstein, J.: Inferring Phylogenies. Sinauer Associates, Sunderland, Massachusetts (2004)
13. Francis, A.R., Steel, M.: Which phylogenetic networks are merely trees with additional arcs? Syst. Biol. **64**(5), 768–777 (2015)
14. Gluck-Thaler, E., Slot, J.C.: Dimensions of horizontal gene transfer in eukaryotic microbial pathogens. PLoS Pathog. **11**(10), e1005156 (2015)
15. Green, R.E., et al.: A draft sequence of the Neandertal genome. Science **328**(5979), 710–722 (2010)
16. Hein, J., Schierup, M., Wiuf, C.: Gene Genealogies, Variation and Evolution: A Primer in Coalescent Theory. Oxford University Press, Oxford (2004)
17. Hejase, H.A., Liu, K.J.: A scalability study of phylogenetic network inference methods using empirical datasets and simulations involving a single reticulation. BMC Bioinform. **17**(1), 422 (2016)
18. Hudson, R.R.: Generating samples under a wright-fisher neutral model of genetic variation. Bioinformatics **18**(2), 337–338 (2002)
19. Huelsenbeck, J.P., Hillis, D.M.: Success of phylogenetic methods in the four-taxon case. Syst. Biol. **42**(3), 247–264 (1993)
20. Hurvich, C.M., Tsai, C.-L.: Regression and time series model selection in small samples. Biometrika **76**(2), 297–307 (1989)
21. Huson, D.H., Rupp, R., Scornavacca, C.: Phylogenetic Networks: Concepts Algorithms and Applications. Cambridge University Press, Cambridge, United Kingdom (2010)

22. Jukes, T.H., Cantor, C.R.: Evolution of Protein Molecules, p. 132. Academic Press, New York (1969)
23. Keeling, P.J., Palmer, J.D.: Horizontal gene transfer in eukaryotic evolution. Nat. Rev. Genet. **9**(8), 605–618 (2008)
24. Kingman, J.F.C.: The coalescent. Stoch. Process. Appl. **13**(3), 235–248 (1982)
25. Leaché, A.D., Harris, R.B., Rannala, B., Yang, Z.: The influence of gene flow on species tree estimation: a simulation study. Syst. Biol. **63**, 17–30 (2013)
26. Liu, K., Raghavan, S., Nelesen, S., Linder, C.R., Warnow, T.: Rapid and accurate large-scale coestimation of sequence alignments and phylogenetic trees. Science **324**(5934), 1561–1564 (2009)
27. Liu, K., et al.: SATé-II: Very fast and accurate simultaneous estimation of multiple sequence alignments and phylogenetic trees. Syst. Biol. **61**(1), 90–106 (2012)
28. Liu, K.J., Steinberg, E., Yozzo, A., Song, Y., Kohn, M.H., Nakhleh, L.: Interspecific introgressive origin of genomic diversity in the house mouse. Proc. Nat. Acad. Sci. **112**(1), 196–201 (2015)
29. McInerney, J.O., Cotton, J.A., Pisani, D.: The prokaryotic tree of life: past, present... and future? Trends Ecol. Evol. **23**(5), 276–281 (2008)
30. Metzker, M.L.: Sequencing technologies - the next generation. Nat. Rev. Genet. **11**(1), 31–46 (2010)
31. Mirarab, S., Warnow, T.: ASTRAL-II: coalescent-based species tree estimation with many hundreds of taxa and thousands of genes. Bioinformatics **31**(12), i44–i52 (2015)
32. Mirarab, S., Reaz, R., Bayzid, M.S., Zimmermann, T., Swenson, M.S., Warnow, T.: ASTRAL: genome-scale coalescent-based species tree estimation. Bioinformatics **30**(17), i541–i548 (2014)
33. Mirarab, S., Nguyen, N., Guo, S., Wang, L.-S., Kim, J., Warnow, T.: PASTA: ultra-large multiple sequence alignment for nucleotide and amino-acid sequences. J. Comput. Biol. **22**(5), 377–386 (2015)
34. Nakhleh, L.: Computational approaches to species phylogeny inference and gene tree reconciliation. Trends Ecol. Evol. **28**(12), 719–728 (2013)
35. Nakhleh, L., Sun, J., Warnow, T., Linder, C.R., Moret, B.M., Tholse, A.: Towards the development of computational tools for evaluating phylogenetic network reconstruction methods. In: Pacific Symposium on Biocomputing, vol. 8, pp. 315–326. World Scientific (2003)
36. Neafsey, D.E.: Highly evolvable malaria vectors: the genomes of 16 Anopheles mosquitoes. Science **347**(6217), 1258522 (2015)
37. Price, M., Dehal, P., Arkin, A.: FastTree 2 - approximately maximum-likelihood trees for large alignments. PLoS ONE **5**(3), e9490 (2010)
38. Rambaut, A., Grassly, N.C.: Seq-Gen: an application for the Monte Carlo simulation of DNA sequence evolution along phylogenetic trees. Comput. Appl. Biosci. **13**, 235–238 (1997)
39. Reich, D., et al.: Genetic history of an archaic hominin group from Denisova Cave in Siberia. Nature **468**(7327), 1053–1060 (2010)
40. Sanderson, M.J.: r8s: inferring absolute rates of molecular evolution and divergence times in the absence of a molecular clock. Bioinformatics **19**(2), 301–302 (2003)
41. Schwarz, G.: Estimating the dimension of a model. Annal. Stat. **6**(2), 461–464 (1978)
42. Solís-Lemus, C., Ané, C.: Inferring phylogenetic networks with maximum pseudolikelihood under incomplete lineage sorting. PLoS Genet. **12**(3), 1–21 (2016)

43. Than, C., Ruths, D., Nakhleh, L.: PhyloNet: a software package for analyzing and reconstructing reticulate evolutionary relationships. BMC Bioinform. **9**(1), 322 (2008)

44. The Heliconious Genome Consortium: Butterfly genome reveals promiscuous exchange of mimicry adaptations among species. Nature **487**(7405), 94–98 (2012)

45. Yun, Y., Nakhleh, L.: A maximum pseudo-likelihood approach for phylogenetic networks. BMC Genomics **16**(Suppl 10), S10 (2015)

46. Yu, Y., Cuong, T., Degnan, J.H., Nakhleh, L.: Coalescent histories on phylogenetic networks and detection of hybridization despite incomplete lineage sorting. Syst. Biol. **60**(2), 138–149 (2011)

47. Yu, Y., Degnan, J.H., Nakhleh, L.: The probability of a gene tree topology within a phylogenetic network with applications to hybridization detection. PLoS Genet. **8**(4), pp. e1002660 (2012)

48. Yu, Y., Dong, J., Liu, K.J., Nakhleh, L.: Maximum likelihood inference of reticulate evolutionary histories. Proc. Nat. Acad. Sci. **111**(46), 16448–16453 (2014)

49. Zhang, L.: On tree-based phylogenetic networks. J. Comput. Biol. **23**(7), 553–565 (2016)

# NJMerge: A Generic Technique for Scaling Phylogeny Estimation Methods and Its Application to Species Trees

Erin K. Molloy(iD) and Tandy Warnow(✉)(iD)

Department of Computer Science, University of Illinois at Urbana-Champaign,
201 N Goodwin Ave., Urbana, IL 61801, USA
{emolloy2,warnow}@illinois.edu

**Abstract.** Divide-and-conquer methods, which divide the species set into overlapping subsets, construct trees on the subsets, and then combine the trees using a supertree method, provide a key algorithmic framework for boosting the scalability of phylogeny estimation methods to large datasets. Yet the use of supertree methods, which typically attempt to solve NP-hard optimization problems, limits the scalability of these approaches. In this paper, we present a new divide-and-conquer approach that does not require supertree estimation: we divide the species set into *disjoint* subsets, construct trees on the subsets, and then combine the trees using a distance matrix computed on the full species set. For this merger step, we present a new method, called NJMerge, which is a polynomial-time extension of the Neighbor Joining algorithm. We report on the results of an extensive simulation study evaluating NJMerge's utility in scaling three popular species tree estimation methods: ASTRAL, SVDquartets, and concatenation analysis using RAxML. We find that NJMerge provides substantial improvements in running time without sacrificing accuracy and sometimes even improves accuracy. Furthermore, although NJMerge can sometimes fail to return a tree, the failure rate in our experiments is less than 1%. Together, these results suggest that NJMerge is a valuable technique for scaling computationally intensive methods to larger datasets, especially when computational resources are limited. NJMerge is freely available on Github: https://github.com/ekmolloy/njmerge. All datasets, scripts, and supplementary materials are freely available through the Illinois Data Bank: https://doi.org/10.13012/B2IDB-1424746_V1.

**Keywords:** Phylogenomics · Species trees
Incomplete lineage sorting · Divide-and-conquer · Neighbor Joining
NJst · ASTRAL · SVDquartets

© Springer Nature Switzerland AG 2018
M. Blanchette and A. Ouangraoua (Eds.): RECOMB-CG 2018, LNBI 11183, pp. 260–276, 2018.
https://doi.org/10.1007/978-3-030-00834-5_15

# 1    Introduction

Species trees provide a useful model for many biological analyses, including the estimation of how life evolved on earth, adaptation, the impact of geological events on speciation, and so on. Yet species tree estimation is challenged by multiple biological processes, such as incomplete lineage sorting, gene duplication and loss, and horizontal gene transfer, that create heterogeneous evolutionary histories across genomes [16], i.e., "gene tree discordance". As a result, species tree estimation is also performed using multiple loci and depends on methods designed to address discordance between species trees and gene trees due to various causes. Different approaches have been developed to estimate species trees in the presence of gene tree discord resulting from incomplete lineage sorting (ILS). However, many species tree methods rely on estimated gene trees, and gene tree estimation error is a problem in many phylogenomic datasets (see discussion in [21]). Site-based methods (e.g., [5,6,8,37]), which do not estimate gene trees, provide an alternative approach to species tree estimation. For example, SVDquartets [6] and SVDquest [37] use the input sequence alignments to estimate quartet trees (using statistical properties of the multi-species coalescent or MSC model [24,26]) and then combine the quartet trees into a tree on the full set of species. This approach is $\Omega(n^4)$ time if all possible quartet trees are evaluated (which is best for accuracy [33]), and hence SVDquartets and SVDquest are computationally intensive on datasets with large numbers of species.

In general, constructing any large species tree presents both statistical challenges (in terms of addressing heterogeneity across the genome) and computational challenges (in terms of scaling to large numbers of species and loci). The divide-and-conquer approaches to species tree estimation that have been proposed operate by dividing the species set into overlapping subsets, constructing trees on the subsets, and then merging the subset trees into a tree on the entire species set; an example is the family of Disk Covering Methods [4,10,13,22,38,40]. The last step of this process, called "supertree estimation", can provide good accuracy (i.e., retain much of the accuracy in the subset trees) if good supertree methods are used. However, the better supertree methods are attempts to solve NP-hard optimization problems, and none of the current supertree methods provide both accuracy and scalability to large datasets (see [39] for discussion).

In this paper, we present a new divide-and-conquer approach to scaling species tree estimation methods to large datasets: we divide the dataset into disjoint subsets, construct trees on the subsets, and then assemble the subset trees into a species tree on the entire species set. Because the subsets we create are disjoint, we cannot use supertree methods (as these require that the subset trees be overlapping) to combine the subset trees. Instead, we construct a "dissimilarity matrix" (i.e., a matrix that is symmetric and zero on the diagonal) on the species set, and we use this matrix to combine the subset trees into a species tree on the full species set (the use of a dissimilarity matrix rather than a distance matrix follows because estimated distances between species may not satisfy the triangle inequality, a requirement for distance matrices).

We present a new polynomial-time method, NJMerge, designed to assemble a tree given a set of disjoint subset trees that operate as constraints on the output tree. We evaluate the impact of using NJMerge with three approaches for species tree estimation. Two of these methods (ASTRAL-III [42] and SVDQuartets [6]) are explicitly designed for species tree estimation in the presence of incomplete lineage sorting, and the third method, commonly referred to as "concatenation", is the popular (and traditional) approach in which the alignments for the different loci are concatenated and then a maximum likelihood tree is estimated on the concatenated data matrix. We show, using an extensive simulation study, that NJMerge provides substantial improvements in running time without sacrificing accuracy and sometimes even improves accuracy. NJMerge also *enables* SVDquartets and RAxML to run on large datasets (e.g., 1000 taxa and 1000 genes), on which SVDquartets and RAxML would otherwise fail to run when limited to 64 GB of physical memory. Finally, while NJMerge is not guaranteed to return a tree (i.e., it can fail under some circumstances); the failure rate in our experiments is extremely low – less than 1%. Hence, NJMerge provides a useful tool for large-scale species tree estimation on multi-locus datasets.

## 2    NJMerge, and Its Use in Species Tree Estimation

### 2.1    NJMerge

The input to NJMerge is a dissimilarity matrix $D$ on leafset $S = \{s_1, \ldots, s_n\}$ as well as a set $T = \{T_i\}_{i=1}^k$ of $k$ unrooted binary trees on pairwise disjoint subsets of the leaf set $S$. The objective is to output a tree $T$ on the species set $S$ that agrees with every tree in $T$ (i.e., $T$ must be a compatibility supertree), and to do so in polynomial time. Thus, when $T$ is restricted to the leaves of $T_i$, after suppressing the internal nodes of degree two, it induces a binary tree that is isomorphic to $T_i$.

Because the trees in $T$ are disjoint, a compatibility supertree always exists (e.g., consider the tree formed by adding a single node $v$ and making it adjacent to an internal node in each of the trees in $T$, and then adding the remaining species arbitrarily). Thus, the objective is to find a tree that is close to the true (but unknown) species tree from the set of compatibility supertrees, and NJMerge tries to do this through the use of the dissimilarity matrix $D$. Note that the trees in $T$ are not required to form clades in $T$. For example, the caterpillar tree on $\{a, b, c, d, e, f, g, h\}$ (i.e., the tree obtained by making a path with the leaves hanging off it in the order $a, b, \ldots, h$) is a compatibility supertree for $T = \{ac|eg, bd|fh\}$, and yet the trees in $T$ do not form clades within the caterpillar tree. Of course, there are also other compatibility supertrees, and in some of these, the input trees will form clades. NJMerge is a heuristic for constructing a compatibility supertree (for the set $T$) on large datasets that can sometimes fail to return a tree. That is, given $T$ and an associated dissimilarity matrix $D$, NJMerge will either return a binary tree $T$ on the leaf set $S$ such that $T$ is a compatibility supertree for $T$ or else NJMerge will fail.

NJMerge is a polynomial-time modification of Neighbor Joining [29], a method that computes a tree given a dissimilarity matrix. Neighbor Joining (NJ) is perhaps the most widely used polynomial-time method for phylogeny estimation and has been proven to be statistically consistent [3] under standard models of sequence evolution (e.g., the Generalized Time Reversible (GTR) Model [35], which contains other sequence evolution models, including Jukes-Cantor [12]). Hence, NJ can be used to estimate gene trees when distances between species are corrected under the appropriate model. NJ can also be used to estimate species trees in a multi-locus setting using a set of gene trees. For example, NJst [15] estimates a species tree by running NJ on an internode distance matrix (i.e., a distance matrix calculated by averaging topological distances between pairs of species in the input set of gene trees). As shown in [2], the internode distance matrix converges to an additive matrix for the species tree, and so NJst and some other methods (e.g., ASTRID [36]) that estimate species trees from internode distance matrices are statistically consistent under the MSC model.

NJ has an iterative design that builds the tree from the bottom up, producing a rooted tree that is then unrooted. Initially, all $n$ species are in separate components. When a pair of species is selected to be siblings, the pair of species is effectively replaced by a rooted tree on two leaves, and the number of components is reduced by one. This process repeats until there is only one component: a tree on the full leaf set. At each iteration, NJ derives a new matrix $Q$ from $D$ and uses it to determine which pair of the remaining nodes to join; specifically, nodes $n_i, n_j$ such that $Q[i,j]$ is minimized are made siblings. Updating the distances in $Q$ at each iteration can be varied to some extent without loss of statistical consistency; see [3] for details.

NJMerge is a modification of NJ to ensure that the constraints implied by the input set $\mathcal{T}$ are upheld in the output tree $T$. When two leaves are made siblings, they are replaced by a new leaf, and the constraint trees are then relabeled. Thus, siblinghood decisions change the set of leaves in the constraint trees and can result in the constraint trees no longer being disjoint. As a result, siblinghood decisions have the potential to make the constraint trees incompatible. Since determining if a set of unrooted phylogenetic trees is compatible is NP-complete [31,41], NJMerge uses a polynomial-time heuristic: it checks every pair of constraint trees to see if they are compatible after being modified based on the proposed siblinghood.

In each iteration, NJMerge sorts the entries of the $Q$ from least to greatest and accepts the first join $x, y$ (i.e., the proposal to make $x, y$ siblings) that satisfies the following properties: (1) if $x$ and $y$ are both in some constraint tree $T_i$, then they are siblings in $T_i$ and (2) the join does not cause any pair of constraint trees to become incompatible (i.e., a compatibility supertree exists for every pair of constraint trees). Note that to determine if some pair of constraint trees becomes incompatible after making $x$ and $y$ siblings, it suffices to check only those pairs of constraint trees that contain at least one of $x$ and $y$; all other pairs of trees are unchanged by the join and are pairwise compatible by induction. Then, the leaves in the two trees labeled $x$ or $y$ are relabeled by the new leaf $(x, y)$. Since the two trees have a leaf in common, they can be treated as rooted trees by rooting them at the common leaf $(x, y)$; testing the compatibility of rooted trees is easily accomplished in polynomial time using [1]. However, pairwise compatibility of unrooted trees does not guarantee that the set of trees is compatible, and hence it is possible for NJMerge to accept a siblinghood decision that will eventually cause the algorithm to fail when none of the remaining leaves can be joined without violating the constraint trees. Although this heuristic can fail, it is easy to see that any tree returned by NJMerge is a compatibility supertree for the input set $\mathcal{T}$ of constraint trees.

### 2.2   Using NJMerge for Multi-locus Species Tree Estimation

Given an input multi-locus dataset and a selected species tree method $M$:

1. We compute gene trees on each locus using a preferred phylogeny estimation method (e.g., FastTree [25] or RAxML [30]).
2. We compute the internode distance matrix $D$ (i.e., the matrix of average topological distances, see [2, 15, 36]).
3. We decompose the taxon set into disjoint subsets, as follows. We compute a "starting species tree" for the multi-locus dataset (e.g., using NJst on the estimated gene trees). We then use the centroid tree decomposition (described in PASTA [18]) to create disjoint subsets of taxa from the starting species tree of the desired maximum size.
4. We apply the selected species tree method $M$ to each such subset, thus producing the set $\mathcal{T}$ of disjoint constraint trees. Subtrees can be estimated in serial or in parallel, depending on the computational resources available.
5. Finally, we apply NJMerge to the input pair $(\mathcal{T}, D)$.

Note that there are several algorithmic parameters that can be selected by the user, for example, the method for computing subset trees, the method for computing the starting tree used to define the subset decomposition, and the maximum subset size. The choice of method for computing subset trees can vary from faster methods (such as ASTRAL) to slower methods (such as SVDquartets) and will impact the choice of the maximum subset size.

## 3   Performance Study

Our study evaluated the effectiveness of using NJMerge to estimate species trees on large datasets. All species tree methods were evaluated in terms of their species tree estimation error, computed using normalized Robinson-Foulds (RF) distances [27], and their running time. We used datasets simulated under a variety of model conditions, described by two numbers of taxa (100 and 1000), two levels of ILS (moderate and very high), each with 20 replicate datasets. Datasets included exon-like sequences and intron-like sequences. Exon-like sequences ("exons") were characterized by slower rates of evolution across sites (less phylogenetic signal), resulting in higher levels of gene tree estimation error, whereas intron-like sequences ("introns") were characterized by faster rates of evolution across sites (greater phylogenetic signal), resulting in lower levels of gene tree estimation error. The 100-taxon datasets were analyzed using 25, 100, and 1000 genes, and the 1000-taxon datasets were analyzed using 1000 genes; note that exons and introns were analyzed separately. For each of these 320 datasets, we constructed constraint trees using true species trees as well as three different species tree methods: ASTRAL, SVDquartets, and RAxML (on the concatenated alignment). This provided 1280 different tests in which to evaluate the impact of using NJMerge. NJMerge failed only twice on these 1280 tests, indicating a failure rate of 0.2%.

## 3.1 Simulated Datasets

Datasets were simulated for this study using the protocol presented in [20] and described below.

*True species and true gene trees.* Datasets, each with a true species tree and 2000 true gene trees, were simulated using SimPhy Version 1.0.2 [17]; see Supplementary Materials (https://doi.org/10.13012/B2IDB-1424746_V1) for command. All model conditions had deep speciation (towards the root) and 20 replicate datasets. By holding the effective population size constant (200 K) and varying the species tree height (in generations), model conditions with different levels of ILS were generated. For species tree heights of 10 M and 500 K generations, the average distance between the true species tree and the true gene trees (as measured by the normalized RF distance) was 8–10% and 68–69% respectively. Thus, we referred to these levels of ILS as "moderate" and "very high" respectively.

*True sequence alignments.* Sequence alignments were simulated for each gene tree using INDELible Version 1.03 [9] under the GTR+$\Gamma$ model of evolution without insertions or deletions. For each gene, the parameters for the GTR+$\Gamma$ model of evolution (base frequencies, substitution rates, and alpha) were drawn from distributions based on estimates of these parameters from the Avian Phylogenomics Dataset [11]; see Supplementary Materials for details. Distributions were fitted for exons and introns, separately (Table S2). For each dataset (with 2000 genes), 1000 gene sequences were simulated with parameters drawn from the exon distributions, and 1000 gene sequences were simulated with parameters drawn from the intron distributions. Note that exons and introns were analyzed separately. The sequence lengths were also drawn from a distribution and varied from 300 to 1500.

*Estimated Gene Trees.* For all gene sequences, a maximum likelihood gene tree was estimated using FastTree-2 [25] under the GTR+CAT model of evolution; see Supplementary Materials for command. The average gene tree estimation error (computed using normalized RF distances between true and estimated gene trees) across all replicates ranged from 26% to 51% for introns and 38% to 64% for exons; thus gene tree estimation error was higher for the exon datasets (Table S1).

## 3.2  Estimated Species Trees

For each model condition (described by number of taxa and level of ILS), species trees estimation methods were run on the exons and the introns genes, separately. Species trees were estimated on 25, 100, or 1000 genes for the 100-taxon datasets and 1000 genes for the 1000-taxon datasets; see Supplementary Materials for commands. Specifically, we ran ASTRAL [19,20,42] Version 5.6.1 (the most recent version of the ASTRAL method, called ASTRAL-III, but henceforth referred to as "ASTRAL"), SVDquartets [6] (as implemented in PAUP* Version 4a161 [34]), and concatenation using unpartitioned maximum likelihood under the GTR+$\Gamma$ model of evolution (as implemented in RAxML [30] Version 8.2.12 with pthreads and SSE3), as the multi-locus species tree estimation methods. ASTRAL is provably statistically consistent under the MSC model and uses estimated gene trees as input. SVDquartets and RAxML both use concatenated multiple sequence alignments (one alignment per gene) as input. SVDquartets is expected to be statistically consistent under the MSC model [7], but unpartitioned maximum likelihood on concatenated alignments is not [28].

*Running NJMerge.* In our experiments, we set the algorithmic parameters as follows. For the distance matrix, we computed the internode distance matrix from the estimated gene trees using ASTRID [36] Version 1.4. For the decomposition of the taxon set into subsets, we built an NJst starting tree by running NJ (as implemented within FastME [14] Version 2.1.5) on the internode distance matrix. The centroid tree decomposition (described in PASTA [18]) was then used to create disjoint subsets of taxa from the NJst tree. The 100-taxon datasets were decomposed into 4–6 subsets with a maximum subset size of 30 taxa, and the 1000-taxon datasets were decomposed into 10–15 subsets with a maximum subset size of 120 taxa.

We ran NJMerge using estimated subset trees and true subset trees as the constraints. Thus, we explored the following uses of NJMerge:

– *NJMerge+True*: NJMerge with the true species tree on each subset
– *NJMerge+ASTRAL*: NJMerge with subset trees estimated using ASTRAL
– *NJMerge+SVDquartets*: NJMerge with subset trees estimated using SVDquartets
– *NJMerge+RAxML*: NJMerge with subset trees estimated using RAxML

Finally, note that running NJst is the same as running NJMerge without any constraint trees.

## 3.3  Evaluation

*Species Tree Estimation Error.* Species tree estimation error was measured as the RF error rate, i.e., the normalized RF distance between the true and the estimated species trees both on the full species set (Table S3). Since both trees were fully resolved or binary, the RF error rate is the proportion of edges in the true tree that are missing in the estimated tree. RF error rates were computed using Dendropy [32]; see Supplementary Materials for command.

*Running Time.* All computational experiments were run on the Blue Waters supercomputer, specifically, the XE6 dual-socket nodes with 64 GB of physical memory and two AMD Interlagos model 6276 CPU processors (i.e., one per socket each with 8 floating point cores). All methods were given access to 16 threads with 1 thread per bulldozer (floating-point) core. SVDquartets and RAxML were explicitly run with 16 threads; however, ASTRAL and NJMerge currently are not implemented with multi-threading. All methods were restricted to a maximum wall-clock time of 48 h.

Running time was measured as the wall-clock time and recorded in seconds for all methods (Table S4). For ASTRAL, SVDquartets, and RAxML, the timing data was recorded for running the method on the full dataset as well as running the method on subsets of the dataset (to produce constraint trees for NJMerge). RAxML did not complete (within the maximum wall-clock time of 48 h) on datasets with 1000 taxa; however, we used the last checkpoint in evaluating species tree estimation error and running time. Specifically, running time was measured as the time between the info file being written and the last checkpoint file being written.

Because NJMerge was run as a pipeline, we approximated total running time by combining the timing data from several different stages of the pipeline. If a study only had access to one compute node, then subset trees would need to be estimated in serial. In this case, the running time of NJMerge would be approximated as

$$T_{serial} = \sum_{i=1}^{N} T_i^{Method} + T^{NJMerge} \tag{1}$$

where $N$ is the number of subsets, $T_i^{Method}$ is the running time of using the given method (i.e., ASTRAL, SVDquartets or RAxML) to compute a species tree on subset $i$, and $T_{NJMerge}$ is the running time of using NJMerge to combine the subset trees into a single tree on the full taxon set. However, if a study had access to multiple compute nodes (specifically at least 6 for the 100-taxon datasets and at least 15 for the 1000-taxon datasets), then the subset trees could be estimated in parallel. In this case, the running time of NJMerge would be approximated as

$$T_{parallel} = \max_{1 \le i \le N} T_i^{Method} + T_{NJMerge} \tag{2}$$

where $N$ is the number of subsets, $T_i^{Method}$ is the running time of using the given method (i.e., ASTRAL, SVDquartets or RAxML) to compute a tree on subset $i$, and $T_{NJMerge}$ is the running time of using NJMerge to combine the subset trees into a single tree on the full taxon set.

The approximate running time for the NJMerge pipeline could have included other timing data; for example, the running time to estimate gene trees or the running time to build the NJst tree. We did not include the running time for estimating gene trees, because it is becoming common for phylogenomic studies to include one or more summary methods when estimating species trees. We did not include the running time for building the NJst tree, because this step is very fast (see [36]), and distance-based methods are typically faster than ASTRAL,

SVDquartets, and RAxML on large datasets. For example, we built the NJst tree in less than two minutes even on datasets with 1000 taxa and 1000 genes; in contrast, running ASTRAL, SVDquartets, or RAxML on these same datasets took hundreds to thousands of minutes.

## 4    Results and Discussion

Results on intron datasets are shown in the main text, and results on exon datasets are shown in the Supplementary Materials. Unless otherwise noted, results were similar for both sequence types; however, species trees estimated on the exon datasets had slightly higher error rates than those estimated on the intron datasets. This is as expected, since exons had slower rates of evolution (and so had less phylogenetic signal) than the introns, which increased gene tree estimation error and subsequent species tree error.

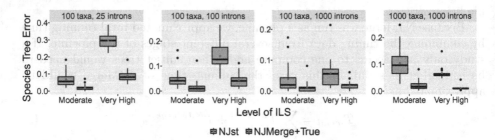

**Fig. 1.** Comparison of NJst (i.e., NJMerge without any subset trees) and NJMerge+True (i.e., NJMerge given subset trees defined by the true species tree as input) on intron datasets. Species tree estimation error is defined as the normalized RF distance between the true and the estimated species tree; bars show medians and red dots show means across replicate datasets. Both NJst and NJMerge were run on the internode distance matrix computed using estimated gene trees. NJMerge failed to return a tree on one replicate dataset with 100 taxa, 25 genes, very high ILS, and intron-like sequences.

*Impact of Constraint Tree Error on NJMerge.* NJMerge is intended to scale highly accurate (yet computationally intensive) species tree estimation methods to larger datasets. Ideally, a highly accurate method would return the true species tree on subsets of taxa, and then NJMerge would be used to combine these subset trees into a tree on the full taxon set using an estimated distance matrix. The error in an NJMerge tree is thus impacted by the error in the estimated constraint trees and the deviation from additivity in the input dissimilarity matrix. To determine the relative contributions of these two factors, we explored the accuracy of NJMerge given true subset trees (i.e., subset trees built from true species trees) but an estimated distance matrix (i.e., an internode distance matrix computed from estimated gene trees) as the input. This

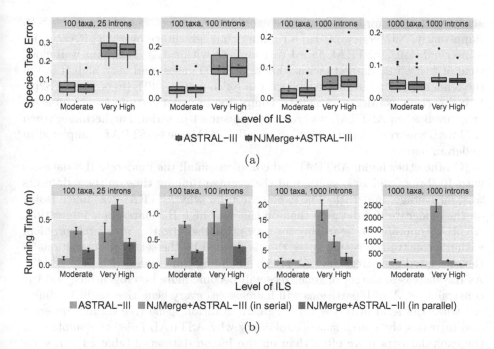

**Fig. 2.** Comparison of ASTRAL and NJMerge+ASTRAL (i.e., NJMerge given the ASTRAL subset trees as input) on intron datasets. Subplot (a) shows species tree estimation error (defined as the normalized RF distance between true and estimated species trees); bars represent medians and red dots represent means, across replicate datasets. Subplot (b) shows running time (in minutes); bars represent means and error bars represent standard deviations, across replicate datasets. For NJMerge+ASTRAL, "in serial" or "in parallel" refers to whether subset trees could be estimated in serial or in parallel; see Eqs. (1) and (2) for more information. ASTRAL did not complete within the maximum wall-clock time of 48 h on four out of the 20 replicate datasets with 1000 taxa and very high ILS.

process produced highly accurate species trees for both introns and exons; for example, the mean species tree error was always less than 5% (regardless of the level of ILS) when given 1000 genes (Figs. 1 and S1). Importantly, the error rate for NJMerge+True (i.e., NJMerge given true constraint trees) was substantially lower than the error rate for NJst (i.e., the equivalent of running NJMerge without constraint trees). We then examined the accuracy of NJMerge given estimated constraint trees (Table S3). Note that the accuracy of the final tree was also very close to the accuracy of the estimated subset trees (often within 1–2%), especially when the subset trees were highly accurate (Table S3). Overall, this study shows that NJMerge loses only a small amount of accuracy when used to combine highly accurate subset trees with an estimated distance matrix.

*NJMerge+ASTRAL.* ASTRAL and NJMerge+ASTRAL had very similar accuracy: the average species tree error across all replicates was often the same and always within 2% for both intron and exon datasets (Figs. 2a and S2a, Table S3). Importantly, NJMerge provided a running time advantage over ASTRAL under some conditions. ASTRAL failed to complete within the maximum wall-clock time of 48 h on 23 datasets with 1000 taxa, 1000 genes, and very high ILS. On the remaining 17 replicates, ASTRAL ran for more than 40 h (Figs. 2b and S2b). In comparison, NJMerge+ASTRAL completed in under 250 min (~4 h) on average – even when ASTRAL was run on each subset in serial. Furthermore, when ASTRAL was run on the subsets in parallel, NJMerge+ASTRAL completed in under an hour.

On the other hand, ASTRAL did complete on all the moderate ILS datasets with 1000 taxa and 1000 genes, and the average running time on these datasets was, on average, under 9 h. The explanation for why ASTRAL was much faster on the moderate ILS datasets than on the very high ILS datasets is interesting. ASTRAL operates by searching for an optimal solution to its search problem within a constrained search space that is defined by the set $\mathcal{X}$ of bipartitions in the estimated gene trees, and ASTRAL's running time scales with $|\mathcal{X}|^{1.726}$ [42]. As ILS increases, the set of gene trees will become more heterogeneous, and the constraint set $\mathcal{X}$ of bipartitions will increase, as every gene tree could be different when the level of ILS is very high. In addition, gene tree estimation error also increases the search space, explaining why ASTRAL failed to complete on the exon datasets more often than on the intron datasets (Table S4). In summary, NJMerge+ASTRAL provided a substantial running time advantage over ASTRAL when datasets were large and had high amounts of gene tree heterogeneity; furthermore, NJMerge+ASTRAL produced species trees of comparable accuracy.

*NJMerge+SVDquartets.* On datasets with 100 taxa, SVDquartets was run using all $\binom{n}{4}$ possible quartets for optimal accuracy. In these experiments, SVDquartets and NJMerge+SVDquartets produced species trees with similar amounts of error, and in some cases, running NJMerge+SVDquartets was even more accurate than running SVDquartets on the full dataset. For example, SVDquartets had, on average, 50% error on intron datasets with 100 taxa, 25 genes, and very high ILS, whereas NJMerge+SVDquartets had, on average, 39% error (Figs. 3a and S3a). NJMerge+SVDquartets also provided running time improvements even on datasets with 100 taxa and 1000 genes; for example, SVDquartets used, on average, 22–64 min, whereas NJMerge+SVDquartets completed in less than 5 min (Figs. 3b and S3b). However, this running time comparison does not take into account the time needed to compute gene trees, so that a fair comparison of running times depends on whether the gene trees would have been computed anyway.

On datasets with 1000 taxa, SVDquartets was run using a random subset of quartets, because the maximum number of quartets allowed by the most recent implementation of SVDquartets inside PAUP* was $4.15833 \times 10^{10}$. However, this resulted in a segmentation fault for all 1000-taxon datasets. In contrast, NJMerge+SVDquartets was able to analyze these datasets, achieving an average error of 5–6% for datasets with moderate ILS and 10–11% for datasets with very

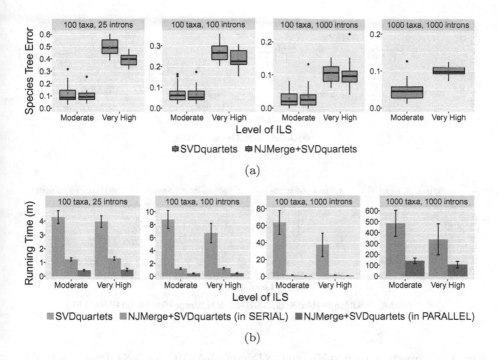

**Fig. 3.** Comparison of SVDquartets and NJMerge+SVDquartets (i.e., NJMerge given the SVDquartets subset trees as input) on intron datasets. Subplot (a) shows species tree estimation error (defined as the normalized RF distance between true and estimated species trees); bars represent medians and red dots represent means, across replicate datasets. Subplot (b) shows running time (in minutes); bars represent means and error bars represent standard deviations, across replicate datasets. For NJMerge+SVDquartets, "in serial" or "in parallel" refers to whether subset trees could be estimated in serial or in parallel; see Eqs. 1 and 2 for more information. SVDquartets did not run any datasets with 1000 taxa due to segmentation faults.

high ILS (Table S3). Hence, NJMerge enabled SVDquartets to be run on large datasets and improved the accuracy of SVDquartets on the smaller datasets.

*NJMerge+RAxML.* Our analyses using RAxML on the 100-taxon datasets were without problems; however, the analyses using RAxML on the 1000-taxon datasets presented some challenges. The first problem was that RAxML required more memory than the 64 GB available (indicated by an "Out of Memory", or OOM, error) on approximately half the datasets (see discussion below). The other problem was that RAxML never converged to a good local optimum within the maximum allowed wall-clock time of 48 h. Therefore, we modified the RAxML command to use checkpointing and used the last checkpoint file written before the job was killed at 48 h.

RAxML and NJMerge+RAxML produced species trees with similar levels of error (within 1–3% on average) on datasets with moderate ILS. On datasets

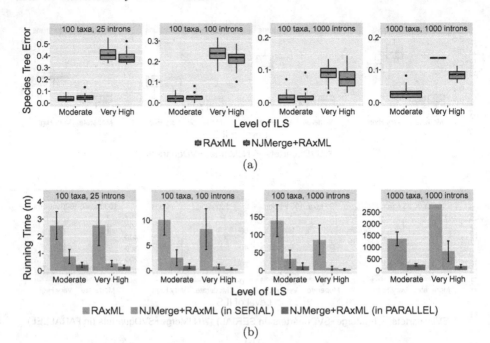

Fig. 4. Comparison of RAxML and NJMerge+RAxML (i.e., NJMerge given the RAxML subset trees as input) on intron datasets. Subplot (a) shows species tree estimation error (defined as the normalized RF distance between true and estimated species trees); bars represent medians and red dots represent means, across replicate datasets. Subplot (b) shows running time (in minutes); bars represent means and error bars represent standard deviations, across replicate datasets. For NJMerge+RAxML, "in serial" or "in parallel" refers to whether subset trees could be estimated in serial or in parallel; see Eqs. 1 and 2 for more information. RAxML was only able to run on one replicate dataset with 1000 taxa due to "Out of Memory" errors.

with very high ILS, NJMerge+RAxML produced better species trees (i.e., lower error) than RAxML (Figs. 4a and S4a, Table S3). This trend was observed for all numbers of taxa and all numbers of genes.

NJMerge+RAxML failed to run on 1 out of the 320 datasets tested; the model condition was challenging with 100 taxa, 25 genes, very high ILS, and exon-like sequences. RAxML failed to run on 42 of the 320 datasets tested due to "Out of Memory" errors. Of these 42 failures, 39 were on the intron datasets and 3 were on the exon datasets, both with 1000 taxa. The explanation for this distinction between introns and exons is interesting. RAxML uses redundancy in site patterns to store the input alignment compactly, so that the memory scales with the number of unique site patterns. We noted that the intron datasets had a substantially larger number of unique site patterns than the exon datasets, which explains why RAxML required more memory when analyzing introns.

Hence, NJMerge enabled RAxML to analyze large data matrices, even when the alignment patterns could not be compressed effectively.

For both the 100-taxon and 1000-taxon datasets, NJMerge+RAxML reduced the running time by more than half (Figs. 4b and S4b) – even when RAxML was run on the subset trees in serial. For the 1000-taxon datasets, the final checkpoint was written by RAxML after more than 2250 min (~37.5 h), on average. In comparison, when RAxML was run on subsets *in parallel* and NJMerge was used to combine the subset trees, NJMerge+RAxML completed in less than 250 min (~4 h), on average. Even when RAxML was run on subsets *in serial*, the average running time of NJMerge+RAxML was less than 1500 min (~25 h). Thus, NJMerge+RAxML substantially reduced the running time of RAxML on large datasets (by tens of hours) and enabled RAxML to run on large intron datasets using a single 64 GB node.

# 5 Conclusions

Our divide-and-conquer approach using NJMerge provided several benefits to large-scale species tree estimation. First, this divide-and-conquer technique ranged from neutral to beneficial with respect to species tree accuracy and even produced substantial gains in accuracy for some methods under some model conditions. Second, the technique often dramatically reduced the running time required for species tree estimation, even when subset trees were estimated in serial. Third, the technique enabled some methods to run on datasets that were too large and/or too heterogeneous for methods to analyze efficiently (e.g., SVDQuartets on 1000-species datasets, ASTRAL on some very high ILS datasets, RAxML on some intron datasets). Fourth, NJMerge had a very low failure rate in these experiments. Hence, this divide-and-conquer technique using NJMerge is an effective approach for scaling species tree estimation methods.

This study suggests several different directions for future research. Since NJMerge uses a heuristic (which can fail) to test for tree compatibility (in deciding whether to accept a siblinghood proposal), a modification to NJMerge to use an exact method for this problem would reduce the failure rate and – if sufficiently fast – would still enable scalability to large datasets. In addition, all aspects of the divide-and-conquer strategy could be modified. The robustness of NJMerge to the starting tree and initial subset decomposition should be explicitly tested, and the potential for improved accuracy through iteration should be evaluated. Other agglomerative techniques for merging disjoint subset trees should be developed, and NJMerge should be compared to such techniques when they are available (e.g., the agglomerative technique described in [43] for gene tree estimation has good theoretical properties but has not yet been implemented).

NJMerge could be further tested using different species tree estimation methods designed for ILS (e.g., STARBEAST2 [23] and SVDquest [37]) as well as using methods that estimate species trees from multi-locus datasets in the presence of duplication/loss. NJMerge could also be used to potentially scale methods that construct trees based on genome rearrangements, fissions, and fusions.

Finally, although we studied NJMerge in the context of species tree estimation, NJMerge could also be tested in the context of gene tree estimation, including cases where the input is a set of unaligned sequences.

**Acknowledgments.** The authors with to thank the anonymous reviewers, whose feedback led to improvements in the paper.

**Funding.** This work was supported by the National Science Foundation (award CCF-1535977) to TW. EKM was supported by the NSF Graduate Research Fellowship (award DGE-1144245) and the Ira and Debra Cohen Graduate Fellowship in Computer Science. Computational experiments were performed on Blue Waters, which is supported by the NSF (awards OCI-0725070 and ACI-1238993) and the state of Illinois. Blue Waters is a joint effort of the University of Illinois at Urbana-Champaign and its National Center for Supercomputing Applications.

# References

1. Aho, A.V., Sagiv, Y., Szymanski, T.G., Ullman, J.D.: Inferring a tree from lowest common ancestors with an application to the optimization of relational expressions. SIAM J. Comput. **10**(3), 405–421 (1981). https://doi.org/10.1137/0210030

2. Allman, E.S., Degnan, J.H., Rhodes, J.A.: Species tree inference from gene splits by unrooted STAR methods. IEEE/ACM Trans. Comput. Biol. Bioinform. **15**(1), 337–342 (2018). https://doi.org/10.1109/TCBB.2016.2604812

3. Atteson, K.: The performance of neighbor-joining methods of phylogenetic reconstruction. Algorithmica **25**(2–3), 251–278 (1999). https://doi.org/10.1007/PL00008277

4. Bayzid, M.S., Hunt, T., Warnow, T.: Disk covering methods improve phylogenomic analyses. BMC Genomics **15**(6), S7 (2014). https://doi.org/10.1186/1471-2164-15-S6-S7

5. Bryant, D., Bouckaert, R., Felsenstein, J., Rosenberg, N.A., RoyChoudhury, A.: Inferring species trees directly from biallelic genetic markers: bypassing gene trees in a full coalescent analysis. Mol. Biol. Evol. **29**(8), 1917–1932 (2012). https://doi.org/10.1093/molbev/mss086

6. Chifman, J., Kubatko, L.: Quartet inference from SNP data under the coalescent model. Bioinformatics **30**(23), 3317–3324 (2014). https://doi.org/10.1093/bioinformatics/btu530

7. Chifman, J., Kubatko, L.: Identifiability of the unrooted species tree topology under the coalescent model with time-reversible substitution processes, site-specific rate variation, and invariable sites. J. Theor. Biol. **374**, 35–47 (2015). https://doi.org/10.1016/j.jtbi.2015.03.006

8. Dasarathy, G., Nowak, R., Roch, S.: Data requirement for phylogenetic inference from multiple loci: a new distance method. IEEE/ACM Trans. Comput. Biol. Bioinform. **12**(2), 422–432 (2015). https://doi.org/10.1109/TCBB.2014.2361685

9. Fletcher, W., Yang, Z.: INDELible: a flexible simulator of biological sequence evolution. Mol. Biol. Evol. **26**(8), 1879–1888 (2009). https://doi.org/10.1093/molbev/msp098

10. Huson, D.H., Vawter, L., Warnow, T.: Solving large scale phylogenetic problems using DCM2. In: Proceedings of the Seventh International Conference on Intelligent Systems for Molecular Biology, pp. 118–129. AAAI Press (1999)

11. Jarvis, E.D., Mirarab, S., et al.: Whole-genome analyses resolve early branches in the tree of life of modern birds. Science **346**(6215), 1320–1331 (2014). https://doi.org/10.1126/science.1253451

12. Jukes, T.H., Cantor, C.R.: Evolution of protein molecules. In: Munro, H. (ed.) Mammalian Protein Metabolism, vol. 3, pp. 21–132. Academic Press, New York (1969)

13. Lagergren, J.: Combining polynomial running time and fast convergence for the disk-covering method. J. Comput. Syst. Sci. **65**(3), 481–493 (2002). https://doi.org/10.1016/S0022-0000(02)00005-3

14. Lefort, V., Desper, R., Gascuel, O.: FastME 2.0: a comprehensive, accurate, and fast distance-based phylogeny inference program. Mol. Biol. Evol. **32**(10), 2798–2800 (2015). https://doi.org/10.1093/molbev/msv150

15. Liu, L., Yu, L.: Estimating species trees from unrooted gene trees. Syst. Biol. **60**(5), 661–667 (2011). https://doi.org/10.1093/sysbio/syr027

16. Maddison, W.P.: Gene trees in species trees. Syst. Biol. **46**(3), 523–536 (1997). https://doi.org/10.1093/sysbio/46.3.523

17. Mallo, D., De Oliveira Martins, L., Posada, D.: SimPhy: phylogenomic simulation of gene, locus, and species trees. Systematic Biol. **65**(2), 334–344 (2016). https://doi.org/10.1093/sysbio/syv082

18. Mirarab, S., Nguyen, N., Guo, S., Wang, L.S., Kim, J., Warnow, T.: PASTA: ultra-large multiple sequence alignment for nucleotide and amino-acid sequences. J. Comput. Biol. **22**(5), 377–386 (2015). https://doi.org/10.1089/cmb.2014.0156

19. Mirarab, S., Reaz, R., Bayzid, M.S., Zimmermann, T., Swenson, M.S., Warnow, T.: ASTRAL: genome-scale coalescent-based species tree estimation. Bioinformatics **30**(17), i541–i548 (2014). https://doi.org/10.1093/bioinformatics/btu462

20. Mirarab, S., Warnow, T.: ASTRAL-II: coalescent-based species tree estimation with many hundreds of taxa and thousands of genes. Bioinformatics **31**(12), i44–i52 (2015). https://doi.org/10.1093/bioinformatics/btv234

21. Molloy, E.K., Warnow, T.: To include or not to include: the impact of gene filtering on species tree estimation methods. Syst. Biol. **67**(2), 285–303 (2018). https://doi.org/10.1093/sysbio/syx077

22. Nelesen, S., Liu, K., Wang, L.S., Linder, C.R., Warnow, T.: DACTAL: divide-and-conquer trees (almost) without alignments. Bioinformatics **28**(12), i274–i282 (2012). https://doi.org/10.1093/bioinformatics/bts218

23. Ogilvie, H.A., Bouckaert, R.R., Drummond, A.J.: StarBEAST2 brings faster species tree inference and accurate estimates of substitution rates. Mol. Biol. Evol. **34**(8), 2101–2114 (2017). https://doi.org/10.1093/molbev/msx126

24. Pamilo, P., Nei, M.: Relationships between gene trees and species trees. Mol. Biol. Evol. **5**(5), 568–583 (1988)

25. Price, M.N., Dehal, P.S., Arkin, A.P.: FastTree 2 - approximately maximum-likelihood trees for large alignments. PLOS ONE **5**(3), 1–10 (2010). https://doi.org/10.1371/journal.pone.0009490

26. Rannala, B., Yang, Z.: Bayes estimation of species divergence times and ancestral population sizes using dna sequences from multiple loci. Genetics **164**(4), 1645–1656 (2003)

27. Robinson, D., Foulds, L.: Comparison of phylogenetic trees. Math. Biosci. **53**(1), 131–147 (1981). https://doi.org/10.1016/0025-5564(81)90043-2

28. Roch, S., Steel, M.: Likelihood-based tree reconstruction on a concatenation of aligned sequence data sets can be statistically inconsistent. Theor. Popul. Biol. **100**, 56–62 (2015). https://doi.org/10.1016/j.tpb.2014.12.005

29. Saitou, N., Nei, M.: The neighbor-joining method: a new method for reconstructing phylogenetic trees. Mol. Biol. Evol. **4**(4), 406–425 (1987). https://doi.org/10.1093/oxfordjournals.molbev.a040454

30. Stamatakis, A.: RAxML version 8: a tool for phylogenetic analysis and post-analysis of large phylogenies. Bioinformatics **30**(9), 1312–1313 (2014). https://doi.org/10.1093/bioinformatics/btu033

31. Steel, M.: The complexity of reconstructing trees from qualitative characters and subtrees. J. Classif. **9**(1), 91–116 (1992). https://doi.org/10.1007/BF02618470

32. Sukumaran, J., Holder, M.T.: DendroPy: a python library for phylogenetic computing. Bioinformatics **26**(12), 1569–1571 (2010). https://doi.org/10.1093/bioinformatics/btq228

33. Swenson, M.S., Suri, R., Linder, C.R., Warnow, T.: An experimental study of Quartets MaxCut and other supertree methods. Algorithm. Mol. Biol. **6**(1), 7 (2011). https://doi.org/10.1186/1748-7188-6-7

34. Swofford, D.L.: PAUP* (*Phylogenetic Analysis Using PAUP), Version 4a161 (2018). http://phylosolutions.com/paup-test/

35. Tavaré, S.: Some probabilistic and statistical problems in the analysis of DNA sequences. Lect. Math. Life Sci. **17**(2), 57–86 (1986)

36. Vachaspati, P., Warnow, T.: ASTRID: accurate species trees from internode distances. BMC Genomics **16**(10), S3 (2015). https://doi.org/10.1186/1471-2164-16-S10-S3

37. Vachaspati, P., Warnow, T.: SVDquest: improving SVDquartets species tree estimation using exact optimization within a constrained search space. Mol. Phylogenet. Evol. **124**, 122–136 (2018). https://doi.org/10.1016/j.ympev.2018.03.006

38. Warnow, T.: Computational Phylogenetics: An Introduction to Designing Methods for Phylogeny Estimation. Cambridge University Press, Cambridge UK (2017)

39. Warnow, T.: Supertree Construction: Opportunities and Challenges. ArXiv e-prints, May 2018. https://arxiv.org/abs/1805.03530

40. Warnow, T., Moret, B.M.E., St. John, K.: Absolute convergence: true trees from short sequences. In: Proceedings of the Twelfth Annual ACM-SIAM Symposium on Discrete Algorithms, SODA 2001, Society for Industrial and Applied Mathematics, Philadelphia, PA, USA, pp. 186–195 (2001)

41. Warnow, T.: Tree compatibility and inferring evolutionary history. J. Algorith. **16**(3), 388–407 (1994). https://doi.org/10.1006/jagm.1994.1018

42. Zhang, C., Rabiee, M., Sayyari, E., Mirarab, S.: ASTRAL-III: polynomial time species tree reconstruction from partially resolved gene trees. BMC Bioinform. **19**(6), 153 (2018). https://doi.org/10.1186/s12859-018-2129-y

43. Zhang, Q.R., Rao, S., Warnow, T.: New absolute fast converging phylogeny estimation methods with improved scalability and accuracy. In: Parida, L., Ukkonen, E. (eds.) 18th International Workshop on Algorithms in Bioinformatics (WABI 2018). Leibniz International Proceedings in Informatics (LIPIcs), vol. 113, pp. 8:1–8:12. Schloss Dagstuhl–Leibniz-Zentrum fuer Informatik, Dagstuhl, Germany (2018). https://doi.org/10.4230/LIPIcs.WABI.2018.8

# On the Non-uniqueness of Solutions
# to the Perfect Phylogeny Mixture Problem

Dikshant Pradhan[1] and Mohammed El-Kebir[2]([⊠])

[1] Department of Bioengineering, University of Illinois at Urbana-Champaign,
Urbana, IL 61801, USA
[2] Department of Computer Science, University of Illinois at Urbana-Champaign,
Urbana, IL 61801, USA
melkebir@illinois.edu

**Abstract.** Tumors exhibit extensive intra-tumor heterogeneity, the presence of groups of cellular populations with distinct sets of somatic mutations. This heterogeneity is the result of an evolutionary process, described by a phylogenetic tree. The problem of reconstructing a phylogenetic tree $T$ given bulk sequencing data from a tumor is more complicated than the classic phylogeny inference problem. Rather than observing the leaves of $T$ directly, we are given mutation frequencies that are the result of mixtures of the leaves of $T$. The majority of current tumor phylogeny inference methods employ the perfect phylogeny evolutionary model. In this work, we show that the underlying PERFECT PHYLOGENY MIXTURE combinatorial problem typically has multiple solutions. We provide a polynomial-time computable upper bound on the number of solutions. We use simulations to identify factors that contribute to and counteract non-uniqueness of solutions. In addition, we study the sampling performance of current methods, identifying significant biases.

## 1 Introduction

Cancer is characterized by somatic mutations that accumulate in a population of cells, leading to the formation of genetically distinct *clones* within the same tumor [19]. This *intra-tumor heterogeneity* is the main cause of relapse and resistance to treatment [24]. The evolutionary process that led to the formation of a tumor can be described by a *phylogenetic tree* whose leaves correspond to tumor cells at the present time and whose edges are labeled by somatic mutations. To elucidate the mechanisms behind tumorigenesis [22,24] and identify treatment strategies [6,28], we require algorithms that accurately infer a phylogenetic tree from DNA sequencing data of a tumor.

Most cancer sequencing studies, including those from The Cancer Genome Atlas [12] and the International Cancer Genome Consortium [8], use bulk DNA sequencing technology, where samples are a mixture of millions of cells. While in classic phylogenetics, one is asked to infer a phylogenetic tree given its leaves, with bulk sequencing data we are asked to infer a phylogenetic tree given mixtures of its leaves in the form of mutation frequencies. More specifically, one

© Springer Nature Switzerland AG 2018
M. Blanchette and A. Ouangraoua (Eds.): RECOMB-CG 2018, LNBI 11183, pp. 277–293, 2018.
https://doi.org/10.1007/978-3-030-00834-5_16

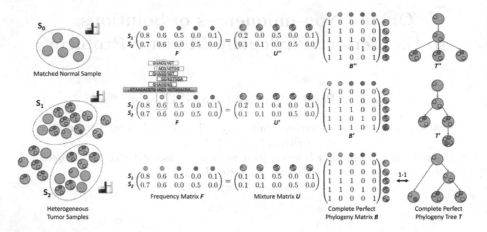

**Fig. 1. Overview of the PERFECT PHYLOGENY MIXTURE (PPM) problem.**
By comparing the aligned reads obtained from bulk DNA sequencing data of a matched
normal sample and $m$ tumor samples, we identify $n$ somatic mutations and their frequencies $F = [f_{p,c}]$. In the PPM problem, we are asked to factorize $F$ into a mixture
matrix $U$ and a complete perfect phylogeny matrix $B$, explaining the composition of
the $m$ tumor samples and the evolutionary history of the $n$ mutations present in the
tumor, respectively. Typically, an input frequency matrix admits multiple distinct solutions. Here, matrix $F$ has three solutions: $(U, B)$, $(U', B')$ and $(U'', B'')$, where only
$(U, B)$ is the correct solution.

first identifies a set of loci containing somatic mutations present in the tumor by
sequencing and comparing the aligned reads of a matched normal sample and
one or more tumor samples. Based on the number reads of each mutation locus
in a sample, we obtain *mutation frequencies* indicating the fraction of cells in
the tumor sample that contain each mutation. From these frequencies, the task
is to infer the phylogenetic tree under an appropriate evolutionary model that
generated the data.

The most commonly used evolutionary model in cancer phylogenetics is the
*two-state perfect phylogeny* model, where mutations adhere to the infinite sites
assumption [2,3,10,11,16,17,20,23,29]. That is, for each mutation locus the
actual mutation occurred exactly once in the evolutionary history of the tumor
and was subsequently never lost. The underlying combinatorial problem of the
majority of current methods is the PERFECT PHYLOGENY MIXTURE (PPM)
problem. Given an $m \times n$ frequency matrix $F$, we are asked to explain the composition of the $m$ tumor samples and the evolutionary history of the $n$ mutations.
More specifically, we wish to factorize $F$ into a mixture matrix $U$ and a perfect
phylogeny matrix $B$. Not only is this problem NP-complete [3], but multiple
perfect phylogeny trees may be inferred from the same input matrix $F$ (Fig. 1).
Tumor phylogenies have been used to identify mutations that drive cancer progression [9,18], to assess the interplay between the immune system and the clonal
architecture of a tumor [15,30] and to identify common evolutionary patterns

in tumorigenesis and metastasis [25,26]. To avoid any bias in such downstream analyses, all possible solutions must be considered. While non-uniqueness of solutions to PPM has been recognized in the field [4,17], a rigorous analysis of its extent and consequences on sampling by current methods has been missing.

In this paper, we study the non-uniqueness of solutions to the PPM problem. We give a upper bound on the number of solutions that can be computed in polynomial time. Using simulations, we identify the factors that contribute to non-uniqueness. In addition, we empirically study how, in addition to bulk sequencing, incorporating single-cell and long-read sequencing technologies affects non-uniqueness. Upon finding that current Markov chain Monte Carlo methods fail to sample uniformly from the solution space, we describe a simple rejection sampling algorithm that is able to sample uniformly for modest numbers $n$ of mutations.

## 2   Preliminaries

In this section, we review the PERFECT PHYLOGENY MIXTURE problem, as introduced in [3] (where it was the called the VARIANT ALLELE FREQUENCY FACTORIZATION PROBLEM or VAFFP). As input, we are given a frequency matrix $F = [f_{p,c}]$ composed of allele frequencies of $n$ single-nucleotide variants (SNVs) measured in $m$ bulk DNA sequencing samples. In the following, we refer to SNVs as mutations.

**Definition 1.** *An $m \times n$ matrix $F = [f_{p,c}]$ is a* frequency matrix *provided $f_{p,c} \in [0,1]$ for all samples $p \in [m]$ and mutations $c \in [n]$.*

Each frequency $f_{p,c}$ indicates the proportion of cells in sample $p$ that have mutation $c$. The evolutionary history of all $n$ mutations is described by a phylogenetic tree. We assume the absence of homoplasy and define a complete perfect phylogeny tree $T$ as follows.

**Definition 2.** *A rooted tree $T$ on $n$ vertices is a* complete perfect phylogeny tree *provided each edge of $T$ is labeled with exactly one mutation from $[n]$ and no mutation appears more than once in $T$.*

We call the unique mutation $r \in [n]$ that does not label any edge of a complete perfect phylogeny tree $T$ the *founder mutation*. Equivalently, we may represent a complete perfect phylogeny tree by an $n \times n$ binary matrix $B$ subject to the following constraints.

**Definition 3.** *An $n \times n$ binary matrix $B = [b_{c,d}]$ is an* $n$-complete perfect phylogeny matrix *provided:*

1. *There exists exactly one $r \in [n]$ such that $\sum_{c=1}^{n} b_{r,c} = 1$.*
2. *For each $d \in [n] \setminus \{r\}$ there exists exactly one $c \in [n]$ such that $\sum_{e=1}^{n} b_{d,e} - \sum_{e=1}^{n} b_{c,e} = 1$, and $b_{d,e} \geq b_{c,e}$ for all $e \in [n]$.*
3. *$b_{c,c} = 1$ for all $c \in [n]$.*

While the rows of a perfect phylogeny matrix $B$ correspond to the leaves of a perfect phylogeny tree $T$ (as per Definition 1), a *complete* perfect phylogeny matrix $B$ includes all vertices of $T$. The final ingredient is an $m \times n$ mixture matrix $U$ defined as follows.

**Definition 4.** *An* $m \times n$ *matrix* $U = [u_{p,c}]$ *is a* mixture matrix *provided* $u_{p,c} \in [0,1]$ *for all samples* $p \in [m]$ *and mutations* $c \in [n]$, *and* $\sum_{c=1}^{n} u_{p,c} \leq 1$ *for all samples* $p \in [m]$.

The forward problem of obtaining a frequency matrix $F$ from a complete perfect phylogeny matrix $B$ and mixture matrix $U$ is trivial. That is, $F = UB$. We are interested in the inverse problem, which is defined as follows.

*Problem 1 (*PERFECT PHYLOGENY MIXTURE *(PPM))*. Given a frequency matrix $F$, find a complete perfect phylogeny matrix $B$ and mixture matrix $U$ such that $F = UB$.

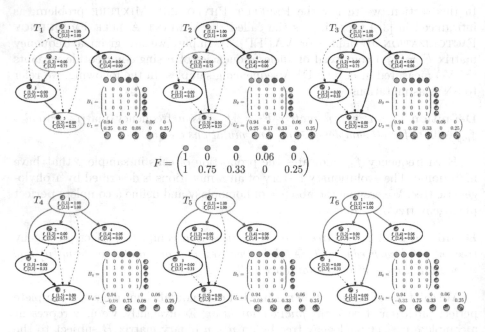

**Fig. 2. Example PPM instance $F$ has three solutions.** Frequency matrix $F$ corresponds to a simulated $n = 5$ instance (#9) and has $m = 2$ samples. The ancestry graph $G_F$ has six spanning arborescences. Among these, only trees $T_1$, $T_2$ and $T_3$ satisfy the sum condition (SC), whereas trees $T_4$, $T_5$ and $T_6$ violate (SC) leading to negative entries in $U_4$, $U_5$ and $U_6$. Tree $T_1$ is the simulated tree of this instance. Trees $T_2$ and $T_3$ differ from $T_1$ by only one edge, and thus each have an edge recall of $3/4 = 0.75$. (Color figure online)

El-Kebir et al. [3] showed that a solution to PPM corresponds to a constrained spanning arborescence of a directed graph $G_F$ obtained from $F$. This directed graph $G_F$ is called the *ancestry graph* and is defined as follows.

**Definition 5.** *The* ancestry graph $G_F$ *obtained from frequency matrix* $F = [f_{p,c}]$ *has* $n$ *vertices* $V(G_F) = \{1, \ldots, n\}$ *and there is a directed edge* $(c, d) \in E(G_F)$ *if and only if* $f_{p,c} \geq f_{p,d}$ *for all samples* $p \in [m]$.

As shown in [3], the square matrix $B$ is invertible and thus matrix $U$ is determined by $F$ and $B$. We denote the set of children of the vertex corresponding to a mutation $c \in [n] \setminus \{r\}$ by $\delta(c)$, and we define $\delta(r) = \{r(T)\}$.

**Proposition 1 (Ref. [3]).** *Given frequency matrix* $F = [f_{p,c}]$ *and complete perfect phylogeny matrix* $B = [b_{c,d}]$, *matrix* $U = [u_{p,c}]$ *where* $u_{p,c} = f_{p,c} - \sum_{d \in \delta(c)} f_{p,d}$ *is the unique matrix* $U$ *such that* $F = UB$.

For matrix $U$ to be a mixture matrix, it is necessary and sufficient to enforce non-negativity as follows.

**Theorem 1 (Ref. [3]).** *Let* $F = [f_{p,c}]$ *be a frequency matrix and* $G_F$ *be the corresponding ancestry graph. Then, complete perfect phylogeny matrix* $B$ *and associated matrix* $U$ *are a solution to* PPM *instance* $F$ *if and only if* $B$ *encodes a spanning arborescence* $T$ *of* $G_F$ *satisfying*

$$f_{p,c} \geq \sum_{d \in \delta_{out}(c)} f_{p,d} \qquad \forall p \in [m], c \in [n]. \tag{SC}$$

The above equation is known as the sum condition (SC), which requires that any mutation with multiple children have a greater frequency than the sum of the frequencies of its children in all samples. In this equation, $\delta_{out}(c)$ denotes the set of children of vertex $c$ in rooted tree $T$. A *spanning arborescence* $T$ of a directed graph $G_F$ is defined as a subset of edges that induce a rooted tree that spans all vertices of $G_F$.

While finding a spanning arborescence in a directed graph can be done in linear time (e.g., using a depth-first or breadth-first search), the problem of finding a spanning arborescence in $G_F$ adhering to (SC) is NP-hard [3,4]. Moreover, the same input frequency matrix $F$ may admit more than one solution (Fig. 2).

## 3   Methods

### 3.1   Characterization of the Solution Space

Let $F$ be a frequency matrix and let $G_F$ be the corresponding ancestry graph. By Theorem 1, we have that solutions to the PPM instance $F$ are spanning arborescences $T$ in the ancestry graph $G_F$ that satisfy (SC). In this section, we describe additional properties that further characterize the solution space. We start with the ancestry graph $G_F$.

**Fact 1.** *If there exists a path from vertex $c$ to vertex $d$ then $(c, d) \in E(G_F)$.*

A pair of mutations that are not connected by a path in $G_F$ correspond to two mutations that must occur on distinct branches in any solution. Such pairs of incomparable mutations are characterized as follows.

**Fact 2.** *Ancestry graph $G_F$ does not contain the edge $(c, d)$ nor the edge $(d, c)$ if and only if there exist two samples $p, q \in [m]$ such that $f_{p,c} > f_{p,d}$ and $f_{q,c} < f_{q,d}$.*

We define the branching coefficient as follows.

**Definition 6.** *The* branching coefficient *$\gamma(G_F)$ is the fraction of unordered pairs $(c, d)$ of distinct mutations such that $(c, d) \notin E(G_F)$ and $(d, c) \notin E(G_F)$.*

In the single-sample case, where frequency matrix $F$ has $m = 1$ sample, we have that $\gamma(G_F) = 0$. This is because either $f_{1,c} \geq f_{1,d}$ or $f_{1,d} \geq f_{1,c}$ for any ordered pair $(c, d)$ of distinct mutations. Since an arborescence is a rooted tree, we have the following fact.

**Fact 3.** *For $G_F$ to contain a spanning arborescence there must exist a vertex in $G_F$ from which all other vertices are reachable.*

Note that $G_F$ may contain multiple source vertices from which all other vertices are reachable. Such source vertices correspond to repeated columns in $F$ whose entries are greater than or equal to every other entry in the same row. In most cases the ancestry graph $G_F$ does not contain any directed cycles because of the following property.

**Fact 4.** *Ancestry graph $G_F$ is a directed acyclic graph (DAG) if and only if $F$ has no repeated columns.*

In the case where $G_F$ is a DAG and contains at least one spanning arborescence, we know that all spanning arborescence $T$ of $G_F$ share the same root vertex. This root vertex $r$ is the unique vertex of $G_F$ with in-degree 0.

**Fact 5.** *If $G_F$ is a DAG and contains a spanning arborescence then there exists exactly one vertex $r$ in $G_F$ from which all other vertices are reachable.*

Figure 2 shows the solutions to a PPM instance $F$ with $m = 2$ tumor samples and $n = 5$ mutations. Since $F$ has no repeated columns, the corresponding ancestry graph $G_F$ is a DAG. Vertex $r = 1$ is the unique vertex of $G_F$ without any incoming edges. There are three solutions to $F$, i.e. $T_1$, $T_2$ and $T_3$ are spanning arborescences of $G_F$, each rooted at vertex $r = 1$ and each satisfying (SC). How do we know that $F$ has three solutions in total? This leads to the following problem.

*Problem 2 (#-PERFECT PHYLOGENY MIXTURE (#PPM)).* Given a frequency matrix $F$, count the number of pairs $(U, B)$ such that $B$ is a complete perfect phylogeny matrix, $U$ is a mixture matrix and $F = UB$.

Since deciding whether a frequency matrix $F$ can be factorized into a complete perfect phylogeny matrix $B$ and a mixture matrix $U$ is NP-complete [3,4], the corresponding counting problem is NP-hard.[1] Since solutions to $F$ correspond to a subset of spanning arborescences of $G_F$ that satisfy (SC), we have the following fact.

**Fact 6.** *The number of solutions to a* PPM *instance $F$ is at most the number of spanning arborescences in the ancestry graph $G_F$.*

Kirchhoff's elegant matrix tree theorem [13] uses linear algebra to count the number of spanning trees in a simple graph. Tutte extended this theorem to count spanning arborescences in a directed graph $G = (V, E)$ [27]. Briefly, the idea is to construct the $n \times n$ Laplacian matrix $L = [\ell_{i,j}]$ of $G$, where

$$\ell_{i,j} = \begin{cases} \deg_{\text{in}}(j), & \text{if } i = j, \\ -1, & \text{if } i \neq j \text{ and } (i,j) \in E \\ 0, & \text{otherwise.} \end{cases} \qquad (1)$$

Then, the number of spanning arborescences $N_i$ rooted at vertex $i$ is $\det(\hat{L}_i)$, where $\hat{L}_i$ is the matrix obtained from $L$ by removing the $i$-th row and column. Thus, the total number of spanning arborescences in $G$ is $\sum_{i=1}^{n} \det(\hat{L}_i)$.

By Fact 4, we have that $G_F$ is a DAG if $F$ has no repeated columns. In addition, by Fact 5, we know that $G_F$ must have a unique vertex $r$ with no incoming edges. We have the following technical lemma.

**Lemma 1.** *Let $G_F$ be a DAG and let $r(G_F)$ be its unique source vertex. Let $\pi$ be a topological ordering of the vertices of $G_F$. Let $L' = [\ell'_{i,j}]$ be the matrix obtained from $L = [\ell_{i,j}]$ by permuting its rows and columns according to $\pi$, i.e. $\ell'_{i,j} = \ell_{\pi(i),\pi(j)}$. Then, $L'$ is an upper triangular matrix and $\pi(1) = r(G_F)$.*

*Proof.* Assume for a contradiction that $L'$ is not upper triangular. Thus, there must exist vertices $i, j \in [n]$ such that $j > i$ and $\ell'_{j,i} \neq 0$. By definition of $L$ and $L'$, we have that $\ell'_{j,i} = -1$. Thus $(\pi(j), \pi(i)) \in E(G_F)$, which yields a contradiction with $\pi$ being a topological ordering of $G_F$. Hence, $L'$ is upper triangular. From Fact 5 it follows that $\pi(1) = r(G_F)$. □

Since the determinant of an upper triangular matrix is the product of its diagonal entries, it follows from the previous lemma that $\det(\hat{L}'_1) = \prod_{i=1}^{n-1} \ell'_{i,i}$. Combining this fact with Tutte's directed matrix-tree theorem, yields the following result.

**Theorem 2.** *Let $F$ be a frequency matrix without any repeated columns and let $r$ be the unique mutation such that $f_{p,r} \geq f_{p,c}$ for all mutations $c$ and samples $p$. Then the number of solutions to $F$ is at most the product of the in-degrees of all vertices $c \neq r$ in $G_F$.*

---

[1] We expect the counting problem #PPM to be #P-complete, as to date no NP-complete problem has been found whose counting version is not NP-complete [14]. To prove that #PPM is #P-complete, we need to give a parsimonious reduction from a known #P-complete problem to #PPM.

In Fig. 2, the number of spanning arborescences in $G_F$ is $\deg_{in}(2) \cdot \deg_{in}(3) \cdot \deg_{in}(4) \cdot \deg_{in}(5) = 1 \cdot 2 \cdot 1 \cdot 3 = 6$. To compute the number of spanning arborescences of $G_F$ that satisfy (SC), we can simply enumerate all spanning arborescences using, for instance, the Gabow-Myers algorithm [7] and only output those that satisfy (SC). El-Kebir et al. [4] extended this algorithm such that it maintains (SC) as an invariant while growing arborescences. Applying both algorithms on the instance in Fig. 2 reveals that trees $T_1$, $T_2$ and $T_3$ comprise all solutions to $F$. We note that the enumeration algorithm in [4] has not been shown to be an output-sensitive algorithm.

## 3.2    Additional Constraints on the Solution Space

*Long-read sequencing.* Most cancer sequencing studies are performed using next-generation sequencing technology, producing short reads containing between 100 and 1000 basepairs. Due to the small size of short reads, it is highly unlikely to observe two mutations that occur on the same read (or read pair). With (synthetic) long read sequencing technology, including 10X Genomics, Pacbio and Oxford Nanopore, one is able to obtain reads with millions of basepairs. Thus, it becomes possible to observe long reads that contain more than one mutation.

As described in [1], the key insight is that a pair $(c, d)$ of mutations that occur on the same read orginate from a single DNA molecule of a single cell, and thus $c$ and $d$ must occur on the same path in the phylogenetic tree. Such mutation pairs provide very strong constraints to the PPM problem. For example in Fig. 2, in addition to frequency matrix $F$, we may be given that mutations 2 and 5 have been observed on a single read. Thus, in $T_1$ and $T_2$ the pair is highlighted in green because it is correctly placed on the same path from the root on the inferred trees. However, the two mutations occur on distinct branches on $T_3$, which is therefore ruled out as a possible solution.

*Single-cell Sequencing.* With single-cell sequencing, we are able to identify the mutations that are present in a single tumor cell. If in addition to bulk DNA sequencing samples, we are given single cell DNA sequencing data from the same tumor, we can constrain the solution space to PPM considerably. In particular, each single cell imposes that its comprising mutations must correspond to a connected path in the phylogenetic tree. These constraints have been described recently in [16].

For an example of these constraints, consider frequency matrix $F$ described in Fig. 2. In addition to frequency matrix $F$, we may observe a single cell with mutations $\{1, 2, 3, 5\}$. $T_1$ is the only potential solution as this is the only tree which places all four mutations on a single path, highlighted in blue. Trees $T_2$ and $T_3$ would be ruled out because the mutation set $\{1, 2, 3, 5\}$ does not induce a connected path in these two trees.

We note that the constraints described above for single-cell sequencing and long-read sequencing assume error-free data. In practice, one must incorporate an error model and adjust the constraints accordingly. However, the underlying principles will remain the same.

### 3.3   Uniform Sampling of Solutions

For practical PPM problem instances, the number $n$ of mutations ranges from 10 to 1000. In particular, for solid tumors in adults we typically observe thousands of point mutations in the genome. As such, exhaustive enumeration of solutions is infeasible in practice. To account for non-uniqueness of solutions and to identify common features shared among different solutions, it would be desirable to have an algorithm that samples uniformly from the solution space. However, as the underlying decision problem is NP-complete, the problem of sampling uniformly from the solution space for arbitrary frequency matrices $F$ is NP-hard. Thus, one must resort to heuristic approaches.

One class of such approaches employs Markov chain Monte Carlo (MCMC) for sampling from the solution space [2,10,11]. Here, we describe an alternative method based on rejection sampling. This method is guaranteed to sample uniformly from the solution space. Briefly, the idea is to generate a spanning arborescence $T$ from $G_F$ uniformly at random and then test whether $T$ satisfies (SC). In the case where $T$ satisfies (SC), we report $T$ as a solution and otherwise reject $T$.

For the general case where $G_F$ may have a directed cycle, we use the cycle-popping algorithm of Propp and Wilson [21]. This algorithm generates a uniform spanning arborescence in time $O(\tau(\tilde{G}_F))$ where $\tau(\tilde{G}_F)$ is the expected hitting time of $\tilde{G}_F$. More precisely, $\tilde{G}_F$ is the multi-graph obtained from $G_F$ by including self-loops such that the out-degrees of all its vertices are identical.

For the case where $G_F$ is a DAG with a unique source vertex $r$, there is a much simpler sampling algorithm. We simply assign each vertex $c \neq r$ to a parent $\pi(c) \in \delta_{in}(c)$ uniformly at random. It is easy to verify that the resulting function $\pi$ encodes a spanning arborescence of $G_F$. Thus, the running time of this procedure is $O(E(G_F))$. In both cases, the probability of success equals the fraction of spanning arborescences of $G_F$ that satisfy (SC) among all spanning arborescences of $G_F$.

An implementation of the rejection sampling for the case where $G_F$ is a DAG is available on https://github.com/elkebir-group/OncoLib.

## 4   Results

Figures 1 and 2 show anecdotal examples of non-uniqueness of solutions to the PERFECT PHYLOGENY MIXTURE problem. The following questions arise: Is non-uniqueness a widespread phenomenon in PPM instances? Which factors contribute to non-uniqueness and how does information from long-read sequencing and single-cell sequencing reduce non-uniqueness? Finally, are current MCMC methods able to sample uniformly from the space of solutions?

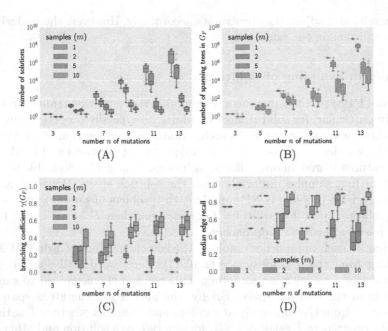

**Fig. 3. Factors that contribute to non-uniqueness.** (A) The number of solutions increased with increasing number $n$ of mutations, but decreased with increasing number $m$ of bulk samples. (B) Every solution of an PPM instance $F$ is a spanning arborescence in the ancestry graph $G_F$. The number of spanning arborescences in $G_F$ also increased with increasing $n$ and decreased with increasing $m$. (C) The decrease in the number of solutions and spanning arborescences with increasing $m$ is explained by the branching coefficient of $\gamma(G_F)$, which is the fraction of distinct pairs of mutations that occur on distinct branches in $G_F$. The fraction of such pairs increased with increasing $m$. (D) The median edge recall of the inferred trees $T$ increased with increasing $m$.

To answer these questions, we used simulated data generated by a previously published tumor simulator [5]. For each number $n \in \{3, 5, 7, 9, 11, 13\}$ of mutations, we generated 10 complete perfect phylogeny trees $T^*$. The simulator assigned each vertex $v \in V(T^*)$ a frequency $f(v) \geq 0$ such that $\sum_{v \in V(T^*)} f(v) = 1$. For each simulated complete perfect phylogeny tree $T^*$, we generated $m \in \{1, 2, 5, 10\}$ bulk samples by partitioning the vertex set $V(T^*)$ into $m$ disjoint parts followed by normalizing the frequencies in each sample. This yielded a frequency matrix $F$ for each combination of $n$ and $m$. In total, we generated $10 \cdot 6 \cdot 4 = 240$ instances. The raw data and scripts to generate the results are available on https://github.com/elkebir-group/PPM-NonUniq.

## 4.1 What Contributes to Non-uniqueness?

The two main factors that influence non-uniqueness are the number $n$ of mutations and the number $m$ of samples taken from the tumor. The former contributes to non-uniqueness while the latter reduces it. As we increased the number $n$ of mutations from 3 to 13, we observed that the number of solutions increased exponentially (Fig. 3A). On the other hand, the number $m$ of samples had an opposing effect: with increasing $m$ the number of solutions decreased.

To understand why we observed these two counteracting effects, we computed the number of spanning arborescences in each ancestry graph $G_F$. Figure 3B shows that the number of spanning arborescences exhibited an exponential increase with increasing number $n$ of mutations, whereas increased number $m$ of samples decreased the number of spanning arborescences. The latter can be explained by studying the effect of the number $m$ of samples on the branching coefficient $\gamma(G_F)$. Figure 3C shows that the branching coefficient increased with increasing $m$, with branching coefficient $\gamma(G_F) = 0$ for all $m = 1$ instances $F$. This finding illustrates that additional samples reveal branching of mutations. That is, in the case where $m = 1$ one does not observe branching in $G_F$, whereas as $m \rightarrow \infty$ each sample will be composed of a single cell with binary frequencies and the ancestry graph $G_F$ will be a rooted tree.

Adding mutations increases the complexity of the problem, as reflected by the number of solutions. To quantify how distinct each solution $T$ is to the simulated tree $T^*$, we computed the edge recall of $T$ defined as $|E(T) \cap E(T^*)|/|E(T^*)|$ (note that $|E(T^*)| = n - 1$ by definition). A recall value of 1 indicates that the inferred tree $T$ is identical to the true tree $T^*$. Figure 3D shows that the median recall decreased with increasing number $n$ of mutations. However, as additional samples provide more information, the recall increased with increasing number $m$ of samples.

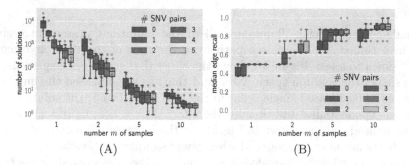

**Fig. 4. Long-read sequencing reduces the size of the solution space.** (A) The number of solutions decreased with increasing pairs of mutations that occurred on the same read. (B) The median edge recall increased with increasing pairs of mutations that co-occur on a read.

## 4.2   How to Reduce Non-uniqueness?

As discussed in Sect. 3.2, the non-uniqueness of solutions can be reduced through various sequencing techniques such as single-cell sequencing and long-read sequencing. We considered the effect of both technologies on the $n = 9$ instances.

By taking longer reads of the genome, long-read sequencing can identify mutations which coexist in a clone if they appear near one another on the genome. If two mutations are observed together on a long read, then one mutation is ancestral to the other. That is, on the true phylogenetic tree $T^*$ there must exist a path from the root to a leaf containing both mutations. We varied the number of mutation pairs observed together from 0 to 5 and observed that increasing this number reduced the size of the solution space (Fig. 4A). In addition, incorporating more simulated long-read information resulted in increased recall of the inferred trees (Fig. 4B).

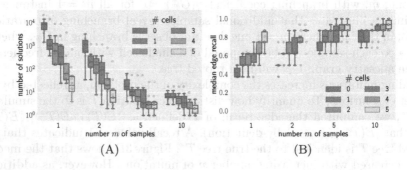

(A)                                     (B)

**Fig. 5. Joint bulk and single-cell sequencing reduces the size of the solution space.** (A) The number of solutions decreased with increasing number of single cells. (B) The median edge recall increased with increasing number of single cells.

Single-cell sequencing illuminates all of the mutations present in a single clone in a tumor. This reveals a path from the root of the true phylogenetic tree $T^*$ down to a leaf. Figure 5A shows the effect that single-cell sequencing has on the size of the solution space. We found that, as we increased the number of known paths (sequenced single cells) in the tree from 0 to 5, the solution space decreased exponentially. Additionally, the inferred trees were more accurate with more sequenced cells, as shown in Fig. 5B by the increase in median edge recall. These effects are more pronounced when fewer samples are available.

In summary, while both single-cell and long-read sequencing reduce the extent of non-uniqueness in the solution space, single-cell sequencing achieves a larger reduction than long-read sequencing.

## 4.3    How Does Non-uniqueness Affect Current Methods?

To study the effect of non-uniqueness, we considered two current methods, PhyloWGS [2] and Canopy [10], both of which use Markov chain Monte Carlo to sample solutions from the posterior distribution. Rather than operating from frequencies $F = [f_{p,c}]$, these two methods take as input two integers $a_{p,c}$ and $d_{p,c}$ for each mutation $c$ and sample $p$. These two integers are, respectively, the number of reads with mutation $c$ and the total number of reads. Given $A = [a_{p,c}]$

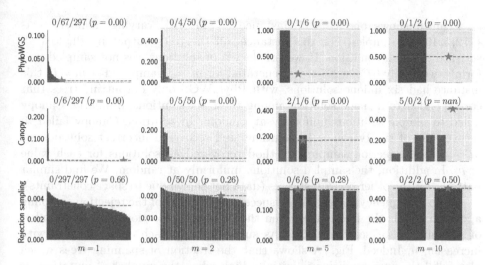

**Fig. 6. PhyloWGS and Canopy do not sample uniformly from the solution space.** We consider an $n = 7$ instance (#81) with varying number $m \in \{1, 2, 5, 10\}$ of bulk samples (columns), from which we sample solutions using different methods (rows). Each plot shows the relative frequency ($y$-axis) of identical trees ($x$-axis) output by each method, with the simulated tree indicated by '$\star$'. While blue bars are correct solutions (satisfying (SC)), red bars correspond to incorrect solutions (violating (SC)). Dashed line indicates the expected relative frequency in the case of uniformity. The title of each plot lists the number of incorrect solutions, the number of recovered correct solutions, the total number of correct solutions and the $p$-value of the chi-squared test of uniformity (null hypothesis is that the samples come from a uniform distribution). (Color figure online)

and $D = [d_{p,c}]$, PhyloWGS and Canopy aim to infer a frequency matrix $\hat{F}$ and phylogenetic tree $T$ with maximum data likelihood $\Pr(D, A \mid \hat{F})$ such that $T$ satisfies (SC) for matrix $\hat{F}$. In addition, the two methods cluster mutations that are inferred to have similar frequencies across all samples. To use these methods in our error-free setting, where we are given matrix $F = [f_{p,c}]$, we set the total number of reads for each mutation $c$ in each sample $p$ to a large number, i.e. $d_{p,c} = 1,000,000$. The number of variant reads is simply set as $a_{p,c} = f_{p,c} \cdot d_{p,c}$. Since both PhyloWGS and Canopy model variant reads $a_{p,c}$ as draws from a binomial distribution parameterized by $d_{p,c}$ and $\hat{f}_{p,c}$, the data likelihood is maximized when $\hat{F} = F$. We also discard generated solutions where mutations are clustered. Hence, we can use these methods in the error-free case.

We ran PhyloWGS, Canopy, and our rejection sampling method (Sect. 3.3) on all $n = 7$ instances. We used the default settings for PhyloWGS (2500 MCMC samples, burnin of 1000) and Canopy (burnin of 100 and 1 out of 5 thinning), with 20 chains per instance for PhyloWGS and 15 chains per instance for Canopy. For each instance, we ran the rejection sampling algorithm until it generated 10,000 solutions that satisfy (SC).

Figure 6 shows one $n = 7$ instance (#81) with varying number $m \in \{1, 2, 5, 10\}$ of samples. For this instance, all the trees output by PhyloWGS satisfied the sum condition. However, the set of solutions was not sampled uniformly, with only 67 out 297 trees generated for $m = 1$ samples. For $m = 5$, this instance had six unique solutions, with PhyloWGS only outputting trees that corresponded to a single solution among these six solutions. Similarly, Canopy failed to sample solutions uniformly at random. In addition, Canopy failed to recover any of the two $m = 10$ solutions and recovered incorrect solutions for $m = 5$. The rejection sampling method recovered all solutions for each value of $m$. In addition, the sampled solutions uniformly at random. We find similar patterns for the other $n = 7$ instances (data not shown due to space constraints).

Given a frequency matrix $F$, the success probability of the rejection sampling approach equals the fraction between the number of solutions and the number of spanning arborescences in $G_F$. As such, this approach does not scale with increasing $n$. Indeed, Fig. 7A shows that the fraction of spanning trees which also fulfill the sum condition is initially high when the number of mutations is low. With $n = 11$ mutations, the fraction is approximately $10^{-2}$ and rejection sampling can be considered to be feasible. However, as the number of mutations is increased further, rejection sampling become infeasible as the fraction can drop to $10^{-10}$ for $n = 21$ mutations (Fig. 7B). Therefore, a better sampling approach is required.

## 5    Discussion

In this work, we studied the problem of non-uniqueness of solutions to the PERFECT PHYLOGENY MIXTURE (PPM) problem. In this problem, we are given a frequency matrix $F$ that determines a directed graph $G_F$ called the ancestry

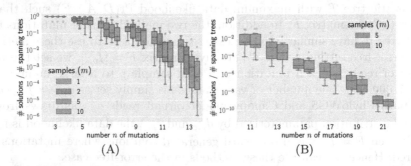

(A)                                          (B)

**Fig. 7. Although rejection sampling achieves uniformity, it becomes impractical with increasing number $n$ of mutations.** (A) Plot shows the ratio of the number of solutions to spanning arborescences. Observe that the number of spanning trees increased with the number $n$ of mutations far more rapidly than the number of solutions. (B) With further increases in $n$, the ratio rapidly decreased and the odds of randomly sampling a solution from the space of spanning arborescences becomes infeasible.

graph. The task is to identify a spanning arborescence $T$ of $G_F$ whose internal vertices satisfy a linear inequality whose terms are entries of matrix $F$. We formulated the #PPM problem of counting the number of solutions to an PPM instance. We showed that the number of solutions is at most the number of spanning arborescences in $G_F$, a number that can be computed in polynomial time. For the case where $G_F$ is a directed acyclic graph, we gave a simple algorithm for counting the number of spanning arborescences. This algorithm formed the basis of a rejection sampling scheme that samples solutions to a PPM instance uniformly at random.

Using simulations, we showed that the number of solutions increases with increasing number $n$ of mutations but decreases with increasing number $m$ of samples. In addition, we showed that the median recall of all solutions increases with increasing $m$ but decreases with increasing $n$. We showed how constraints from single-cell and long-read sequencing reduce the number of solutions. Finally, we showed that current MCMC methods fail to sample uniformly from the solution space. This is problematic as it leads to biases that propagate to downstream analyses.

There are a couple of avenues for future research. First, it remains to show that #PPM is #P-complete. Second, while the rejection sampling algorithm achieves uniformity, it does not scale to practical problem instance sizes. Further research is needed to develop sampling algorithms that achieve near-uniformity and have reasonable running time for practical problem instances. Third, in terms of practical applications, the problem of sampling solutions uniformly at random in the case of noisy frequencies must be studied. Fourth, just as single-cell sequencing and long-read sequencing impose constraints on the solution space of PPM, it will be worthwhile to include additional prior knowledge to further constrain the solution space. Finally, the PPM problem and the simulations in this paper assumed error-free data. Further research is needed to study the effect of sequencing, sampling and mapping errors. It is to be expected that the problem of non-uniqueness is further exacerbated with additional sources uncertainty.

**Acknowledgements.** This research is part of the Blue Waters sustained-petascale computing project, which is supported by the National Science Foundation (awards OCI-0725070 and ACI-1238993) and the state of Illinois. Blue Waters is a joint effort of the University of Illinois at Urbana-Champaign and its National Center for Supercomputing Applications. The authors thank the anonymous referees for insightful comments that have improved the manuscript.

# References

1. Deshwar, A.G., et al.: Abstract B2-59: PhyloSpan: Using multi-mutation reads to resolve subclonal architectures from heterogeneous tumor samples. Cancer Res. **75**(22 Suppl. 2), B2-59–B2-59 (2015)
2. Deshwar, A.G., et al.: PhyloWGS: reconstructing subclonal composition and evolution from whole-genome sequencing of tumors. Genome Biol. **16**(1), 35 (2015)

3. El-Kebir, M., Oesper, L., Acheson-Field, H., Raphael, B.J.: Reconstruction of clonal trees and tumor composition from multi-sample sequencing data. Bioinformatics **31**(12), i62–i70 (2015)
4. El-Kebir, M., Satas, G., Oesper, L., Raphael, B.J.: Inferring the mutational history of a tumor using multi-state perfect phylogeny mixtures. Cell Syst. **3**(1), 43–53 (2016)
5. El-Kebir, M., Satas, G., Raphael, B.J.: Inferring parsimonious migration histories for metastatic cancers. Nature Genetics **50**(5), 718–726 (2018)
6. Fisher, R., Pusztai, L., Swanton, C.: Cancer heterogeneity: implications for targeted therapeutics. Br. J. Cancer **108**(3), 479–485 (2013)
7. Gabow, H.N., Myers, E.W.: Finding all spanning trees of directed and undirected graphs. SIAM J. Comput. **7**(3), 280–287 (1978)
8. Gerstung, M., et al.: PCAWG Evolution, Heterogeneity Working Group, and PCAWG network. The evolutionary history of 2,658 cancers. bioRxiv, p. 161562, July 2017
9. Jamal-Hanjani, M., et al.: Trackingthe evolution of non-small-cell lung cancer. New Engl. J. Med. **376**(22), 2109–2121 (2017)
10. Jiang, Y., Qiu, Y., Minn, A.J., Zhang, N.R.: Assessing intratumor heterogeneity and tracking longitudinal and spatial clonal evolutionary history by next-generation sequencing. Proc. National Acad. Sci. United States Am. **113**(37), E5528–37 (2016)
11. Jiao, W., Vembu, S., Deshwar, A.G., Stein, L., Morris, Q.: Inferring clonal evolution of tumors from single nucleotide somatic mutations. BMC Bioinform. **15**, 35 (2014)
12. Kandoth, C., et al.: Mutational landscape and significance across 12 major cancer types. Nature **502**(7471), 333–339 (2013)
13. Kirchhoff, G.: Ueber die Auflösung der Gleichungen, auf welche man bei der Untersuchung der linearen Vertheilung galvanischer Ströme geführt wird. Annalen der Physik **148**, 497–508 (1847)
14. Livne, N.: A note on #P-completeness of NP-witnessing relations. Inf. Process. Lett. **109**(5), 259–261 (2009)
15. Łuksza, M., et al.: A neoantigen fitness model predicts tumour response to checkpoint blockade immunotherapy. Nature **551**(7681), 517 (2017)
16. Malikic, S., Jahn, K., Kuipers, J., Sahinalp, C., Beerenwinkel, N.: Integrative inference of subclonal tumour evolution from single-cell and bulk sequencing data. bioRxiv, p. 234914, December 2017
17. Malikic, S., McPherson, A.W., Donmez, N., Sahinalp, C.S.: Clonality inference in multiple tumor samples using phylogeny. Bioinformatics **31**(9), 1349–1356 (2015)
18. McGranahan, N., et al.: Clonal status of actionable driver events and the timing of mutational processes in cancer evolution. Sci. Trans. Med. **7**(283), 283ra54 (2015)
19. Nowell, P.C.: The clonal evolution of tumor cell populations. Science **194**(4260), 23–8 (1976)
20. Popic, V., Salari, R., Hajirasouliha, I., Kashef-Haghighi, D., West, R.B., Batzoglou, S.: Fast and scalable inference of multi-sample cancer lineages. Genome Biol. **16**(1), 91 (2015)
21. Propp, J.G., Wilson, D.B., James Gary Propp and David Bruce Wilson: How to get a perfectly random sample from a generic Markov chain and generate a random spanning tree of a directed graph. J. Algorithms **27**(2), 170–217 (1998)
22. Schwartz, R., Schäffer, A.A., Russell Schwartz and Alejandro: The evolution of tumour phylogenetics: principles and practice. Nature Rev. Genet. **18**(4), 213–229 (2017)

23. Strino, F., Parisi, F., Micsinai, M., Kluger, Y.: Trap: a tree approach for finger-printing subclonal tumor composition. Nucleic Acids Res. **41**(17), e165 (2013)
24. Tabassum, D.P., Polyak, K.: Tumorigenesis: it takes a village. Nature Rev. Cancer **15**(8), 473–483 (2015)
25. Turajlic, S., et al.: Tracking cancer evolution reveals constrained routes to metastases: TRACERx Renal. Cell **173**(3), 581–594 (2018)
26. Turajlic, S., et al.: Deterministic evolutionary trajectories influence primary tumor growth: TRACERx renal. Cell **173**(3), 581–594 (2018)
27. Tutte, W.T.: The dissection of equilateral triangles into equilateral triangles. Math. Proc. Camb. Philos. Soc. **44**(4), 463–482 (1948)
28. Venkatesan, S., Swanton, C.: Tumor evolutionary principles: how intratumor heterogeneity influences cancer treatment and outcome. Am. Soc. Clin. Oncol. Educ. Book. **35**, e141–9 (2016). American Society of Clinical Oncology. Meeting
29. Yuan, K., Sakoparnig, T., Markowetz, F., Beerenwinkel, N.: BitPhylogeny: a probabilistic framework for reconstructing intra-tumor phylogenies. Genome Biol. **16**(1), 1 (2015)
30. Zhang, A.W., et al.: Interfaces of malignant and immunologic clonal dynamics in ovarian cancer. Cell **173**(7), 1755–1769.e22 (2018)

# Non-parametric and Semi-parametric Support Estimation Using SEquential RESampling Random Walks on Biomolecular Sequences

Wei Wang[1], Jack Smith[1], Hussein A. Hejase[2], and Kevin J. Liu[1(✉)]

[1] Department of Computer Science and Engineering, Michigan State University, East Lansing, MI 48824, USA
kjl@msu.edu
[2] Simons Center for Quantitative Biology, Cold Spring Harbor Laboratory, Cold Spring Harbor, NY 11724, USA

**Abstract.** Non-parametric and semi-parametric resampling procedures are widely used to perform support estimation in computational biology and bioinformatics. Among the most widely used methods in this class is the standard bootstrap method, which consists of random sampling with replacement. While not requiring assumptions about any particular parametric model for resampling purposes, the bootstrap and related techniques assume that sites are independent and identically distributed (i.i.d.). The i.i.d. assumption can be an over-simplification for many problems in computational biology and bioinformatics. In particular, sequential dependence within biomolecular sequences is often an essential biological feature due to biochemical function, evolutionary processes such as recombination, and other factors.

To relax the simplifying i.i.d. assumption, we propose a new non-parametric/semi-parametric sequential resampling technique that generalizes "Heads-or-Tails" mirrored inputs, a simple but clever technique due to Landan and Graur. The generalized procedure takes the form of random walks along either aligned or unaligned biomolecular sequences. We refer to our new method as the SERES (or "SEquential RESampling") method.

To demonstrate the performance of the new technique, we apply SERES to estimate support for the multiple sequence alignment problem. Using simulated and empirical data, we show that SERES-based support estimation yields comparable or typically better performance compared to state-of-the-art methods.

**Electronic supplementary material** The online version of this chapter (https://doi.org/10.1007/978-3-030-00834-5_17) contains supplementary material, which is available to authorized users.

M. Blanchette and A. Ouangraoua (Eds.): RECOMB-CG 2018, LNBI 11183, pp. 294–308, 2018.
https://doi.org/10.1007/978-3-030-00834-5_17

# 1   Introduction

Resampling methods are widely used throughout computational biology and bioinformatics as a means for assessing statistical support. At a high level, resampling-based support estimation procedures consist of a methodological pipeline: resampled replicates are generated, inference/analysis is performed on each replicate, and results are then compared across replicates. Among the most widely used resampling methods are non-parametric approaches including the standard bootstrap method [5], which consists of random sampling with replacement. We will refer to the standard bootstrap method as the bootstrap method for brevity. Unlike parametric methods, non-parametric approaches need not assume that a particular parametric model is applicable to a problem at hand. However, the bootstrap and other widely used non-parametric approaches assume that observations are independent and identically distributed (i.i.d.).

In the context of biomolecular sequence analysis, there are a variety of biological factors that conflict with this assumption. These include evolutionary processes that cause intra-sequence dependence (e.g., recombination) and functional dependence among biomolecular sequence elements and motifs. Felsenstein presciently noted these limitations when he proposed the application of the bootstrap to phylogenetic inference: "A more serious difficulty is lack of independence of the evolutionary processes in different characters. . . . For the purposes of this paper, we will ignore these correlations and assume that they cause no problems; in practice, they pose the most serious challenge to the use of bootstrap methods." (reproduced from p. 785 of [6]).

To relax the simplifying assumption of i.i.d. observations, Landan and Graur [10] introduced the Heads-or-Tails (HoT) technique for the specific problem of multiple sequence alignment (MSA) support estimation. The idea behind HoT is simple but quite powerful: inference/analysis should be repeatable whether an MSA is read either from left-to-right or from right-to-left – i.e., in either heads or tails direction, respectively. While HoT resampling preserves intra-sequence dependence, it is limited to two replicates, which is far fewer than typically needed for reasonable support estimation; often, hundreds of resampled replicates or more are used in practice. Subsequently developed support estimation procedures increased the number of possible replicates by augmenting HoT with bootstrapping, parametric resampling, and domain-specific techniques (e.g., progressive MSA estimation) [11,15,17]. The combined procedures were shown to yield comparable or improved support estimates relative to the original HoT procedure [17] as well as other state-of-the-art parametric and domain-specific methods [9,13], at the cost of some of the generalizability inherent to non-parametric approaches. In this study, we revisit the central question that HoT partially addressed: how can we resample many non-parametric replicates that account for dependence within a sequence of observations, and how can such techniques be used to derive improved support estimates for biomolecular sequence analysis?

## 2     Methods

In our view, a more general statement of HoT's main insight is the following, which we refer to as the "neighbor preservation property": a neighboring observation is still a neighbor, whether reading an observation sequence from the left or the right. In other words, the key property needed for non-parametric resampling is preservation of neighboring bases within the original sequences, where any pair of bases that appear as neighbors in a resampled sequence must also be neighbors in the corresponding original sequence. To obtain many resampled replicates that account for intra-sequence dependence while retaining the neighbor preservation property, we propose a random walk procedure which generalizes a combination of the bootstrap method and the HoT method. We refer to the new resampling procedure as SERES ("SEquential RESampling"). Note that the neighbor preservation property is necessary but not sufficient for statistical support estimation. Other important properties include computational efficiency of the resampling procedure and unbiased sampling of observations within the original observation sequence.

SERES walks can be performed on both aligned and unaligned sequence inputs. We discuss the case of aligned inputs first, since it is simpler than the case of unaligned inputs.

### 2.1     SERES Walks on Aligned Sequences

Detailed pseudocode for a non-parametric SERES walk on a fixed MSA is shown in the Appendix's Supplementary Methods section: Algorithm 1.

The random walk is performed on the sequence of aligned characters (i.e., MSA sites). The starting point for the walk is chosen uniformly at random from the alignment sites, and the starting direction is also chosen uniformly at random. The random walk then proceeds in the chosen direction with non-deterministic reversals, or direction changes, that occur with probability $\gamma$; furthermore, reversals occur with certainty at the start and end of the fixed MSA. Aligned characters are sampled during each step of the walk. The random walk ends once the number of sampled characters is equal to the fixed MSA length.

The long-term behavior of an infinitely long SERES random walk can be described by a second-order Markov chain. Certain special cases (e.g., $\gamma = 0.5$) can be described using a first-order Markov chain.

In theory, a finite-length SERES random walk can exhibit biased sampling of sites since reversal occurs with certainty at the start and end of the observation sequence, whereas reversal occurs with probability $\gamma$ elsewhere. However, for practical choices of walk length and reversal probability $\gamma$, sampling bias is expected to be minimal.

### 2.2     SERES Walks on Unaligned Sequences

Detailed pseudocode for SERES resampling of unaligned sequences is shown in the Appendix's Supplementary Methods section: Algorithm 2. Figure 1 provides an illustrated example.

The procedure begins with estimating a set of anchors – sequence regions that exhibit high sequence similarity – which enable resampling synchronization across unaligned sequences. A conservative approach for identifying anchors would be to use highly similar regions that appear in the strict consensus of multiple MSA estimation methods. In practice, we found that highly similar regions within a single guide MSA produced reasonable anchors. We used the average normalized Hamming distance (ANHD) as our similarity measure, where indels are treated as mismatches.

Unaligned sequence indices corresponding to the start and end of each anchor serve as "barriers" in much the same sense as in parallel computing: asynchronous sequence reads occur between barrier pairs along a current direction (left or right), and a random walk is conducted on barrier space in a manner similar to a SERES walk on a sequence of aligned characters. The set of barriers also includes trivial barriers at the start and end of the unaligned sequences. The random walk concludes once the unaligned sequences in the resampled replicate have sufficient length; our criterion requires that the longest resampled sequence has minimum length that is a multiple maxReplicateLengthFactor of the longest input sequence length.

Technically, the anchors in our study make use of parametric MSA estimation and the rest of the SERES walk is non-parametric. The overall procedure is therefore semi-parametric (although see Conclusions for an alternative).

## 2.3   Performance Study

Our study evaluated the performance of SERES-based support estimation in the context of MSA support estimation. Of course, there are many other applications for non-parametric/semi-parametric support estimation – too many to investigate in one study. We focus on this application since it is considered to be a classical problem in computational biology and bioinformatics and its outputs are useful for studying a range of topics (e.g., phylogenetics and phylogenomics, proteomics, comparative genomics, etc.).

**Computational Methods.** We examined the problem of evaluating support in the context of MSA estimation. The problem input consists of an estimated MSA $A$ which has a corresponding set of unaligned sequences $S$. The problem output consists of support estimates for each nucleotide-nucleotide homology in $A$, where each support estimate is on the unit interval. Note that this computational problem is distinct from the full MSA estimation problem.

There are a variety of existing methods for MSA support estimation. The creators of HoT and their collaborators subsequently developed alignment-specific parametric resampling techniques [11] and then combined the two to obtain two new semi-parametric approaches: GUIDANCE [15] (which we will refer to as GUIDANCE1) and GUIDANCE2 [17]. Other parametric MSA support estimation methods include PSAR [9] and T-Coffee [13].

We focus on GUIDANCE1 and GUIDANCE2, which subsume HoT and have been demonstrated to have comparable or better performance relative to other state-of-the-art methods [17]. We used MAFFT for re-estimation on resampled replicates, since it has been shown to be among the most accurate progressive MSA methods to date [8,12].

We then used SERES to perform resampling in place of the standard bootstrap that is used in the first step of GUIDANCE1/GUIDANCE2. Re-estimation was performed on 100 SERES replicates – each consisting of a set of unaligned sequences – using MAFFT with default settings, which corresponds to the FFT-NS-2 algorithm for progressive alignment. The SERES resampling procedure used a reversal probability $\gamma = 0.5$, which is equivalent to selecting a direction uniformly at random (UAR) at each step of the random walk; each SERES replicate utilized a total of $\lfloor \frac{k}{20} \rfloor$ anchors with anchor size of 5 bp and a minimum distance between neighboring anchors of 25 bp, where $k$ is the length of the input alignment $A$. All downstream steps of GUIDANCE1/GUIDANCE2 were then performed using the re-estimated alignments as input.

**(a)** Estimate consensus alignment on input set of unaligned sequences.

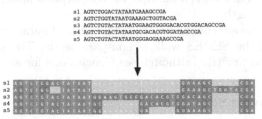

**(b)** Obtain anchors on consensus alignment. Barriers (dashed lines) consist of anchor boundaries plus trivial start/end barriers.

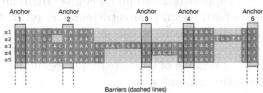

**Fig. 1. Illustrated example of SERES resampling random walk on unaligned sequences.** Detailed pseudocode is provided in the Appendix's Supplementary Methods section (Algorithm 2 in Appendix). (a) The resampling procedure begins with the estimation of a consensus alignment on the input set of unaligned sequences. (b) A set of conservative anchors is then obtained using the consensus alignment, and anchor boundaries define a set of barriers (including two trivial barriers – one at the start of the sequences and one at the end of the sequences). (c) The SERES random walk is conducted on the set of barriers. The walk begins at a random barrier and proceeds in a random direction to the neighboring barrier. The walk reverses with certainty when the trivial start/end barriers are encountered; furthermore, the walk direction can randomly reverse with probability $\gamma$. As the walk proceeds from barrier to barrier, unaligned sequences are sampled between neighboring barrier pairs. (d) The resampling procedure terminates when the resampled sequences meet a specified sequence length threshold.

**(c)** Choose an initial barrier and walk direction at random.
Begin random walk (red arrow) from first barrier to neighboring barrier.
As walk proceeds from one barrier to neighboring barrier,
sample unaligned sequences between barrier pairs.

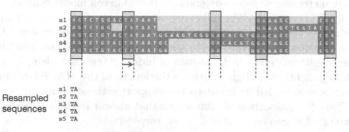

Resampled
sequences

```
s1  TA
s2  TA
s3  TA
s4  TA
s5  TA
```

**(d)** Random walk terminates when resampled sequences reach required length.

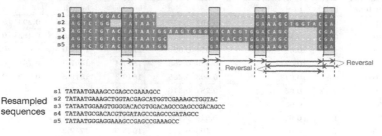

Resampled
sequences

```
s1  TATAATGAAAGCCGAGCCGAAAGCC
s2  TATAATGAAAGCTGGTACGAGCATGGTCGAAAGCTGGTAC
s3  TATAATGGAAGTGGGGGACACGTGGACAGCCGAGCCGACAGCC
s4  TATAATGCGACACGTGGATAGCCGAGCCGATAGCC
s5  TATAATGGGAGGAAAGCCGAGCCGAAAGCC
```

**Fig. 1.** (*continued*)

**Simulated Datasets.** Model trees and sequences were simulated using INDELible [7]. First, non-ultrametric model trees with either 10 or 50 taxa were sampled using the following procedure. Model trees were generated under a birth-death process [18], branch lengths were chosen UAR from the interval $(0, 1)$, and the model tree height was re-scaled from its original height $h_0$ to a desired height $h$ by multiplying all branch lengths by the factor $h/h_0$. Next, sequences were evolved down each model tree under the General Time-Reversible (GTR) model of substitution [16] and the indel model of Fletcher and Yang [7], where the root sequence had length of 1 kb. We used the substitution rates and base frequencies from the study of Liu et al. [12], which were based upon empirical analysis of the nematode Tree of Life. Sequence insertions/deletions occurred at rate $r_i$, and we used the medium gap length distribution from the study of Liu et al. [12]. The model parameter values used for simulation are shown in Table 1, and each combination of model parameter values constitutes a model condition. Model conditions are enumerated in order of generally increasing sequence divergence, as reflected by average pairwise ANHD. For each model condition, the simulation procedure was repeated to generate twenty replicate datasets. Summary statistics for simulated datasets are shown in Table 1.

We evaluated performance based upon receiver operating characteristic (ROC) curves, precision-recall curves (PR), and area under ROC and PR curves (ROC-AUC and PR-AUC, respectively). Consistent with other studies of MSA

**Table 1. Model condition parameter values and summary statistics.** The simulation study parameters consist of the number of taxa, model tree height, and insertion/deletion probability. Each model condition corresponds to a distinct set of model parameter values. The 10-taxon model conditions are named 10.A through 10.E in order of generally increasing sequence divergence; the 50-taxon model conditions are named 50.A through 50.E similarly. The following table columns list average summary statistics for each model condition ($n = 20$). "NHD" is the average normalized Hamming distance of a pair of aligned sequences in the true alignment. "Gappiness" is the percentage of true alignment cells which consists of indels. "True align length" is the length of the true alignment. "Est align length" is the length of the MAFFT-estimated alignment [8] which was provided as input to the support estimation methods. "SP-FN" and "SP-FP" are the proportion of homologies that appear in the true alignment but not in the estimated alignment and vice versa, respectively.

| Model condition | Number of taxa | Tree height | Insertion/deletion probability | NHD | Gappiness | True align length | Est align length | SP-FN | SP-FP |
|---|---|---|---|---|---|---|---|---|---|
| 10.A | 10 | 0.4 | 0.13 | 0.297 | 0.474 | 1965.3 | 1552.3 | 0.294 | 0.341 |
| 10.B | 10 | 0.7 | 0.1 | 0.394 | 0.512 | 2165.1 | 1563.5 | 0.483 | 0.533 |
| 10.C | 10 | 1 | 0.06 | 0.514 | 0.526 | 2162.8 | 1554.0 | 0.657 | 0.684 |
| 10.D | 10 | 1.6 | 0.031 | 0.599 | 0.485 | 1874.4 | 1507.5 | 0.747 | 0.752 |
| 10.E | 10 | 4.3 | 0.013 | 0.693 | 0.465 | 1849.3 | 1612.8 | 0.945 | 0.943 |
| 50.A | 50 | 0.45 | 0.06 | 0.281 | 0.516 | 2043.5 | 1785.7 | 0.086 | 0.088 |
| 50.B | 50 | 0.7 | 0.03 | 0.398 | 0.475 | 1935.5 | 1714.2 | 0.105 | 0.102 |
| 50.C | 50 | 1 | 0.02 | 0.514 | 0.498 | 2047.6 | 1703.1 | 0.245 | 0.230 |
| 50.D | 50 | 1.8 | 0.012 | 0.594 | 0.471 | 1945.0 | 1712.2 | 0.455 | 0.419 |
| 50.E | 50 | 4.3 | 0.004 | 0.688 | 0.459 | 1890.2 | 2319.2 | 0.963 | 0.948 |

support estimation techniques [15,17], the MSA support estimation problem in our study entails annotation of nucleotide-nucleotide homologies in the estimated alignment; thus, homologies that appear in the true alignment but not the estimated alignment are not considered. For this reason, the confusion matrix quantities used for ROC and PR calculations are defined as follows. True positives (TP) are the set of nucleotide-nucleotide homologies that appear in the true alignment and the estimated alignment with support value greater than or equal to a given threshold, false positives (FP) are the set of nucleotide-nucleotide homologies that appear in the estimated alignment with support value greater than or equal to a given threshold but do not appear in the true alignment, false negatives (FN) are the set of nucleotide-nucleotide homologies that appear in the true alignment but appear in the estimated alignment with support value below a given threshold, and true negatives (TN) are the set of nucleotide-nucleotide homologies that do not appear in the true alignment and appear in the estimated alignment with support value below a given threshold. The ROC curve plots the true positive rate ($|TP|/(|TP| + |FN|)$) versus the false positive rate ($|FP|/(|FP| + |TN|)$). The PR curve plots the true positive rate versus precision ($|TP|/(|TP| + |FP|)$). Varying the support threshold yields different points along these curves. Custom scripts were used to perform confusion matrix calculations. ROC curve, PR curve, and AUROC calculations were performed using the scikit-learn Python library [14].

**Table 2. Empirical dataset summary statistics.** The empirical study made use of reference alignments ("Ref align") from the CRW database [2]. The reference alignments were curated using heterogeneous data including secondary structure information. The column description is identical to Table 1, where the empirical study made use of reference alignments in lieu of the simulation study's true alignments.

| Dataset | Number of taxa | NHD | Gappiness | Ref align length | Est align length | SP-FP | SP-FN |
|---------|------|-------|-------|-------|-------|-------|-------|
| IGIA | 110 | 0.606 | 0.915 | 10368 | 6675 | 0.734 | 0.784 |
| IGIB | 202 | 0.579 | 0.910 | 10633 | 7379 | 0.825 | 0.864 |
| IGIC2 | 32 | 0.533 | 0.700 | 4243 | 3514 | 0.689 | 0.715 |
| IGID | 21 | 0.719 | 0.782 | 5061 | 3023 | 0.874 | 0.904 |
| IGIE | 249 | 0.451 | 0.838 | 2751 | 2775 | 0.393 | 0.376 |
| IGIIA | 174 | 0.668 | 0.814 | 6406 | 7005 | 0.816 | 0.800 |
| PA23 | 142 | 0.293 | 0.267 | 3991 | 3552 | 0.078 | 0.077 |
| PE23 | 117 | 0.300 | 0.612 | 9436 | 10083 | 0.202 | 0.213 |
| PM23 | 102 | 0.361 | 0.797 | 10999 | 8803 | 0.262 | 0.288 |
| SA16 | 132 | 0.212 | 0.205 | 1866 | 1673 | 0.031 | 0.028 |
| SA23 | 144 | 0.304 | 0.460 | 4048 | 3678 | 0.077 | 0.081 |

**Empirical Datasets.** We downloaded empirical benchmarks from the Comparative RNA Web (CRW) Site database, which can be found at www.rna.icmb. utexas.edu [2]. In brief, the CRW database includes ribosomal RNA sequence datasets than span a range of dataset sizes and evolutionary divergence. We focused on datasets where high-quality reference alignments are available; the reference alignments were produced using intensive manual curation and analysis of heterogeneous data, including secondary structure information. We selected primary 16 S rRNA, primary 23 S rRNA, primary intron, and seed alignments with at most 250 sequences. Aligned sequences with 99% or more missing data and/or indels were omitted from analysis. Summary statistics for the empirical benchmarks are shown in Table 2.

## 2.4   Computational Resources Used and Software/Data Availability

All computational analyses were run on computing facilities in Michigan State University's High Performance Computing Center. We used compute nodes in the intel16-k80 cluster, each of which had a 2.4 GHz 14-core Intel Xeon E5-2680v4 processor. All replicates completed with memory usage less than 10 GiB. Open-source software and open data can be found at https://gitlab.msu.edu/liulab/SERES-Scripts-Data.

**Table 3. Simulation study results.** Results are shown for five 10-taxon model conditions (named 10.A through 10.E in order of generally increasing sequence divergence) and five 50-taxon model conditions (similarly named 50.A through 50.E). We evaluated the performance of two state-of-the-art methods for MSA support estimation – GUIDANCE1 [15] and GUIDANCE2 [17] – versus re-estimation on SERES and parametrically resampled replicates (using parametric techniques from either GUIDANCE1 or GUIDANCE2). (See Methods section for details.) We calculated each method's precision-recall (PR) and receiver operating characteristic (ROC) curves. Performance is evaluated based upon aggregate area under curve (AUC) across all replicates for a model condition ($n = 20$). The top rows show AUC comparisons of GUIDANCE1 ("GUIDANCE1") vs. SERES combined with parametric techniques from GUIDANCE1 ("SERES+GUIDANCE1"), and the bottom rows show AUC comparisons of GUIDANCE2 ("GUIDANCE2") vs. SERES combined with parametric techniques from GUIDANCE2 ("SERES+GUIDANCE2"); for each model condition and pairwise comparison, the best AUC is shown in bold. Statistical significance of PR-AUC or AUC-ROC differences was assessed using a one-tailed pairwise t-test or DeLong et al. [4] test, respectively, and multiple test correction was performed using the method of Benjamini and Hochberg [1]. Corrected q-values are reported ($n = 20$) and all were significant ($\alpha = 0.05$).

| Model condition | PR-AUC (%) | | Pairwise t-test corrected q-value | ROC-AUC (%) | | DeLong et al. test corrected q-value |
|---|---|---|---|---|---|---|
| | GUID-ANCE1 | SERES+ GUID-ANCE1 | | GUID-ANCE1 | SERES+ GUID-ANCE1 | |
| 10.A | 88.74 | **91.17** | $5.4 \times 10^{-7}$ | 80.22 | **85.57** | $< 10^{-10}$ |
| 10.B | 82.21 | **86.26** | $1.5 \times 10^{-6}$ | 84.83 | **88.66** | $< 10^{-10}$ |
| 10.C | 76.23 | **83.49** | $1.9 \times 10^{-4}$ | 86.98 | **91.23** | $< 10^{-10}$ |
| 10.D | 74.65 | **85.81** | $1.9 \times 10^{-4}$ | 88.55 | **93.72** | $< 10^{-10}$ |
| 10.E | 42.61 | **59.20** | $3.1 \times 10^{-4}$ | 82.24 | **87.40** | $< 10^{-10}$ |
| 50.A | 98.22 | **98.92** | $5.3 \times 10^{-10}$ | 83.09 | **90.64** | $< 10^{-10}$ |
| 50.B | 97.84 | **98.69** | $2.8 \times 10^{-9}$ | 82.85 | **90.39** | $< 10^{-10}$ |
| 50.C | 95.08 | **96.80** | $5.6 \times 10^{-8}$ | 85.54 | **90.64** | $< 10^{-10}$ |
| 50.D | 90.79 | **95.75** | $5.3 \times 10^{-6}$ | 88.89 | **94.56** | $< 10^{-10}$ |
| 50.E | 62.47 | **79.14** | $8.0 \times 10^{-10}$ | 91.02 | **93.23** | $< 10^{-10}$ |
| Model condition | PR-AUC (%) | | Pairwise t-test corrected q-value | ROC-AOC (%) | | DeLong et al. test corrected q-value |
| | GUID-ANCE2 | SERES+ GUID-ANCE2 | | GUID-ANCE2 | SERES+ GUID-ANCE2 | |
| 10.A | 92.55 | **93.33** | $7.4 \times 10^{-6}$ | 87.17 | **88.34** | $< 10^{-10}$ |
| 10.B | 88.08 | **89.31** | $8.4 \times 10^{-4}$ | 89.45 | **90.56** | $< 10^{-10}$ |
| 10.C | 84.28 | **86.86** | $3.1 \times 10^{-4}$ | 91.36 | **92.88** | $< 10^{-10}$ |
| 10.D | 86.03 | **88.75** | $1.9 \times 10^{-4}$ | 93.34 | **94.69** | $< 10^{-10}$ |
| 10.E | 51.17 | **62.30** | $1.3 \times 10^{-3}$ | 86.00 | **88.28** | $< 10^{-10}$ |
| 50.A | 98.98 | **99.14** | $5.3 \times 10^{-6}$ | 91.17 | **92.50** | $< 10^{-10}$ |
| 50.B | 98.79 | **98.96** | $1.5 \times 10^{-6}$ | 91.24 | **92.44** | $< 10^{-10}$ |
| 50.C | 96.86 | **97.45** | $3.2 \times 10^{-7}$ | 90.81 | **92.31** | $< 10^{-10}$ |
| 50.D | 94.04 | **96.23** | $1.5 \times 10^{-5}$ | 92.67 | **95.09** | $< 10^{-10}$ |
| 50.E | 72.61 | **81.47** | $1.5 \times 10^{-8}$ | 92.94 | **94.22** | $< 10^{-10}$ |

# 3   Results

## 3.1   Simulation Study

For all model conditions, SERES-based resampling and re-estimation yielded improved MSA support estimates compared to GUIDANCE1 and GUIDANCE2, two state-of-the-art methods, where performance was measured by PR-AUC or ROC-AUC (Table 3). In all cases, PR-AUC or ROC-AUC improvements were statistically significant (corrected pairwise t-test or DeLong et al. [4] test, respectively; $n = 20$ and $\alpha = 0.05$). The observed performance improvement was robust to several experimental factors: dataset size, increasing sequence divergence due to increasing numbers of substitutions, insertions, and deletions, and the choice of alignment-specific parametric support estimation techniques (i.e., the parametric approaches used by either GUIDANCE1 or GUIDANCE2) that were used in combination with SERES-based support estimation.

Compared to dataset size, sequence divergence had a relatively greater quantitative impact on each method's performance. For each dataset size (10 or 50 taxa), PR-AUC differed by at most 3% on the least divergent model condition. The SERES-based method's performance advantage grew as sequence divergence increased – to as much as 28% – and the largest performance advantages were seen on the most divergent datasets in our study. The most divergent datasets were also the most challenging. For each method, PR-AUC generally degraded as sequence divergence increased; however, the SERES-based method's PR-AUC degraded more slowly compared to the non-SERES-based method. Consistent with the study of Sela et al. [17], GUIDANCE2 consistently outperformed GUIDANCE1 on each model conditions and using either AUC measure. The performance improvement of SERES+GUIDANCE1 over GUIDANCE1 was generally greater than that seen when comparing SERES+GUIDANCE2 and GUIDANCE2; furthermore, the PR-AUC-based corrected q-values were more significant for the former compared to the latter in all cases except for the 10.D model condition, where the corrected q-values were comparable. Finally, while the SERES-based method consistently yielded performance improvements over the corresponding non-SERES-based method regardless of the choice of performance measure (either PR-AUC or ROC-AUC), the PR-AUC difference was generally larger than the ROC-AUC difference, especially on more divergent model conditions. On average across all replicates of all model conditions with a given dataset size, the runtime overhead contributed by SERES was minimal – amounting to just a few minutes per replicate dataset – and all methods in the simulation study completed analysis of each replicate dataset in less than half an hour (Supplementary Table S1 in Appendix).

## 3.2   Empirical Study

Relative to GUIDANCE1 or GUIDANCE2, SERES-based support estimates consistently returned higher AUC on all datasets – primary, seed, and intronic – with a single exception: the comparison of SERES+GUIDANCE2 and

**Table 4. Empirical study results.** The empirical study made use of benchmark RNA datasets and curated reference alignments from the CRW database [2]. Results are shown for intronic ("IG" prefix) and non-intronic datasets ("P" prefix and "S" prefix, following "primary" and "seed" nomenclature from the CRW database). For each dataset, we report each method's PR-AUC and ROC-AUC. For each dataset and pairwise method comparison, the best AUC is shown in bold. Methods, performance measures, table layout, and table description are otherwise identical to Table 3.

| Dataset | PR-AUC (%) | | ROC-AUC (%) | |
|---------|-----------|-----------|-----------|-----------|
| | GUIDANCE1 | SERES+ GUIDANCE1 | GUIDANCE1 | SERES+ GUIDANCE1 |
| IGIA | 62.67 | **69.28** | 89.50 | **91.62** |
| IGIB | 73.60 | **87.47** | 94.49 | **97.39** |
| IGIC2 | 72.67 | **75.36** | 82.25 | **83.87** |
| IGID | 63.74 | **76.30** | 95.10 | **96.73** |
| IGIE | 93.56 | **95.42** | 90.08 | **93.30** |
| IGIIA | 73.03 | **83.06** | 86.49 | **96.45** |
| PA23 | 98.54 | **99.41** | 82.59 | **93.63** |
| PE23 | 98.44 | **99.27** | 94.75 | **97.41** |
| PM23 | 97.53 | **98.48** | 94.20 | **96.44** |
| SA16 | 99.72 | **99.86** | 91.07 | **95.57** |
| SA23 | 98.35 | **99.24** | 81.76 | **92.18** |

| Dataset | PR-AUC (%) | | ROC-AUC (%) | |
|---------|-----------|-----------|-----------|-----------|
| | GUIDANCE2 | SERES+ GUIDANCE2 | GUIDANCE2 | SERES+ GUIDANCE2 |
| IGIA | 67.4 | **68.49** | 91.38 | **91.94** |
| IGIB | 80.66 | **86.72** | 96.47 | **97.38** |
| IGIC2 | **74.44** | 73.27 | **84.63** | 82.51 |
| IGID | 75.15 | **78.38** | 96.44 | **97.09** |
| IGIE | 94.6 | **95.44** | 91.84 | **93.49** |
| IGIIA | 78.16 | **85.09** | 94.50 | **96.82** |
| PA23 | 99.24 | **99.53** | 91.48 | **94.88** |
| PE23 | 99.07 | **99.34** | 96.72 | **97.63** |
| PM23 | 98.68 | **98.85** | 96.93 | **97.28** |
| SA16 | 99.88 | **99.91** | 96.22 | **97.22** |
| SA23 | 99.04 | **99.33** | 89.93 | **93.18** |

GUIDANCE2 on the intronic IGIC2 dataset, where the PR-AUC and ROC-AUC differences were 1.17% and 2.12%, respectively. For each pairwise comparison of methods (i.e., SERES+GUIDANCE1 vs. GUIDANCE1 or SERES+GUIDANCE2 vs. GUIDANCE2), the SERES-based method returned relatively larger PR-AUC improvements on datasets with greater sequence divergence, as measured by ANHD and gappiness. In particular, PR-AUC improvements were less than 1% on seed and primary non-intronic datasets. Intronic datasets yielded PR-AUC improvements of as much as 13.87%. Observed AUC improvements of SERES+GUIDANCE1 over GUIDANCE1 were relatively greater than those seen for SERES+GUIDANCE2 in comparison to GUID-ANCE2. Finally, GUIDANCE2 consistently returned higher AUC relative to GUIDANCE1, regardless of whether PR or ROC curves were the basis for AUC comparison.

## 4  Discussion

Re-estimation using SERES resampling resulted in comparable or typically improved support estimates for the applications in our study. We believe that this performance advantage is due to the ability to generate many distinct repli-cates while enforcing the neighbor preservation principle. The latter is critical for retaining sequence dependence which is inherent to the application in our study.

On all model conditions, SERES+GUIDANCE1 support estimation resulted in significant improvements in AUC-PR and AUC-ROC com-pared to GUIDANCE1. A similar outcome was observed when comparing SERES+GUIDANCE2 and GUIDANCE2. The main difference in each com-parison is the resampling technique – either SERES or standard bootstrap. Our findings clearly demonstrate the performance advantage of the former over the latter. SERES accounts for intra-sequence dependence due to insertion and dele-tion processes, while the bootstrap method assumes that sites are independent and identically distributed. Regarding comparisons involving GUIDANCE2 ver-sus GUIDANCE1, a contributing factor may have been the greater AUC of GUIDANCE2 over GUIDANCE1. We used SERES to perform semi-parametric support estimation in conjunction with the parametric support techniques of GUIDANCE1 or GUIDANCE2. The latter method's relatively greater AUC may be more challenging to improve upon (Table 4).

The performance comparisons on empirical benchmarks were consistent with the simulation study. In terms of ANHD and gappiness, the non-intronic datasets in our empirical study were more like the low divergence model conditions in our simulation study, and the intronic datasets were more like the higher diver-gence model conditions. Across all empirical datasets, SERES-based support estimation consistently yielded comparable or better AUC versus GUIDANCE1 or GUIDANCE2 alone. The SERES-based method's AUC advantage generally increased as datasets became more divergent and challenging to align – partic-ularly when comparing performance on non-intronic versus intronic datasets. We found that the support estimation methods returned comparable AUC (within a few percentage points) on datasets with 1–2 dozen sequences and low sequence divergence relative to other datasets. In particular, IGIC2 was the only dataset where SERES+GUIDANCE2 did not return an improved AUC rel-ative to GUIDANCE2. IGIC2 was the second-smallest dataset – about an order of magnitude smaller than all other datasets except the IGID dataset – and IGIC2 also had the second-lowest ANHD and lowest gappiness among intronic datasets. IGID was the smallest dataset, but had higher ANHD and gappiness compared to the IGIC2 dataset. Compared to the other empirical datasets, SERES+GUIDANCE2 returned a small AUC improvement over GUIDANCE2 on the IGID dataset – at most 3.2%.

On simulated and empirical datasets, greater sequence divergence generally resulted in increased inference error for all methods. However, the SERES-based method's performance tended to degrade more slowly than the corresponding non-SERES-based method as sequence divergence increased, and the greatest performance advantage was seen on the most divergent model conditions and empirical datasets.

Finally, we note that non-parametric/semi-parametric resampling techniques are orthogonal to parametric alternatives. Consistent with previous studies [15, 17], we found that combining two different classes of methods yielded better performance than either by itself.

## 5 Conclusions

This study introduced SERES, which consists of new non-parametric and semi-parametric techniques for resampling biomolecular sequence data. Using simulated and empirical data, we explored the use of SERES resampling for support estimation involving a classical problem in computational biology and bioinformatics. We found that SERES-based support estimation yields comparable or typically better performance compared to state-of-the-art approaches.

We conclude with possible directions for future work. First, the SERES algorithm in our study made use of a semi-parametric resampling procedure on unaligned inputs, since anchors were constructed using progressive multiple sequence alignment. While this approach worked well in our experiments, non-parametric alternatives could be substituted (e.g., unsupervised $k$-mer clustering using alignment-free distances [3]) to obtain a purely non-parametric resampling procedure. Second, the unaligned input application focused on nucleotide-nucleotide homologies to enable direct comparison against existing MSA support estimation procedures (i.e., GUIDANCE1 and GUIDANCE2). The SERES framework can be extended in a straightforward manner to estimate support for nucleotide-indel pairs. Third, SERES resampling can be used to perform full MSA inference. One approach would be to analyze homologies that appeared in re-estimated inferences across resampled replicates, without regard to any input alignment. Fourth, in the case where biomolecular sequences evolved under insertion/deletion processes, we consider the distinction between aligned and unaligned inputs to be an unnecessary dichotomy. In theory, the latter subsumes the former. We can apply this insight using a two-phase approach: (1) perform SERES-based re-estimation on unaligned sequences to estimate support for aligned homologies (from either an input MSA or the de novo procedure proposed above), and (2) perform support-weighted SERES walks on the annotated MSA from the previous stage to obtain support estimates on downstream inference. Alternatively, we can simultaneously address both problems using co-estimation. Finally, we envision many other SERES applications. Examples in computational biology and bioinformatics include protein structure prediction, detecting genomic patterns of natural selection, and read mapping and assembly. Non-parametric resampling for support estimation is widely used throughout science and engineering, and SERES resampling may similarly prove useful in research areas outside of computational biology and bioinformatics.

**Acknowledgments.** This work has been supported in part by the National Science Foundation (grant nos. CCF-1565719, CCF-1714417, and DEB-1737898 to KJL) and MSU faculty startup funds (to KJL). Computational experiments were performed using the High Performance Computing Center (HPCC) at MSU.

# References

1. Benjamini, Y., Hochberg, Y.: Controlling the false discovery rate: a practical and powerful approach to multiple testing. J. Royal Stat. Soc. Ser. B (Methodol) **57**(1), 289–300 (1995)
2. Cannone, J.J., et al.: The Comparative RNA Web (CRW) site: an online database of comparative sequence and structure information for Ribosomal, Intron and Other RNAs. BMC Bioinform. **3**(15) (2002). http://www.rna.ccbb.utexas.edu
3. Daskalakis, C., Roch, S.: Alignment-free phylogenetic reconstruction. In: Berger, B. (ed.) RECOMB 2010. LNCS, vol. 6044, pp. 123–137. Springer, Heidelberg (2010). https://doi.org/10.1007/978-3-642-12683-3_9
4. DeLong, E.R., DeLong, D.M., Clarke-Pearson, D.L.: Comparing the areas under two or more correlated receiver operating characteristic curves: a nonparametric approach. Biometrics **44**(3), 837–845 (1988)
5. Efron, B.: Bootstrap methods: another look at the jackknife. Ann. Stat. **7**(1), 1–26 (1979)
6. Felsenstein, J.: Confidence limits on phylogenies: an approach using the bootstrap. Evolution **39**(4), 783–791 (1985)
7. Fletcher, W., Yang, Z.: INDELible: a flexible simulator of biological sequence evolution. Mol. Biol. Evol. **26**(8), 1879–1888 (2009)
8. Katoh, K., Standley, D.M., Kazutaka Katoh and Daron: MAFFT multiple sequence alignment software version 7: improvements in performance and usability. Mol. Biol. Evol. **30**(4), 772–780 (2013)
9. Kim, J., Ma, J.: PSAR: measuring multiple sequence alignment reliability by probabilistic sampling. Nucleic Acids Res. **39**(15), 6359–6368 (2011)
10. Landan, G., Graur, D.: Heads or tails: a simple reliability check for multiple sequence alignments. Mol. Biol. Evol. **24**(6), 1380–1383 (2007)
11. Landan, G., Graur, D.: Local reliability measures from sets of co-optimal multiple sequence alignments. In: Biocomputing, pp. 15–24. World Scientific (2008)
12. Liu, K., et al.: SATé-II: very fast and accurate simultaneous estimation of multiple sequence alignments and phylogenetic trees. Syst. Biol. **61**(1), 90–106 (2012)
13. Notredame, C., Higgins, D.G., Heringa, J.: T-Coffee: a novel method for fast and accurate multiple sequence alignment. J. Mol. Biol. **302**, 205–217 (2000)
14. Pedregosa, F., et al.: Scikit-learn: machine learning in Python. J. Mach. Learn. Res. **12**, 2825–2830 (2011)

15. Penn, O., Privman, E., Landan, G., Graur, D., Pupko, T.: An alignment confidence score capturing robustness to guide tree uncertainty. Mol. Biol. Evol. **27**(8), 1759–1767 (2010)

16. Rodriguez, F., Oliver, J.L., Marin, A., Medina, J.R.: The general stochastic model of nucleotide substitution. J. Theor. Biol. **142**, 485–501 (1990)

17. Sela, I., Ashkenazy, H., Katoh, K., Pupko, T.: GUIDANCE2: accurate detection of unreliable alignment regions accounting for the uncertainty of multiple parameters. Nucleic Acids Res. **43**(W1), W7–W14 (2015)

18. Yang, Z., Rannala, B.: Bayesian phylogenetic inference using DNA sequences: a Markov chain Monte Carlo method. Mol. Biol. Evol. **14**(7), 717–724 (1997)

# Linear-Time Tree Containment in Phylogenetic Networks

Mathias Weller$^{(\boxtimes)}$ (iD)

CNRS, LIGM, Université Paris Est, Marne-la-Vallée, France
`mathias.weller@u-pem.fr`

**Abstract.** We consider the NP-hard TREE CONTAINMENT problem that has important applications in phylogenetics. The problem asks if a given single-rooted leaf-labeled network ("phylogenetic network") $N$ "contains" a given leaf-labeled tree ("phylogenetic tree") $T$. We develop a fast algorithm for the case that $N$ is a phylogenetic tree in which multiple leaves might share a label. Generalizing a previously known decomposition scheme lets us leverage this algorithm, yielding linear-time algorithms for so-called "reticulation visible" networks and "nearly stable" networks. While these are special classes of networks, they rank among the most general of the previously considered cases. We also present a dynamic programming algorithm that solves the general problem in $O(3^{t^*} \cdot |N| \cdot |T|)$ time, where the parameter $t^*$ is the maximum number of "tree components with unstable roots" in any block of the input network. Notably, $t^*$ is stronger (that is, smaller on all networks) than the previously considered parameter "number of reticulations" and even the popular parameter "level" of the input network.

## 1 Introduction

The quest to find the famous "tree of life" has been popular in life sciences since the widespread adoption of evolution as the source of biodiversity on earth. With the discovery of DNA, the task of constructing a history of the evolution of a set of species has become both a blessing and a curse. A blessing because we no longer rely on phenotypical characteristics to distinguish between species and a curse because we are being overwhelmed with data that has to be cleaned, interpreted and visualized in order to draw conclusions. The use of DNA also gave strong support to the realization that trees are not always suited to display ancestral relations, as they fail to model recombination events such as hybridization (occurring frequently in plants) and horizontal gene transfer (a dominating factor in bacterial evolution) [8,25]. Thus, researchers are more and more interested in evolutionary networks and algorithms dealing with them (see the monographs by Gusfield [20] and Huson et al. [22]).

The particular task that we consider in this work is to tell whether a given evolutionary network $N$ "displays" an evolutionary tree $T$, that is, whether the tree-like information that we might have come to believe in the past is consistent with a proposed recombinant evolution. This problem is known as TREE

© Springer Nature Switzerland AG 2018
M. Blanchette and A. Ouangraoua (Eds.): RECOMB-CG 2018, LNBI 11183, pp. 309–323, 2018.
https://doi.org/10.1007/978-3-030-00834-5_18

CONTAINMENT and it has been studied extensively. As it is NP-hard for general binary $N$ and $T$ [23,26], research focuses on moderately exponential time algorithms [18] and biologically relevant special cases of networks [6,14,15,17,23,26]. Prominent among these special classes are the following:

- nearly-stable networks for which a linear-time algorithm is known [15]
- reticulation-visible networks for which cubic-time [6], quadratic-time [17] and linear-time [16] algorithms are known.

The related task of finding a "cluster" instead of a tree [17,19,26] has also been considered. Allowing high-degree nodes ("polytomies") splits the problem into two variants. In the "soft" variant, loosely speaking, polytomies are compatible with any binary subnetwork (see Bentert et al. [4]). In this work, we consider the "hard" version, where polytomies in $T$ must correspond to polytomies in $N$. Using the decomposition of Gunawan et al. [17] for *general* networks, we show that TREE CONTAINMENT can be solved in $O(|N| \cdot \overleftarrow{\Delta}_N^2 \cdot \overrightarrow{\Delta}_T^2)$ time[1] if each tree vertex with a reticulation parent is stable[2] on some leaf. This running time degenerates to linear time for binary $N$ in which the length of a longest "reticulation chain" (directed path consisting only of reticulations) is constant. This class of networks properly includes both reticulation visible and nearly stable networks and, therefore, subsumes previous work mentioned above. We culminate the ideas that lead to the linear-time algorithms to develop an $O((\overrightarrow{\Delta}_T + 1)^{t^*} \cdot (\overleftarrow{\Delta}_N + \overrightarrow{\Delta}_T^{2.5}) \cdot |V(N)| \cdot |V(T)|)$-time algorithm, where $t^*$ is the maximum number of unstable tree components (see Definition 1) in any biconnected component[3] of $N$. For bifurcating $N$, this degenerates to $O(3^{t^*} \cdot |V(N)| \cdot |V(T)|)$ time.

*Preliminaries.* Let $N$ be a weakly connected, directed acyclic graph (DAG) with a single source $\rho(N)$ called the *root* and each of the sinks $\mathcal{L}(N)$ (called *leaves*) carries a label (its "taxon"). Then, we call $N$ an evolutionary (or phylogenetic) *network* (or "network" for short). We call the vertices of in-degree at least two in $N$ *reticulations* and all other vertices *tree vertices* and we demand that all reticulations have out-degree one (if any vertex has in- and out-degree more than one, it can be "split" into a tree vertex with a reticulation parent without impact on the computational problem). If $N$ has no reticulations, then it is called a *tree*. We denote the number of arcs in $N$ by $|N|$.

　　We denote the maximum in- and out-degree in $N$ by $\overleftarrow{\Delta}_N$ and $\overrightarrow{\Delta}_N$, respectively. Then, we call $N$ *forward-binary* (or *bifurcating*) if $\overrightarrow{\Delta}_N \leq 2$ and *binary* if also $\overleftarrow{\Delta}_N \leq 2$. If each label occurs $\leq k$ times in $N$, we call $N$ *$k$-labeled* or, if $k$ is

---

[1] Herein, $\overrightarrow{\Delta}_T$ is the maximum out-degree in $T$ and $\overleftarrow{\Delta}_N$ is the maximum in-degree in the result of contracting all arcs between reticulations in $N$.

[2] $u$ is stable on $\ell$ if all root-$\ell$-paths contain $u$. The notion of stability is equivalent to the notion of "dominators" in directed graphs [1,24].

[3] A biconnected component (or "block") of a network is a subdigraph induced by the vertices of a biconnected component of its underlying undirected graph, that is, a connected component in the result of removing all bridges.

unknown or inconsequential, *multi-labeled.* We define the relation $\leq_N$ such that $u \leq_N v \iff u$ is a descendant of $v$ (that is, $v$ is an ancestor of $u$) in $N$. Note that $u \leq_N \rho(N)$ for all $u \in V(N)$. For each vertex $v$ of $N$, we define $N_v$ to be the *subnetwork rooted at* $v$, that is, the subnetwork of $N$ that contains exactly the vertices $u$ with $u \leq_N v$ and all arcs of $N$ between those vertices. The subnetwork $N|_U$ of $N$ *restricted* to a set $U$ of vertices is the result of first removing all vertices $v$ with $\forall_{u \in U} u \not\leq_N v$ and then contracting all arcs that are outgoing of vertices $w$ with in-degree and out-degree at most one, unless $w \in U$. Note that the least common ancestor (LCA) of any two vertices of $U$ is also in $N|_U$.

We call any vertex $v$ of $N$ *stable on another vertex* $u$ if all $\rho(N)$-$u$-paths contain $v$ and we call $v$ *stable* if $v$ is stable on a leaf of $N$. Then, $N$ is called *reticulation*

```
┌─ TREE CONTAINMENT (TC) ──────┐
│ Input: a network N, a tree T │
│ Question: Does N display T?  │
└──────────────────────────────┘
```

*visible* if each reticulation $r$ is stable. Further, $N$ is called *nearly stable* if, for each vertex $v$, either $v$ or its direct predecessors (called *parents*) are stable. For all $k$, a $k$-labeled network $N$ is said to *contain* a tree $T$ if $T$ is a subgraph of $N$ (respecting the leaf-labeling). Further, $N$ is said to *display* $T$ if $N$ contains a subdivision of $T$ (that is, the result of a series of arc-subdivisions in $T$). In this work, we consider the TREE CONTAINMENT problem defined on the right. We assume that each arc between two reticulations in $N$ is initially contracted. Since $T$ has no reticulations, this has no influence on whether or not $N$ displays $T$.

**Assumption 1.** *The children and parents of all reticulations are tree vertices.*

## 2    Multi-labeled Tree Containment

The following is a simple dynamic programming deciding if a $k$-labeled tree $\hat{T}$ displays a tree $T$. To this end, we define a table with entries $[u, v]$ where $u \in V(\hat{T})$ and $v \in V(T)$ such that, for all computed entries $[u, v]$, we have

$$[u, v] = 1 \iff u \in \min_{\leq_{\hat{T}}}\{w \mid \hat{T}_w \text{ displays } T_v\}. \tag{1}$$

While the table $[u, v]$ might have $|T| \cdot |\hat{T}|$ cells, we will not compute all of them, but only those with $[u, v] = 1$ which we show to be at most $|\hat{T}| \cdot k$. We represent the table as a sparse set, allowing efficient enumeration, setting, and querying [7].

**Observation 1.** *Given $v$, we can get $U := \{u \mid [u, v] = 1\}$ in $O(|U|)$ time.*

Note that, if $T_v$ is displayed by subtrees $\hat{T}_u$ and $\hat{T}_w$ of $\hat{T}$ and $u$ and $w$ are incomparable wrt. $\leq_{\hat{T}}$, then for each leaf-label $\lambda$ in $T_v$, each of $\hat{T}_u$ and $\hat{T}_w$ contains a different leaf labeled $\lambda$. Thus, there cannot be more than $k$ such subtrees.

**Observation 2.** *For all $v$, we have $|\{u \mid [u, v] = 1\}| \leq k$.*

We compute the table in a bottom-up manner. If $v$ is a leaf, then we find the $\leq k$ leaves of $\hat{T}$ with the same label as $v$ and set $[u, v] = 1$ for them. If $v$ has children $v_1, v_2, \ldots, v_d$, then we first compute $U := \bigcup_i \{w \mid [w, v_i] = 1\}$ and then compute the subtree $\hat{T}|_U$ of $\hat{T}$ that is restricted to $U$. Finally, we find the lowest vertices $u$ of $\hat{T}|_U$ such that there is a matching $M$ between the children of $v$ in $T$ and the children of $u$ in $\hat{T}|_U$ such that each $M(v_i)$ has a descendant $z_i$ in $\hat{T}|_U$ with $[z_i, v_i] = 1$. For all these $u$, we set $[u, v] = 1$.

**Lemma 1.** *The computation is correct, that is, (1) holds for all entries $[u, v]$.*

*Proof.* The proof is by induction on the height of $v$ in $T$. If $v$ is a leaf, then (1) clearly holds. Otherwise, let $v_1, v_2, \ldots, v_d$ be the children of $v$ in $T$ and let $U := \bigcup \{w \mid [w, v_i] = 1\}$.

"$\Leftarrow$": Let $u$ be a lowest vertex in $\hat{T}$ such that $\hat{T}_u$ contains a subdivision $S$ of $T_v$. Then, $\hat{T}_u$ contains lowest $z_1, z_2, \ldots, z_d$ such that $S_{z_i}$ displays $T_{v_i}$. Suppose that $S$ is chosen such as to maximize the sum of the distances between $u$ and $z_i$. By minimality of $u$, we know that $u$ is the LCA of the $z_i$ in $\hat{T}$. Then, by induction hypothesis, $[z_i, v_i] = 1$ for all $i$, implying that $z_i \in U$ and, thus, $z_i \in V(\hat{T}|_U)$. Since $u$ is the LCA of the $z_i$ in $\hat{T}$, we also have $u \in V(\hat{T}|_U)$. Moreover, $u$ has children $w_1, w_2, \ldots, w_d$ in $\hat{T}$ such that $z_i \leq_{\hat{T}} w_i$ for all $i$ and, thus, $u$ also has children $w'_1, w'_2, \ldots, w'_d$ in $\hat{T}|_U$ with $z_i \leq_{\hat{T}|_U} w'_i$ for all $i$. Hence, mapping $v_i$ to $w'_i$ for each $i$ constitutes a matching $M$ as demanded by the above construction, implying $[u, v] = 1$.

"$\Rightarrow$": Suppose that $[u, v] = 1$. By construction, $u$ is a lowest vertex of $\hat{T}|_U$ for which there is a matching $M$ between the children of $v$ in $T$ and the children of $u$ in $\hat{T}|_U$ such that each $M(v_i)$ has a descendant $z_i$ in $\hat{T}|_U$ with $[z_i, v_i] = 1$. By induction hypothesis, each $z_i$ is a minimum wrt. $\leq_{\hat{T}}$ of $\{w \mid \hat{T}_w$ displays $T_{v_i}\}$. Thus, each $T_{v_i}$ has a subdivision $S_i$ in $\hat{T}_{z_i}$ and, since $M$ is a matching, each $z_i$ descends from a different child $u_i$ of $u$ in $\hat{T}$. Thus, the $S_i$ together with the unique $u$-$z_i$-paths in $\hat{T}$ can be merged to form a subdivision of $T_v$ contained in $\hat{T}_u$. Towards a contradiction, assume that $u$ is not minimal wrt. $\leq_{\hat{T}}$ among such vertices, that is, there is a different lowest $u' <_{\hat{T}} u$ such that $\hat{T}_{u'}$ contains a subdivision $S'$ of $T_v$. By the argument in the "$\Leftarrow$"-direction, there is a matching $M'$ between the children of $v$ in $T$ and the children of $u'$ in $\hat{T}|_U$ such that each $M'(v_i)$ has a descendant $z'_i$ in $\hat{T}|_U$ with $[z'_i, v_i] = 1$. But then, $u$ is not minimal among such vertices in $\hat{T}|_U$, and we would not have constructed $[u, v] = 1$.    $\square$

To show the running time, we assume that $\hat{T}$ is preprocessed to allow translating labels into leaves and leaves of $\hat{T}$ into leaves of $T$.

**Assumption 2.** *Given a label $\lambda$, we can get the list $L$ of all leaves of $\hat{T}$ with label $\lambda$ in $O(|L|)$ time. Given a leaf $\ell$ of $\hat{T}$, we can get the leaf of $T$ with the same label in $O(1)$ time.*

We also assume that we can compute the LCA of two vertices in $T$ or in $\hat{T}$ in constant time (see, for example [3]). This helps us compute the restriction of $T$ and $\hat{T}$ to any preordered list $U \subseteq V(T)$ in $O(|U|)$ time (see, for example [10, Section 8]).

**Assumption 3.** *Given vertices $x$ and $y$ in $T$ or in $\hat{T}$, we can find $\text{LCA}_T(xy)$ and $\text{LCA}_{\hat{T}}(xy)$ in $O(1)$ time. Given a preordered list $U$, we can find $T|_U$ and $\hat{T}|_U$ in $O(|U|)$ time.*

**Lemma 2.** *Let $\hat{T}$ be $k$-labeled. Then, we can find the maximal (wrt. $\leq_T$) vertices $v$ such that $\hat{T}$ displays $T_v$ in $O(|\hat{T}| \cdot k^2 \overline{\Delta}_T^2)$ time.*

*Proof.* First, we get the set $Y$ of leaves of $T$ whose label occurs in $\hat{T}$ by scanning all leaves of $\hat{T}$ and translating these leaves to $T$ using Assumption 2. This allows us to set $[u, v]$ for all leaves $v$ of $T$ in $O(|\hat{T}|)$ time. Furthermore, we can compute $T|_Y$ in $O(|Y|)$ time using Assumption 3.

Scanning $T|_Y$ in a bottom-up manner, we compute $[u, v]$ for each $u$ and each $v$ with children $v_1, v_2, \ldots, v_d$ as described. To this end, we construct $U_i := \{w \mid [w, v_i] = 1\}$ in $O(k)$ time by Observation 1 and Observation 2 and $U := \bigcup_i U_i$ in $O(kd)$ time since $i \leq d$. Then, we construct $\hat{T}|_U$ in $O(|U|)$ time using Assumption 3. For each $x \in V(\hat{T}|_U)$, we then compute the set $L_x$ of indices $i$ such that there is some $w <_{\hat{T}|_U} x$ with $[w, v_i] = 1$. With a bottom-up dynamic programming in $\hat{T}|_U$, this can be done in $O(|\hat{T}|_U| \cdot d)$ time since $|L_x| \leq d$ for each $x$. Then, we compute a list $C$ of all vertices $x \in V(\hat{T}|_U)$ with $L_x = \{1, 2, \ldots, d\}$. Since the subtrees of $\hat{T}$ rooted at each minimum wrt. $\leq_{\hat{T}|_U}$ of $C$ are leaf-disjoint, we know that each such minimum has its own private descendant $w$ with $[w, v_1] = 1$ and, thus, there are at most $k$ such minima, implying $|C| \leq 2k - 1$.

For each vertex $u \in C$, we then construct a bipartite graph $B$ whose two partitions are the children of $u$ in $\hat{T}|_U$ and the children of $v$ in $T$, respectively, and $B$ contains an edge $\{x, v_i\}$ if and only if $i \in L_x$ (that is, $x$ has a descendant $w$ in $\hat{T}|_U$ with $[w, v_i] = 1$). If $B$ has a size-$d$ matching, we set $[u, v] = 1$. This can be done in $O(\sqrt{d} \cdot \min\{d^2, kd\}) \subseteq O(kd^{1.5})$ time [9] for each $u$. Note that no vertex $u \notin C$ can have such a matching and, thus, we set $[u, v] = 1$ correctly for all $u$ and $v$.

Summing up the total time spent and noting that $|Y| \leq |\hat{T}|$, and $|U| \leq kd$, and $|\hat{T}|_U| \leq 2|U| - 1$, and $|C| \leq 2k - 1$, we arrive at a total running time of $O(|\hat{T}| \cdot (k\overline{\Delta}_T^2 + k^2\overline{\Delta}_T^{1.5}))$. Since our algorithm runs bottom-up in $T|_Y$, we can retain the highest $v$ for which there is some $w$ in $\hat{T}$ with $[w, v] = 1$ as claimed in the lemma. $\square$

If we are only interested in whether or not $\hat{T}$ displays $T$, then we can prepend a size check and refuse the instance if $|T| > |\hat{T}|$. Thus, we can bound all preprocessing in $O(|\hat{T}|)$ time and Lemma 2 implies the following theorem.

**Theorem 1.** *Let $\hat{T}$ be a $k$-labeled tree and let $T$ be a tree with maximum out-degree $\overline{\Delta}_T$. Then, we can decide if $\hat{T}$ displays $T$ in $O(|\hat{T}| \cdot k^2\overline{\Delta}_T^2)$ time $(O(|\hat{T}| \cdot k^2)$ time if $T$ is binary).*

## 3   Tree Containment in Special Networks

In this section, we move from multi-labeled trees to single-labeled networks, that is, in what follows, each label occurs exactly once (the leaf-labelling function is bijective).

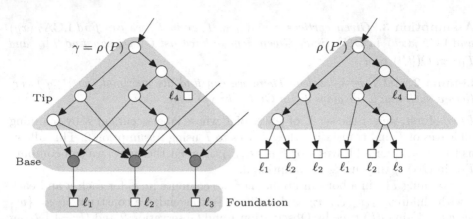

**Fig. 1. Left:** A leaf $\gamma$ of the component DAG $Q$ of $N$ implies a layering of the *pyramid* $P = N_\gamma$ into its *tip* $P^\Delta$ (tree nodes $\bigcirc$), its *base* $P^B$ (reticulations $\bullet$), and its *foundation* $P^F$ (leaves $\square$ below reticulations). Note that leaves may also be in the tip of $P$. **Right:** The multi-labeled tree $P'$ computed from the pyramid on the left in the proof of Lemma 3.

*Network Decomposition.* Gunawan et al. [17] introduced a decomposition for reticulation-visible networks which we apply to arbitrary networks. To this end, we have to do some initial cleanup using the following reduction.

**Rule 1.** *Let $ab$ be a* cherry *(that is, a pair of leaves sharing a common parent) in $N$. If $ab$ is not a cherry in $T$, then reject $(N,T)$ and, otherwise, delete $a$ in both $N$ and $T$ and contract the arc incoming to $b$ in both $N$ and $T$.*

**Definition 1 (See [17]).** *Let $(N,T)$ be reduced wrt. Rule 1 and let $F$ be the forest that results from removing all reticulations from $N$. Then, each tree of $F$ is called* tree component *of $N$. A tree component of $N$ is called trivial if it contains only a leaf of $N$ and stable if its root is stable. Let $\Gamma$ be the set of roots of the non-trivial tree components of $N$. The restriction of "$\leq_N$" to $\Gamma$ forms a DAG $Q$ and we call it the* component DAG *of $N$. More formally, $Q := (\Gamma, (\leq_N) \cap (\Gamma \times \Gamma))$.*

The goal will be to repeatedly find a leaf $\gamma$ of $Q$ and the best possible $v$ of $T$ such that $N_\gamma$ displays $T_v$. Then, we shrink both $N_\gamma$ and $T_v$ to a single leaf and remove $\gamma$ from $Q$. We make use of the special structure of $N_\gamma$, implied by the fact that all tree nodes with a reticulation ancestor in $N_\gamma$ are leaves of $N$ (otherwise, they are in a tree component below $\gamma$, contradicting $\gamma$ being a leaf).

**Definition 2.** *Let $\gamma$ be a leaf of $Q$. Then, $P := N_\gamma$ consists of a tree with root $\rho(P) := \gamma$, some reticulations and some leaves of $N$. Further, $P$ can be divided into "layers" (see Fig. 1) and we call $P$ a* pyramid *with a tip $P^\Delta$ (layer of tree vertices), a base $P^B$ (layer of reticulations) and a foundation $P^F$ (layer of leaves below reticulations).*

*Algorithm.* In this section, we show how Lemma 2 can be applied to pyramids. Given a pyramid $P$ in $N$, our goal is to display as much of $T$ as possible in $P$ and reduce $N$ and $T$ using this information. To this end, we consider only the tip $P^\Delta$ of $P$ and replace each arc $xy$ from the tip to the base by an arc to a copy of the child $\ell$ of $y$. By Definition 2, $\ell$ is a leaf of $N$. Recall that $\overleftarrow{\Delta}_P$ is the maximum in-degree in $P$ and $\overrightarrow{\Delta}_T$ is the maximum out-degree in $T$.

**Lemma 3.** *In $O(|P| \cdot \overleftarrow{\Delta}_P^2 \cdot \overrightarrow{\Delta}_T^2)$ time, we can find all maximal $v$ (wrt. $\leq_T$) s.t. $P$ displays $T_v$.*

*Proof.* Let $P'$ denote the multi-labeled tree that results from $P^\Delta$ by, for each arc $xy \in V(P^\Delta) \times V(P^B)$, hanging a leaf onto $x$ that is labeled with the same label as the unique child $\ell$ of $y$ in $P$ (see Fig. 1). Note that $P'$ is indeed $\overleftarrow{\Delta}_P$-labeled, its size is at most $|P|$, and it can be constructed in $O(|P|)$ time. Having constructed $P'$, we compute the maximal (wrt. $\leq_T$) vertices $v$ such that $P'$ displays $T_v$. By Lemma 2, this can be done in $O(|P'| \cdot \overleftarrow{\Delta}_P^2 \cdot \overrightarrow{\Delta}_T^2)$ time. It remains to show for all $v$ of $T$ that $P$ displays $T_v$ if and only if $P'$ does (see also [17]).

"$\Rightarrow$": Let $P$ contain a subdivision $S$ of $T_v$. Let $S'$ result from $S$ by contracting all arcs that are incoming to a vertex of the base $P^B$ of $P$. Since $S$ is a tree, all vertices of $P^B$ have indegree one and outdegree one in $S$ and, thus, $S'$ is also a subdivision of $T_v$. To show that $P'$ contains $S'$, assume that $S'$ contains an arc $xy$ that is not in $P'$. If $xy$ is in $S$, then neither $x$ nor $y$ is a reticulation in $N$, implying that $xy$ is in $P^\Delta$ and, thus, in $P'$. Otherwise, $S$ contains a path $(x, r, y)$, where $r \in P^B$ and $y$ is a (copy of a) leaf in the foundation of $P$. Then, $xr$ is an arc in $V(P^\Delta) \times V(P^B)$, implying that $P'$ contains a copy of $y$ hanging from $x$.

"$\Leftarrow$": Let $P'$ contain a subdivision $S'$ of $T_v$. Let $x\ell$ be an arc of $S'$ that is not in $P$. Then, $\ell$ is a leaf of $P$ and its parent $r$ is in $P^B$. Let $S$ result from $S'$ by replacing each such arc $x\ell$ by the path $(x, r, \ell)$. Clearly, $S$ is a subdivision of $S'$ and, thus, of $T_v$. To show that $P$ contains $S$, it suffices to show that none of the new paths $p$ introduces vertices that were already in $S'$ or in any previously added path. For the first claim, note that all newly added vertices are in $P^B$ and, thus, not in $P'$. For the second claim, note that each label of $P'$ occurs at most once in $S'$ and each vertex of $P^B$ is parent of a unique leaf in $P$. Thus, $P$ contains $S$ and, therefore, $P$ displays $T_v$. □

It is noteworthy that Lemma 3 might return many vertices $v$ such that $T_v$ is displayed by $P$ and, without any more assumptions regarding $N$, the number of possible combinations grows exponentially. Thus, we restrict the class of networks that we are considering by demanding that each tree vertex of $N$ that has a reticulation parent is stable. Hence, $\rho(P)$ is stable for all tree components $P^\Delta$ which form the tips of the pyramids $P$ that we are seeing in the algorithm. In the following, let $c$ be a leaf that $\rho(P)$ is stable on and observe that the set of all vertices $v$ such that $P$ displays $T_v$ and $c \leq_T v$ has a unique maximum wrt. $\leq_T$. Thus, at most one of the maxima obtained by Lemma 3 is an ancestor of $c$ in $T$ and we can find it in $O(|P|)$ time. We then apply the following reduction that places $T_v$ into $P$ and removes all arcs that disagree (see Fig. 2).

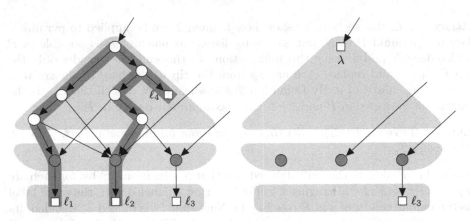

**Fig. 2.** An example of an application of Rule 2 to the network $N$ of Fig. 1 with a subdivision of $T_v$ shown in dark gray on the left.

**Rule 2.** *Let $\rho(P)$ be stable on a leaf $c$ and let $v$ be the unique maximum wrt. $\leq_T$ such that $c \leq_T v$ and $P$ displays $T_v$. Then, remove all leaves of $N$ whose label occurs in $T_v$, remove all vertices (with their incident arcs) in the tip of $P$ except $\rho(P)$, remove all arcs outgoing of $\rho(P)$, remove all vertices of $T_v$ except $v$, and label $v$ and $\rho(P)$ with the same new label $\lambda$.*

For correctness of Rule 2, see [17, Proposition 5] or our proof in the appendix. To apply Rule 2 in $O(|P|)$ time, we have to find a leaf $c$ that $\rho(P)$ is stable on, in $O(|P|)$. This is easy if $P^\triangle$ contains a leaf of $N$. Otherwise, we mark all arcs between the tip and the base of $P$ and check if any vertex in the base has all its incoming arcs marked. For a vertex $r$ with $m$ incoming marked arcs, this check can be done in $O(m)$ time. Thus, we can find $c$ in $O(|P|)$ time.

**Observation 3.** *We can find a leaf $c$ that $\rho(P)$ is stable on in $O(|P|)$ time.*

Further, note that Rule 2 might leave former reticulations as isolated vertices or pending leaves without label. Clearly, such a vertex is created by the deletion of an incoming or outgoing arc. To remove them, we mark such vertices as garbage upon removal of this incident arc in $O(1)$ time per removed arc. Then, we run a cleanup phase after Rule 2 that removes garbage in constant time per removed vertex, that is $O(|N|)$ overall. Thus, we can assume that $N$ does not contain isolated vertices or unlabeled leaves.

Each time Rule 2 is applied to a leaf $\gamma$ of the component DAG $Q$, it will replace the tip of $N_\gamma$ by a single leaf in $N$. To keep $Q$ up to date we just need to delete $\gamma$ from $Q$ at that point (since the tree component of $\gamma$ is no longer non-trivial), but none of the other tree component roots are affected.

**Observation 4.** *We can find a leaf of $Q$ in constant time.*

The algorithm terminates when Rule 2 has been applied to the last pyramid of $N$ and we return yes if and only if $\rho(T)$ is a leaf whose label occurs in $N$.

By Lemma 3, the overall running time can be bounded by $O(\sum_i |P_i| \cdot \overleftarrow{\Delta}_N^2 \cdot \overrightarrow{\Delta}_T^2)$, where the summation is over all applications of Rule 2. Since no arc outgoing of $P^\Delta$ survives an application of Rule 2 to $P$, we conclude $\sum_i |P_i| \le |N|$.

**Theorem 2.** *Let $T$ be a tree with maximum out-degree $\overrightarrow{\Delta}_T$, let $N$ be a network with maximum in-degree $\overleftarrow{\Delta}_N$ (after contraction of arcs between reticulations) and let each tree vertex of $N$ that has a reticulation parent be stable. Then, we can determine if $N$ displays $T$ in $O(|N| \cdot \overleftarrow{\Delta}_N^2 \cdot \overrightarrow{\Delta}_T^2)$ time.*

Consider the special case that $T$ and $N$ are binary. If $N$ is reticulation-visible, it already verifies Assumption 1, implying $\overleftarrow{\Delta}_N \le 2$ and, as each reticulation is stable, each tree vertex with a reticulation parent is also stable. If $N$ is nearly-stable, it cannot have three consecutive reticulations, implying $\overleftarrow{\Delta}_N \le 4$ after the contraction operation of Assumption 1 and, as each node is either stable or has a stable parent, each tree vertex with a reticulation parent is stable.

**Corollary 1.** *Let $T$ be a binary tree and let $N$ be forward-binary and reticulation-visible or nearly stable. Then, we can decide if $N$ displays $T$ in $O(|N|)$ time.*

See also Fig. 3 for an estimation of the probability to encounter the three discussed types of networks when simulating recombinant evolution.

# 4   Tree Containment in General Networks

In this section, we present an algorithm, based on the ideas of the previous section, that solves TREE CONTAINMENT in $O((\overrightarrow{\Delta}_T + 1)^t \cdot (\overleftarrow{\Delta}_N + \overrightarrow{\Delta}_T^{2.5}) \cdot |V(N)| \cdot |V(T)|)$, where $\overleftarrow{\Delta}_N$ and $\overrightarrow{\Delta}_T$ are the respective maximal out-degrees of $N$ and $T$ and $t$ is the number of unstable tree components of $N$ (see Definition 1). For bifurcating $N$ and $T$, this simplifies to $O(3^t \cdot |V(N)| \cdot |V(T)|)$ time. Remarkably, as each root of a tree component (except $\rho(N)$) has its own, distinct reticulation parent, we know that $t$ is always smaller than the number of reticulations in $N$ (plus 1) which has been considered as parameter [18]. Indeed, we prove that we can check all biconnected components of $N$ independently, so the parameter can be improved to the maximum number $t^*$ of unstable tree components in any biconnected component of $N$. For large classes of networks, $t^*$ is arbitrarily small compared to the number of reticulations or even the level[4] of $N$. Figure 3 shows a preliminary comparison of these parameters in networks generated by simulating recombinant evolution.

The main difficulty of applying the presented algorithm to general networks is that the roots of the tree components are not necessarily stable on any leaf in $N$. Upon finding such a root $\gamma$, we thus have to keep track of all the (maximal) vertices $v$ of $T$ such that $N_\gamma$ displays $T_v$. This brings further difficulties: picture two roots $\gamma_1$ and $\gamma_2$ of tree components of $N$ with $\gamma <_N \gamma_1, \gamma_2$. When computing

---

[4] The *level* of a phylogenetic network is the largest number of reticulations in any biconnected component (of its underlying undirected graph).

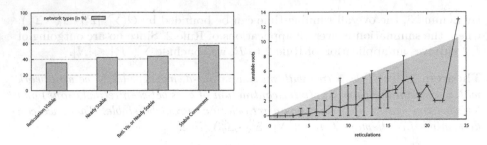

**Fig. 3.** Comparison of 250 networks generated under the coalescent with recombination model (10 taxa, recombination rate 4, see [2]). **Left:** Percentages of network types. Here "Stable Component" refers to the condition that every tree component is stable. **Right:** Comparison of the parameter $t$ to the number $r$ of reticulations. For data points in the gray area, our algorithm has smaller exponential dependance of the asymptotic running time when compared to the one proposed by Gunawan et al. [18] (that is, $1.618^r > 3^t$). The refined parameter $t^*$ relates similarly to the level.

the possible vertices $v_1$ and $v_2$ in $T$ whose subtrees are displayed by $N_{\gamma_1}$ and $N_{\gamma_2}$, respectively, we have to make sure that we are not using $N_\gamma$ to display subtrees in both $N_{\gamma_1}$ and $N_{\gamma_2}$. In the previous algorithm this was not necessary because, if $\gamma$ is stable, then $N_\gamma$ cannot display a subtree of $T_{v_1}$ as well as a subtree of $T_{v_2}$.

Recall that $\Gamma$ is the set of roots of non-trivial tree components in $N$ and let $\Gamma^*$ be the set of roots of unstable tree components in $N$. For any $u \in V(N)$, we call a subnetwork $S$ of $N_u$ *nice* for $N_u$ if $u \in V(S)$ and, for each $v \in V(S)$ and each leaf $\ell$ that $v$ is stable on in $N$, $\ell \in V(S)$. Note that, if $S$ is nice for $N_u$, then $S_v$ is nice for $N_v$ for all $v \in V(S)$. For technical reasons, we use a slightly extended notion of subdivisions that allow adding an arc incoming to the root before subdividing arcs. Then, all subdivisions of $T$ in $N$ containing the root of $N$ are nice. The dynamic programming table has an entry for each triple $(u, v, R) \in V(N) \times V(T) \times 2^{\Gamma^*}$ with the following semantics:

$$[u, v, R] := 1 \iff N_u \text{ contains a nice subdivision } S \text{ of } T_v \text{ with } V(S) \cap \Gamma^* = R \tag{2}$$

Note that $N$ displays $T$ if and only if $[\rho(N), \rho(T), R] = 1$ for some $R \subseteq \Gamma^*$. We give some special cases of $[u, v, R]$ for which (2) can be easily verified:

**Case 1.** If $u \in \Gamma^* \setminus R$, then we set $[u, v, R] = 0$ since all nice subdivisions of $T$ in $N_u$ contain $u$, and, thus, $u \in V(S) \cap \Gamma^*$ but $u \notin R$.

**Case 2.** If $u$ is a reticulation with child $w$, then we set $[u, v, R] = [w, v, R]$ for all $v \in V(T)$ and $R \subseteq \Gamma^*$, since $N_u$ cannot display any more of $T$ than $N_w$.

**Case 3.** If $u$ is stable on a leaf $\ell \notin \mathcal{L}(T_v)$, then we set $[u, v, R] = 0$ for all $R \subseteq \Gamma^*$.

**Case 4.** If $u$ and $v$ are leaves, then we set $[u, v, R] = 1$ if and only if $u$ and $v$ have the same label and $R = \varnothing$.

**Case 5.** If $u$ is a leaf and $v$ is not a leaf, then we set $[u, v, R] = 0$ for all $R \subseteq \Gamma^*$.

**Case 6.** If $u$ is a tree vertex of $N$ and $[w, v, R \setminus \{u\}] = 1$ for any child $w$ of $u$ in $N$ and Case 3 does not apply, then we set $[u, v, R] = 1$.

We call an entry $[u, v, R]$ *trivial* if it corresponds to any of the above cases. Otherwise, we set $[u, v, R] = 1$ if and only if there is a size-$d$ matching $M$ between the children $v_1, v_2, \ldots, v_d$ of $v$ in $T$ to the children of $u$ in $N$ and pairwise disjoint sets $R_1, R_2, \ldots, R_d$ such that $\forall_i\ [M(v_i), v_i, R_i] = 1$ and $\bigcup_i R_i = R \setminus \{u\}$.

**Lemma 4.** *For $[u, v, R]$ computed as above, (2) holds.*

*Proof.* We prove the lemma by induction on the index of $u$ in any fixed DAG-ordering of $N$. We suppose that $[u, v, R]$ is non-trivial, as the other cases are evident. This also implies the induction base (where $u$ is a leaf of $N$).

"$\Leftarrow$": Suppose that there is a nice subdivision $S$ of $T_v$ in $N_u$ with $V(S) \cap \Gamma^* = R$. First, $u$ is not a leaf of $S$, since all leaves of $S$ are labeled. Second, $u$ does not have degree two in $S$ since, otherwise, $[w, v, R] = 1$ for the child $w$ of $u$ in $S$, contradicting the non-triviality of $[u, v, R]$. Hence, there is a matching $M$ between the children $v_1, v_2, \ldots, v_d$ of $v$ in $T$ and the children of $u$ in $S$ such that, for each child $v_i$ of $v$, $S_i := S_{M(v_i)}$ is a subdivision of $T_{v_i}$ and, by the above observation, niceness of $S$ implies niceness of $S_i$. Also note that $M$ has size $d$. Then, by induction hypothesis, $[M(v_i), v_i, R_i] = 1$ for all $i$, where $R_i := V(S_i) \cap \Gamma^*$. Since the $S_i$ are pairwise disjoint, the sets $R_i$ are pairwise disjoint and, since $\bigcup_i V(S_i) = V(S) \setminus \{u\}$, we have $\bigcup_i R_i = \bigcup_i V(S_i) \cap \Gamma^* = V(S) \cap \Gamma^* \setminus \{u\} = R \setminus \{u\}$. Thus, by construction, $[u, v, R] = 1$.

"$\Rightarrow$": Suppose $[u, v, R] = 1$. Then, by construction, there are $M$ and $R_i$ such that $R_i$ are pairwise disjoint, $\forall_i\ [M(v_i), v_i, R_i] = 1$, and $\bigcup_i R_i = R \setminus \{u\}$. By induction hypothesis, Lemma 4 holds for each $[M(v_i), v_i, R_i]$, implying that, for each $i$, there is a nice subdivision $S_i$ of $T_{v_i}$ in $N_{M(v_i)}$ and $V(S_i) \cap \Gamma^* = R_i$. To show that these subdivisions are pairwise vertex-disjoint, assume that $S_i$ and $S_j$ intersect for some $i \neq j$. Let $w$ be the minimum with respect to $\leq_N$ among the vertices in $V(S_i) \cap V(S_j)$. Then, $w$ is neither a reticulation (otherwise its child is smaller wrt. $\leq_N$) nor a leaf (since $T_{v_i}$ and $T_{v_j}$ cannot share leaves). Hence, $w$ is in a tree-component of $N$ and it has a root $r$. Then, both $S_i$ and $S_j$ contain $r$ as well, as otherwise, $M(v_i) = M(v_j)$ contradicting $M$ being a matching. If $r$ is stable on some leaf $\ell$ then, by niceness of $S_i$ and $S_j$, both contain $\ell$, contradicting again that $T_{v_i}$ and $T_{v_j}$ are leaf-disjoint. If $r$ is not stable, then $r \in V(S_i) \cap R_i$ and $r \in V(S_j) \cap R_j$, contradicting that $R_i$ and $R_j$ are disjoint. $\square$

To compute all $[u, v, R]$ for fixed $u$ and $v$ where $d$ is the out-degree of $v$, we enumerate all partitions of $\Gamma^*$ into $d + 1$ cells, one for each child $v_i$ of $v$, corresponding to the $R_i$, plus one cell corresponding to "$\notin R$". Then, we construct the bipartite graph $B$ whose vertices are the children of $u$ in $N$ and $v$ in $T$, respectively, and the edge set is $\{u_j v_i \mid [u_j, v_i, R_i] = 1\}$. Finally, we set $[u, v, R] = 1$ if $B$ has a size-$d$ matching for any of the partitions of $\Gamma^*$. Since one cell of the bipartition has size $d$, such a matching can be computed in $O(d^{2.5})$ time [9]. Thus, the implied bottom-up dynamic programming runs in $O((\overline{\Delta}_T + 1)^{|\Gamma^*|} \cdot (\overline{\Delta}_N + \overline{\Delta}_T^{2.5}) \cdot |V(N)| \cdot |V(T)|)$ time. If $N$ and $T$ are forward-binary, this simplifies to $O(3^{|\Gamma^*|} \cdot |V(N)| \cdot |V(T)|)$. We can, however, further refine the algorithm by splitting off biconnected components of $N$. To this end, we use the following lemma.

**Lemma 5 (See also** [17]**).** *Let $u \in \Gamma$ such that $N - u$ is disconnected, let $v := \mathrm{LCA}_T(\mathcal{L}(N_u))$, and let $(N', T')$ be the result of contracting $N_u$ and $T_v$, respectively, into a single vertex and giving a new label $\lambda$ to both of them. Then, $N$ displays $T$ if and only if $N_u$ displays $T_v$ and $N'$ displays $T'$.*

*Proof.* First, note that $u$ is stable on all leaves of $\mathcal{L}(N_u)$.

"$\Rightarrow$": Let $S$ be a subdivision of $T$ in $N$. Since $u$ is stable on all leaves of $\mathcal{L}(N_u)$, we know that $N_u$ displays $T_v$ and cannot display $T_w$ for any $w >_T v$. Thus, the result $S'$ of contracting $S_u$ into a single vertex and labeling it $\lambda$ displays $T'$ and it is clearly a subdivision of $N'$.

"$\Leftarrow$": Since $V(N') \cap V(N_u) = \{u\}$, the result of gluing a subdivision of $T'$ in $N'$ (which has to contain $u$ as leaf) and a subdivision of $T_v$ in $N_u$ together at $u$ is contained in $N$ and it is clearly a subdivision of $T$. □

With Lemma 5, we can check tree containment in all biconnected components of $N$ independently.

**Theorem 3.** *Let $T$ be a tree, let $N$ be a network, and let $\vec{\Delta}_N$ and $\vec{\Delta}_T$ be their respective maximum out-degrees. Let $t^*$ be the maximum number of unstable tree components of any biconnected component of $N$ (see Definition 1). Then, we can decide whether $N$ displays $T$ in $O((\vec{\Delta}_T + 1)^{t^*} \cdot (\vec{\Delta}_N + \vec{\Delta}_T^{2.5}) \cdot |V(N)| \cdot |V(T)|)$ time. If $N$ and $T$ are forward-binary, this is $O(3^{t^*} \cdot |V(N)| \cdot |V(T)|)$ time.*

Albeit inconsequential for the algorithm itself, note that $t^*$ can be computed in linear time as biconnected components and tree components can be found in linear time [17,21] and the stability of the roots can be checked in linear time [1].

We finish this section with a note on polynomial-time preprocessing concerning the number $t^*$ of unstable tree components in any biconnected component. Indeed, to show that TREE CONTAINMENT does not admit a polynomial-size kernel (see [11,12] for more details on "kernelization") it suffices to show that instances $(N_i, T_i)$ of TREE CONTAINMENT can be combined to a single instance $(N, T)$ such that 1. the number $t^*$ of unstable tree components in any biconnected component of $N$ is in $O(\max_i |N_i|)$ and 2. $N$ displays $T$ if and only if $N_i$ displays $T_i$ for each $i$ (see [5,13] or [11, Section 15.1.3] for details on "AND compositions").

Let $C_k$ be a caterpillar tree with $k$ leaves labeled with $\{1, 2, \ldots, k\}$. Given $k$ instances $(N_i, T_i)$ of TREE CONTAINMENT with disjoint label-sets, let $N$ denote the result of, for each $i$, replacing the leaf labeled $i$ in $C_k$ by $N_i$. Likewise, let $T$ be the result of, for each $i$, replacing the leaf labeled $i$ in $C_k$ by $T_i$. It is then straightforward to verify that $N$ displays $T$ if and only if, for each $i$, $N_i$ displays $T_i$. Note that the argument above is independent of the actual parameter that we take per block. For example, it holds as well for the "level" of $N$.

**Observation 5.** *Let $\varphi^*$ map networks to integers such that, for all networks $N$ and all cut-vertices $u$ in $N$, we have $\varphi^*(N) = \max\{\varphi^*(N_u), \varphi^*(N^u)\}$ where $N^u$ results from $N$ by contracting $N_u$ into a single vertex (with new label). Then, TREE CONTAINMENT does not admit a polynomial-size kernel with respect to $\varphi^*$, unless $NP \subseteq coNP/\mathrm{poly}$.*

# 5    Conclusion

We developed efficient algorithms for the TREE CONTAINMENT problem in various settings, continuing existing efforts to speed up the process of solving the problem in special types of networks, as well as developing first parameterized algorithms and preliminary results concerning efficient and effective preprocessing. We showed that, if each label occurs at most $k$ times in $N$, the problem can be solved in $O(|N| \cdot \vec{\Delta}_T \cdot k^2)$ time (where $\vec{\Delta}_T$ is the maximum out-degree in $T$). Together with the powerful network decomposition of Gunawan et al. [17], this implies an $O(|N|)$-time algorithm for binary reticulation visible networks and nearly stable networks. We further developed an algorithm that solves the general case in $O((\vec{\Delta}_T + 1)^{t^*} \cdot (\vec{\Delta}_N + \vec{\Delta}_T^{2.5}) \cdot |N| \cdot |T|)$ time where $t^*$ is the maximum number of unstable tree components in any biconnected component of $N$. For binary $N$ and $T$, this simplifies to $O(3^{t^*} \cdot |N| \cdot |T|)$. The discovery of the parameter $t$ (and $t^*$) is interesting in its own regard, as previous algorithms used to study phylogenetic networks focus on the "number $r$ of reticulations" or the "maximum number of reticulations in a biconnected component" (the"level"), but the parameter $t^*$ can be arbitrarily small when compared to these parameters. As there is an implementation of an $O(1.618^r \cdot |N| \cdot |T|)$-time algorithm for TREE CONTAINMENT [18], I am eager to compare our algorithm to it on practical data sets. Preliminary comparisons show its potential on data-sets generated from simulating evolutionary processes (see Fig. 3). Finally, I am highly motivated to research more parameters of phylogenetic networks as we presume that practical networks are likely to be highly structured (since evolution is not a totally random process). The distance of the input network to being reticulation visible or nearly stable seems to be the canonical starting point.

**Acknowledgement.** Thanks to Celine Scornavacca for her thorough proof-reading.

# Appendix

*Proof. (Proof of correctness of Rule 2).* Let $S^v$ be a subdivision of $T_v$ in $P$ and let $(N', T')$ be the result of applying Rule 2 to $(N, T)$.

"$\Leftarrow$": Let $N'$ contain a subdivision $S'$ of $T'$. It suffices to show that the result $S$ of replacing $\rho(P)$ with $S^v$ in $S'$ is contained in $N$ since $S$ is clearly a subdivision of $T$. Since $S^v$ is contained in $P$, it suffices to show that $S'$ and $S^v$ are vertex disjoint (except for $\rho(P)$). Towards a contradiction, assume that $S'$ and $S^v$ both contain a vertex $u \neq \rho(P)$ of $P$. Since $\mathcal{L}(S')$ and $\mathcal{L}(S^v)$ are disjoint, $u$ is ancestor to at least two different leaves in $N$. Thus, $u$ is in the tip of $P$, contradicting that $u$ is in $N'$.

"$\Rightarrow$": Let $N$ contain a subdivision $S$ of $T$ and let $u := \text{LCA}_S(\mathcal{L}(T_v))$. Since $\rho(P)$ is stable on $c$ and $c \in \mathcal{L}(T_v)$, we have $u \leq_N \rho(P)$, implying $\mathcal{L}(S_{\rho(P)}) \supseteq \mathcal{L}(T_v)$. Further, maximality of $v$ implies $\mathcal{L}(S_{\rho(P)}) \subseteq \mathcal{L}(T_v)$. Let $S'$ result from $S$ by contracting $S_{\rho(P)}$ into a single vertex and labeling this vertex $\lambda$. Since $\mathcal{L}(S_{\rho(P)}) = \mathcal{L}(T_v)$, we know that $S'$ is a subdivision of $T'$ and it suffices to show that $N'$ contains $S'$. To do this, we show that all vertices of $S'$ are in $N'$. Assume towards a contradiction that $S'$ contains a vertex $w$ that is not in $N'$. Then, $w$ is in the tip of $P$, implying $\mathcal{L}(S_w) \subseteq \mathcal{L}(S_{\rho(P)})$. Thus, $w$ is a vertex of $S_{\rho(P)}$ contradicting $w$ being in $S'$.

# References

1. Alstrup, S., Harel, D., Lauridsen, P.W., Thorup, M.: Dominators in linear time. SIAM J. Comput. **28**(6), 2117–2132 (1999)
2. Arenas, M., Valiente, G., Posada, D.: Characterization of reticulate networks based on the coalescent with recombination. Mol. Biol. Evol. **25**(12), 2517–2520 (2008)
3. Bender, M.A., Farach-Colton, M.: The LCA problem revisited. In: Gonnet, G.H., Viola, A. (eds.) LATIN 2000. LNCS, vol. 1776, pp. 88–94. Springer, Heidelberg (2000). https://doi.org/10.1007/10719839_9
4. Bentert, M., Malík, J., Weller, M.: Tree containment with soft polytomies. In: Proceedings of the 16th SWAT. LIPIcs, vol. 101, pp. 9:1–9:14. Schloss Dagstuhl (2018)
5. Bodlaender, H.L., Jansen, B.M.P., Kratsch, S.: Kernelization lower bounds by cross-composition. SIAM J. Discrete Math. **28**(1), 277–305 (2014)
6. Bordewich, M., Semple, C.: Reticulation-visible networks. Adv. Appl. Math. **78**, 114–141 (2016)
7. Briggs, P., Torczon, L.: An efficient representation for sparse sets. ACM Lett. Program. Lang. Syst. (LOPLAS) **2**(1–4), 59–69 (1993)
8. Chan, J.M., Carlsson, G., Rabadan, R.: Topology of viral evolution. Proc. Natl. Acad. Sci. **110**(46), 18566–18571 (2013)
9. Chandran, B.G., Hochbaum, D.S.: Practical and theoretical improvements for bipartite matching using the pseudoflow algorithm. CoRR abs/1105.1569 (2011)
10. Cole, R., Farach-Colton, M., Hariharan, R., Przytycka, T., Thorup, M.: An $o(n \log n)$ algorithm for the maximum agreement subtree problem for binary trees. SIAM J. Comput. **30**(5), 1385–1404 (2000)
11. Cygan, M., et al.: Parameterized Algorithms. Springer, Heidelberg (2015). https://doi.org/10.1007/978-3-319-21275-3
12. Downey, R.G., Fellows, M.R.: Fundamentals of Parameterized Complexity. Texts in Computer Science. Springer, London (2013). https://doi.org/10.1007/978-1-4471-5559-1
13. Drucker, A.: New limits to classical and quantum instance compression. SIAM J. Comput. **44**(5), 1443–1479 (2015)
14. Fakcharoenphol, J., Kumpijit, T., Putwattana, A.: A faster algorithm for the tree containment problem for binary nearly stable phylogenetic networks. In: Proceedings of the 12th JCSSE, pp. 337–342. IEEE (2015)
15. Gambette, P., Gunawan, A.D., Labarre, A., Vialette, S., Zhang, L.: Solving the tree containment problem in linear time for nearly stable phylogenetic networks. Discrete Appl. Math. **246**, 62–79 (2018)
16. Gunawan, A.D.M.: Solving the tree containment problem for reticulation-visible networks in linear time. In: Jansson, J., Martín-Vide, C., Vega-Rodríguez, M.A. (eds.) AlCoB 2018. LNCS, vol. 10849, pp. 24–36. Springer, Cham (2018). https://doi.org/10.1007/978-3-319-91938-6_3
17. Gunawan, A.D., DasGupta, B., Zhang, L.: A decomposition theorem and two algorithms for reticulation-visible networks. Inf. Comput. **252**, 161–175 (2017)

18. Gunawan, A.D., Lu, B., Zhang, L.: A program for verification of phylogenetic network models. Bioinformatics **32**(17), i503–i510 (2016)
19. Gunawan, A.D., Lu, B., Zhang, L.: Fast methods for solving the cluster containment problem for phylogenetic networks. CoRR abs/1801.04498 (2018)
20. Gusfield, D.: ReCombinatorics: The Algorithmics of Ancestral Recombination Graphs and Explicit Phylogenetic Networks. MIT Press, Cambridge (2014)
21. Hopcroft, J., Tarjan, R.: Algorithm 447: efficient algorithms for graph manipulation. Commun. ACM **16**(6), 372–378 (1973)
22. Huson, D.H., Rupp, R., Scornavacca, C.: Phylogenetic Networks: Concepts, Algorithms and Applications. Cambridge University Press, New York (2010)
23. Kanj, I.A., Nakhleh, L., Than, C., Xia, G.: Seeing the trees and their branches in the network is hard. Theor. Comput. Sci. **401**(1–3), 153–164 (2008)
24. Lengauer, T., Tarjan, R.E.: A fast algorithm for finding dominators in a flowgraph. ACM Trans. Program. Lang. Syst. **1**(1), 121–141 (1979)
25. Treangen, T.J., Rocha, E.P.: Horizontal transfer, not duplication, drives the expansion of protein families in prokaryotes. PLoS Genet. **7**(1), e1001284 (2011)
26. Van Iersel, L., Semple, C., Steel, M.: Locating a tree in a phylogenetic network. Inf. Process. Lett. **110**(23), 1037–1043 (2010)

# Author Index

# Author Index

Printed in the United States
By Bookmasters